T0332666

umerical Integration
ecent Developments, Software and Applications

NATO ASI Series

Advanced Science Institutes Series

A series presenting the results of activities sponsored by the NATO Science Committee,
which aims at the dissemination of advanced scientific and technological knowledge,
with a view to strengthening links between scientific communities.

The series is published by an international board of publishers in conjunction with the
NATO Scientific Affairs Division

A Life Sciences	Plenum Publishing Corporation
B Physics	London and New York
C Mathematical and Physical Sciences	D. Reidel Publishing Company Dordrecht, Boston, Lancaster and Tokyo
D Behavioural and Social Sciences	Martinus Nijhoff Publishers
E Engineering and Materials Sciences	Dordrecht, Boston and Lancaster
F Computer and Systems Sciences	Springer-Verlag
G Ecological Sciences	Berlin, Heidelberg, New York, London,
H Cell Biology	Paris, and Tokyo

Series C: Mathematical and Physical Sciences Vol. 203

Numerical Integration

Recent Developments, Software and Applications

edited by

Patrick Keast

Department of Mathematics, Statistics and Computing Science,
Dalhousie University, Halifax, Nova Scotia, Canada

and

Graeme Fairweather

Departments of Mathematics and Engineering Mechanics,
University of Kentucky, Lexington, Kentucky, U.S.A.

Reidel Publishing Company

Dordrecht / Boston / Lancaster / Tokyo

Published in cooperation with NATO Scientific Affairs Division

Proceedings of the NATO Advanced Research Workshop on
Numerical Integration
Recent Developments, Software and Applications
Halifax, Canada
August 11-15, 1986

Library of Congress Cataloging in Publication Data

NATO Advanced Research Workshop on Numerical Integration: Recent Developments,
    Software, and Applications (1986: Halifax, N.S.)
    Numerical integration.

    (NATO ASI series. Series C, Mathematical and physical sciences; vol. 203)
    "Proceedings of the NATO Advanced Research Workshop on Numerical Integration:
Recent Developments, Software, and Applications, Halifax, Canada, August 11–August 15,
1986"—T.p. verso.
    "Published in cooperation with NATO Scientific Affairs Division."
    Includes Index.
    1. Numerical integration—Congresses.   2. Numerical integration—Data processing—
Congresses.   I. Keast, Patrick.   II. Fairweather, Greame.   III. North Atlantic Treaty
Organization. Scientific Affairs Division.   IV. Title.   V. Series: NATO ASI series. Series C,
Mathematical and physical sciences; vol. 203.
    QA299.3.N38   1986   511   87–9685
    ISBN 90–277–2514–4

Published by D. Reidel Publishing Company
P.O. Box 17, 3300 AA Dordrecht, Holland

Sold and distributed in the U.S.A. and Canada
by Kluwer Academic Publishers,
101 Philip Drive, Assinippi Park, Norwell, MA 02061, U.S.A.

In all other countries, sold and distributed
by Kluwer Academic Publishers Group,
P.O. Box 322, 3300 AH Dordrecht, Holland

D. Reidel Publishing Company is a member of the Kluwer Academic Publishers Group

Dedicated

to

Philip Rabinowitz

on the occasion of his sixtieth birthday, August 14, 1986

PREFACE

This volume contains refereed papers and extended abstracts of papers presented at the NATO Advanced Research Workshop entitled 'Numerical Integration : Recent Developments, Software and Applications', held at Dalhousie University, Halifax, Canada, August 11-15, 1986. The Workshop was attended by thirty-six scientists from eleven NATO countries. Thirteen invited lectures and twenty-two contributed lectures were presented, of which twenty-five appear in full in this volume, together with extended abstracts of the remaining ten.

It is more than ten years since the last workshop of this nature was held, in Los Alamos in 1975. Many developments have occurred in quadrature in the intervening years, and it seemed an opportune time to bring together again researchers in this area. The development of QUADPACK by Piessens, de Doncker, Uberhuber and Kahaner has changed the focus of research in the area of one dimensional quadrature from the construction of new rules to an emphasis on reliable robust software. There has been a dramatic growth in interest in the testing and evaluation of software, stimulated by the work of Lyness and Kaganove, Einarsson, and Piessens. The earlier research of Patterson into Kronrod extensions of Gauss rules, followed by the work of Monegato, and Piessens and Branders, has greatly increased interest in Gauss-based formulas for one-dimensional integration. The work of Rabinowitz and Mantel, together with work of Grundmann and Moeller, of Lyness and Keast, and of Haegemans and Piessens, has stimulated much research into the structure of optimal quadrature formulas for multiple integrals by Genz and Malik, Espelid, Sorevik, Cools and Haegemans and many others. While extrapolation methods for one dimensional quadrature have received less attention in the last ten years, much work has been done on extrapolation techniques for multiple integrals by Genz, Lyness, Kahaner and Wells, Hollosi and others. But perhaps potentially the most exciting development in quadrature over the past decade has arisen from the new computer architectures. Vector and parallel machines will almost certainly change the way in which one looks at algorithms and their implementation. The workshop included papers on all of the above topics, with lectures presented by most of the researchers named. These talks not only summarised the main developments of the past ten years, but should also serve to stimulate further research, and to forge productive links between various research groups.

The papers in this volume fall into four broad categories, and are listed under the headings: Theoretical aspects of one dimensional quadrature; Theoretical aspects of multiple quadrature; Algorithms, software and applications; and Classification and testing. In addition to containing material on software classification, the paper by David Kahaner in the last section summarises the Workshop, and puts it in perspective with reflections on the state of quadrature in 1975 and the progress that has been made since.

The contributors to this volume unanimously agreed to dedicate the proceedings to Philip Rabinowitz who celebrated his sixtieth birthday on August 14, 1986, during the Workshop. Dr. Rabinowitz' contributions to numerical integration span more than twenty-five years, with publications in almost every aspect of the area. The book "Numerical Integration", which he co-authored with P. J. Davis, was the first comprehensive source-book in one dimensional integration. The second book by these authors, now in its second edition, is already the classic reference for researchers in the field. Almost everyone who attended the Workshop could point to some part of his or her research which had been stimulated by Dr. Rabinowitz' results.

We wish to acknowledge the financial support received from the NATO Science Committee. We are also grateful to the Faculty of Graduate Studies, and the Office of Research Services, Dalhousie University, and to the College of Arts and Science, the College of Engineering and the Graduate School at the University of Kentucky for additional financial support. We also wish to thank Elizabeth Hampton, Brian Moses, Patsy Penrod and Bob Crovo for assistance during the preparation this volume; Carlos Cacola for the photographic record of the highlights of the meeting, from which the photograph of Philip Rabinowitz appearing in this book was taken; and the referees who cooperated so willingly with us in keeping to very strict deadlines : they have helped to ensure that the papers are of a uniformly high technical standard. The original idea for the Workshop came from Julian Gribble while he was a member of the Faculty at Dalhousie University. We gratefully acknowledge Jules' early efforts in organizing the meeting. Finally our thanks go to all contributors and participants who made the Workshop a success.

<table>
<tr><td>Patrick Keast</td><td>Graeme Fairweather</td></tr>
<tr><td>Halifax</td><td>Lexington</td></tr>
</table>

TABLE OF CONTENTS

Software Classification and Testing

THE CONVERGENCE OF NONINTERPOLATORY
PRODUCT INTEGRATION RULES

Philip Rabinowitz
Weizmann Institute of Science
Rehovot 76100
Israel

ABSTRACT. Two types of results on the convergence of noninterpolatory (or nonpolynomial) product integration rules are proved. This first concerns rules based on generalized piecewise polynomial interpolation. The second concerns rules involving modified moments for which there is a double limit process. First, convergence of an iterated limit is proved under assumptions similar to those required for polynomial product integration. Then, convergence of the double limit is shown under more stringent assumptions.

1. INTRODUCTION

Product integration rules have been around for at least forty years since Beard [7] and were used over thirty years ago by Young [49,50] for the numerical solution of integral equations. Since then, these rules have been studied extensively by many authors, including Elliott and Paget [11,12,21], Sloan and Smith [37,38,42,43,44,45,46,52], de Hoog and Weiss [10], Dagnino [8], Lubinsky and Sidi [19], Schneider[34], the author [27,28,32,51] and others. For their application to the numerical solution of integral equations, see Baker [6], Atkinson [2,3,4], Schneider [33,35,36], Sloan [39,40,41], Spence [47], Töpler and Volk [48] and others. Most of the work on product integration has dealt with either polynomial (or interpolatory) product integration or piecewise polynomial product integration. In the present paper we shall deal with a more general approach which includes as special cases the usual implementations of polynomial product integration based on modified moments. This approach was also used by the author in the numerical treatment of Cauchy principal value integrals [30,31]. However, by adopting a more general approach, we end up with weaker theorems. Before discussing the approach based on modified moments, we shall give some results for piecewise polynomial product integration which generalize those in [32] since this also falls in the category of noninterpolatory product integration.

1

P. Keast and G. Fairweather (eds.), Numerical Integration, 1–16.
© *1987 by D. Reidel Publishing Company.*

In our general setting, we are given two functions, $k \in L_1(a,b)$ and $f \in F[a,b]$, where F is a family of functions defined on the finite or infinite interval $[a,b]$ such that

$$I(kf) = \int_a^b k(x)f(x)dx \qquad (1)$$

is defined. The family F will usually be either $C[a,b]$ or $R[a,b]$, the set of all (bounded) Riemann-integrable functions on the finite interval $[a,b]$. However, it may be a more restricted class of functions or a more extended class, as the case may be. In addition we are given a sequence of integration rules $\{I_n f\}$ of the form

$$I_n f = \sum_{i=1}^{m_n} w_{in} f(x_{in}) \qquad (2)$$

such that

$$I_n f \rightarrow I(wf) = \int_a^b w(x)f(x)dx \qquad (3)$$

as $n \rightarrow \infty$ for all $f \in F[a,b]$, where w is an admissible weight function on $[a,b]$, $w \in A[a,b]$, i.e. $w(x) \geq 0$, $w(x) > 0$ on a subinterval of $[a,b]$ and the moments

$$\int_a^b w(x)x^i \, dx < \infty, \qquad i = 0,1,2,\dots \; .$$

Using this information plus some additional information which we shall specify in each particular case, we derive a sequence of integration rules $\{Q_n f\}$ where

$$Q_n f = \sum_{i=1}^{m_n} w_{in}(k)f(x_{in}) . \qquad (4)$$

We now wish to establish conditions on k, f and the sequence of integration rules $\{I_n f\}$ which will ensure that

$$Q_n f \rightarrow I(kf) \text{ as } n \rightarrow \infty. \qquad (5)$$

For some situations in which the x_{in} are the zeros of

$$q_{n+r+s}(x) = (1-x)^r(1+x)^s p_n(x;w), \qquad r,s \in \{0,1\}$$

where $\{p_n(x;w)\}$ is the sequence of orthonormal polynomials with respect to some $w \in A[a,b]$, and the $w_{in}(k)$ are interpolatory, i.e. are chosen so that (4) is exact for all $f \in P_{n+r+s-1}$, the set of polynomials of degree $< n+r+s$, Sloan and Smith [37,42,43,44,45,52] and Rabinowitz [27,28,51] have derived conditions on k ensuring convergence of $\{Q_n f\}$ to $I(kf)$ for all $f \in F[a,b]$ or $f \in C[a,b]$. Furthermore, they have also shown convergence of the companion rule

$$|Q_n|f = \sum_{i=1}^{m_n} |w_{in}(k)| f(x_{in}) \qquad (6)$$

to $I(|k|f)$. We shall show how these rules fit into our framework but as indicated above, we shall not get the same convergence results. Our results will be weaker when the conditions on k and f are similar and will only be the same under more stringent conditions on k.

In Section 2, we shall extend a result in [32] on the convergence of a sequence of generalized piecewise polynomial product integration rules (GP³IR's) which was proved there for $f \in C[a,b]$ or $f \in PC[a,b]$, the set of all piecewise continuous functions on $[a,b]$, to the case $f \in R[a,b]$. We shall also prove a second result which removes one of the restrictions of the GP³IR's but requires that f be much smoother. In Section 3, we shall approximate $I(kf)$ by a linear combination of modified moments, $M_j(k)$, of k with respect to a given sequence of orthonormal polynomials,

$$M_j(k) = \int_a^b k(x)p_j(x;w)dx . \qquad (7)$$

These modified moments can also be expressed as Fourier coefficients of the function

$$K(x) = k(x)/w(x) \qquad (8)$$

since

$$M_j(k) = \int_a^b w(x)K(x)p_j(x;w)dx .$$

Consequently, the quality of our results will depend on the smoothness of $K(x)$ as was observed previously [27,28,44,45,52].

2. GENERALIZED PIECEWISE POLYNOMIAL PRODUCT INTEGRATION RULES

Since in the present section we shall need two subscripts to label our variables, we shall suppress their dependence on n. Consider now the partition

$$\infty < a = t_0 < t_1 < \cdots < t_n = b < \infty$$

and define $h_i = t_i - t_{i-1}$, $i = 1, \ldots, n$, and $H_n = \max h_i$. Now partition each subinterval $[t_{i-1},t_i]$ as follows,

$$t_{i-1} \equiv x_{i0} \leq x_{i1} < \cdots < x_{i,m_i} \leq x_{i,m_i+1} \equiv t_i ,$$

where $m_i \geq 1$. The points x_{ij}, $j = 1, \ldots, m_i$, $i = 1, \ldots, n$ are called grid points, and will be the abscissas of the integration rule. Two grid points coincide if $x_{i,m_i} = t_i = x_{i+1,1}$. We define $h_{ij} = x_{i,j+1} - x_{ij}$, $j = 0, \ldots, m_i$. Note that h_{i0} or h_{i,m_i} vanish if $x_{i1} = t_{i-1}$ or $x_{i,m_i} = t_i$, respectively.

We now define a product integration rule based on the grid points x_{ij},

$$Q_n f = \sum_{i=1}^{n} \sum_{j=1}^{m_i} w_{ij} f_{ij}, \tag{9}$$

where $f_{ij} = f(x_{ij})$, the weights w_{ij} are given by

$$w_{ij} = \int_{t_{i-1}}^{t_i} l_{ij}(x) k(x) dx, \quad i = 1, \ldots, n, \quad j = 1, \ldots, m_i, \tag{10}$$

and

$$l_{ij}(x) = \prod_{\substack{k=1 \\ k \neq j}}^{m_i} \frac{x - x_{ij}}{x_{ij} - x_{ij}}, \quad i = 1, \ldots, n, \quad j = 1, \ldots, m_i, \tag{11}$$

are the fundamental Lagrange polynomials.

An alternative expression for $Q_n f$ may be obtained if we define the Lagrange interpolating polynomial on each subinterval $[t_{i-1}, t_i]$,

$$L_i(x) = \sum_{j=1}^{m_i} l_{ij}(x) f_{ij}. \tag{12}$$

Then it follows from (9) and (10) that

$$Q_n f = \sum_{i=1}^{n} \int_{t_{i-1}}^{t_i} L_i(x) k(x) dx,$$

so that if we define

$$P_n(x) = \begin{cases} L_i(x) & \text{if } t_{i-1} < x \leq t_i, \\ L_1(a) & \text{if } x = a, \end{cases} \tag{13}$$

then we obtain $Q_n f = I(kP_n)$. Clearly P_n is a piecewise polynomial of degree at most $M_n - 1$, where $M_n = \max_{1 \leq i \leq n} m_i$, so that $P_n \in PC[a,b]$. If $x_{i,m_i} = t_i = x_{i+1,1}$ for $1 \leq i \leq n-1$, then $P_n \in C[a,b]$.

Now, for a general distribution of the points x_{ij}, $Q_n f$ may fail to converge to $I(kf)$, even if $f \in C[a,b]$, $k(x) \equiv 1$, $H_n \to 0$ and M_n is bounded. This is shown by Example 1 in the Appendix in [32]. The problem is that the fundamental Lagrange polynomials $l_{ij}(x)$ restricted to the interval $[t_{i-1}, t_i]$ may not be uniformly bounded as $n \to \infty$. To ensure that

$$|l_{ij}(x)| \leq L , \qquad t_{i-1} \leq x \leq t_i , \tag{14}$$

for all i, j, n, we shall require that $h_{ij} \geq dh_i$ for some constant d, with $0 < d < 1$, for all $i = 1, \ldots, n$, $j = 1, \ldots, m_i - 1$, and that $M_n \leq \bar{M}$ for all n. In that case we have

$$|l_{ij}(x)| = \left| \prod_{\substack{k=1 \\ k \neq j}}^{m_i} \frac{x - x_{ik}}{x_{ij} - x_{ik}} \right| \leq \frac{h_i^{m_i - 1}}{(dh_i)^{m_i - 1}} \leq \frac{1}{d^{\bar{M} - 1}} \equiv L , \qquad t_{i-1} \leq x \leq t_i .$$

Note that there is no restriction on h_{i0} or h_{i,m_i}. The condition (14) ensures that the weight w_{ij} is bounded by

$$|w_{ij}| = \left| \int_{t_{i-1}}^{t_i} l_{ij}(x) k(x) dx \right| \leq L \int_{t_{i-1}}^{t_i} |k(x)| dx , \tag{15}$$

a result that we shall frequently find useful.

We shall say that a sequence of rules $\{Q_n f\}$ given by (9) and (10) is a sequence of *generalized piecewise polynomial product integration rules* (GP³IR's) if $h_{ij} \geq dh_i$, where $0 < d < 1$, for $i = 1, \ldots, n$, $j = 1, \ldots, m_i - 1$, $H_n \to 0$ as $n \to \infty$ and $M_n \leq \bar{M}$ for all n. For such a sequence of rules it was shown in [32] that $Q_n f \to I(kf)$ for all $f \in C[a,b]$, and in fact for all $f \in PC[a,b]$.

We now show that $Q_n f \to I(kf)$ for all $f \in R[a,b]$. To this end we use the generalization in [29] of a theorem of Polya [9, p. 131]. This generalization is to $I(wf)$ where $w \in A[a,b]$. However, the proof extends to $I(kf)$ for any $k \in L_1(a,b)$. Before stating the theorem, we need the following definitions [9, p. 131]. Let J designate the union of a finite number of intervals (disjoint or not) located in $[a,b]$ and let $m(J)$ be the sum of lengths of the individual intervals of J. The notation $\sum_J |w_{ij}|$ will designate the sum taken over those w_{ij} for which $x_{ij} \in J$. Set

$$\Delta(J) = \limsup_{n \to \infty} \sum_J |w_{ij}| .$$

The set function $\Delta(J)$ is called semicontinuous if for any sequence $J_1 \supset J_2 \supset \cdots$ with $m(J_n) \to 0$, we have $\lim_{n \to \infty} \Delta(J_n) = 0$. We now have the following lemma.

Lemma 1. If $k \in L_1(a,b)$ and $\lim_{n \to \infty} Q_n f = I(kf)$ holds for all $f \in C[a,b]$ then it holds for all $f \in R[a,b]$ if and only if $\Delta(J)$ is semicontinuous.

The proof of this lemma is the same as that in [29] with $w \in A[a,b]$ replaced by k.

With the help of Lemma 1, we can prove the following theorem.

Theorem 1. Let $k \in L_1(a,b)$ and let $\{Q_n f\}$ be a sequence of GP^3IR's. Then $Q_n f \to I(kf)$ as $H_n \to 0$ for all $f \in R[a,b]$.

Proof: Since $Q_n f \to I(kf)$ for all $f \in C[a,b]$, it suffices to show that $\Delta(J)$ is semicontinuous. Let $\epsilon > 0$. We shall find $\delta > 0$ such that if $m(J_k) < \delta$, $\Delta(J_k) < \epsilon$. For any n, define A_{kn} to be the union of intervals $[t_{i-1}, t_i]$ intersecting J_k. By choosing n sufficiently large, we get $m(A_{kn}) < 2m(J_k)$. Now, by (15), $\Delta(J_k) \leq \bar{M}L \int_{A_{kn}} |k(x)| dx$. Further, since $k \in L_1(a,b)$,

$$\int_{A_{kn}} |k(x)| dx < \epsilon / \bar{M}L \quad \text{if} \quad m(A_{kn}) < \eta .$$

Hence, if $\delta = \eta/2$, $\Delta(J_k) < \epsilon$ proving our theorem.

If we wish to remove the restriction that $H_{ij} \geq dh_i$ in the definition of GP^3IR's, we have to require a much higher degree of smoothness for f to ensure convergence for all $k \in L_1(a,b)$ as in the following theorem.

Theorem 2. Let $k \in L_1(a,b)$ and let $\{Q_n f\}$ be a sequence of GP^3IR's without the restriction that $h_{ij} \geq dh_i$. Then if $f \in C^{\bar{M}}[a,b]$, $Q_n f \to I(kf)$ as $H_n \to 0$.

Proof: We have that

$$I(kf) - Q_n f = \sum_{i=1}^{n} \int_{t_{i-1}}^{t_i} k(x)[f(x) - L_i(x)] dx .$$

Now, by the well known result on Lagrange interpolation,

$$f(x) - L_i(x) = \prod_{k=1}^{m_i} (x - x_{ik}) f^{(m_i)}(\xi_i(x)), \quad t_{i-1} < x, \xi_i(x) < t_i .$$

If we define $B_p = \max_{a \leq x \leq b} |f^{(p)}(x)|$ and $B = \max_{0 \leq p \leq \bar{M}} B_p$, then

$$|I(kf) - Q_n f| \leq \sum_{i=1}^{n} \int_{t_{i-1}}^{t_i} |k(x)| \prod_{k=1}^{m_i} |x - x_{ik}| B_{m_i}$$

$$\leq B \sum_{i=1}^{n} h_i^{m_i} \int_{t_{i-1}}^{t_i} |k(x)|dx \leq B H_n \int_a^b |k(x)|dx \longrightarrow 0.$$

This ends our treatment of piecewise polynomial product integration. In the next section, we discuss an approach to general noninterpolatory product integration.

3. GENERAL PRODUCT INTEGRATION

We now consider a more general situation in which we have a sequence of integration rules $\{I_n f\}$ which we wish to use for product integration. These integration rules are associated with a weight function $w \in A[a,b]$ in the sense that $I_n f \longrightarrow I(wf)$ for all $f \in F[a,b]$. We now proceed as follows: We approximate f by a finite segment of its Fourier expansion in the orthonormal polynomials $p_n(x;w)$,

$$f(x) \simeq \sum_{j=0}^{N} a_j p_j(x;w) \tag{16}$$

where

$$a_j = (f,p_j) = \int_a^b w(x)f(x)p_j(\bar{x};w)dx . \tag{17}$$

The second stage in the approximation is to replace a_j by

$$\hat{a}_{jn} = [f,p_j]_n = I_n(fp_j) = \sum_{i=1}^{m_n} w_{in} f(x_{in})p_j(x_{in};w). \tag{18}$$

Then

$$I(kf) \simeq \sum_{j=0}^{N} \hat{a}_{jn} M_j(k) \equiv Q_n^N f \tag{19}$$

where the $M_j(k)$ are given by (7). If we use orthogonal polynomials $\tilde{p}_j(x;w)$ instead of orthonormal polynomials, then we must multiply a_j by $h_j^{-1} = (\tilde{p}_j,\tilde{p}_j)$ and $\hat{a}_{jn}$ by h_j^{-1} or $[\tilde{p}_j,\tilde{p}_j]_n^{-1}$, depending on whether we have an analytic expression for h_j or not.

The evaluation of $M_j(k)$ for various families of orthogonal polynomials is discussed by many authors. Thus Piessens and Branders [22-25] have given recurrence relations for $M_j(k)$ for a variety of functions k when $[a,b]=[-1,1]$, $w(x)=(1-x^2)^{-1/2}$ and $\tilde{p}_j(x;w)=T_j(x)$, the Chebyshev polynomial of the first kind. Lewanowicz [17,18] has shown how to derive such recurrence relations when k satisfies a linear differential equation with polynomial coefficients and

$w(x)=(1-x^2)^{\lambda-1/2}$, $\lambda>-1/2$, so that $\widetilde{p}_j(x;w)=C_j^\lambda(x)$, the Gegenbauer or ultraspherical polynomial [9, p. 36]. Of course, the modified moments $M_j(k)$ are precisely the Fourier coefficients of $K(x)$, $I(wKp_j)$. In particular, if $[a,b]=[-1,1]$ and $w(x)\equiv1$ so that $\widetilde{p}_j(x;w)=P_j(x)$, the Legendre polynomial, then the modified moments coincide with the Fourier coefficients. For the case $w(x)=(1-x)^\alpha(1+x)^\beta$, the Jacobi weight function, Paget [21] has given formulas for the Fourier coefficients of $K(x;\lambda)=\exp(i\lambda x)$, λ real, $\log|x-\lambda|$, $-1<\lambda<1$, and $|x-\lambda|^s$, $s>-1$, $-1<\lambda<1$ and an algorithm to evaluate the sum $\sum c_j M_j(wK)$. For additional modified moments with respect to Jacobi and Laguerre polynomials, see Gatteschi [15].

The quality of the approximation (19) depends on the smoothness properties of k and f as well as on the accuracy of the integration rules $I_n f$. For some special choices of $Q_n f$, the approximation (19) turns out to be identical with that arrived at by polynomial product integration which has been treated extensively in the literature [11,12,27,28,32,37,42-46,51,52]. This occurs when $N=n-1$ and $I_n f$ is the n-point Gauss integration rule with respect to any $w\in A[a,b]$ or when $I_n f$ is either the n-point Gauss, the $(n+1)$-point left or right Radau or the $(n+1)$-point Lobatto rule with respect to the generalized smooth Jacobi weight function defined as follows:

Definition. w is a generalized smooth Jacobi weight function on $[-1,1]$, $w\in GSJ$, if

$$w(x)=\psi(x)(1-x)^\alpha(1+x)^\beta\prod_{k=1}^m\left|t_k-x\right|^{\gamma_k} \tag{20}$$

where $-1<t_1<\cdots<t_m<1$, $\alpha,\beta,\gamma_k>-1$, $k=1,\ldots,m$ and $\psi>0$, $\psi\in C[-1,1]$ and the modulus of continuity ω of ψ satisfies

$$\int_0^1\omega(\psi;t)t^{-1}dt<\infty.$$

In particular, if $w(x)=(1-x^2)^{-1/2}$ and $I_n f$ is the n-point Gauss-Chebyshev rule, we get the product integration rule studied by Sloan [37,38] while if $I_n f$ is the $(n+1)$-point Lobatto-Chebyshev rule, we get the modified Clenshaw-Curtis rule used extensively in QUADPACK [26] and studied by Sloan and Smith [42,43] and Piessens and Branders [23,25]. In all the above cases, Q_n^N is exact for all $f\in P_{n-1}$ and we have a single limit process. In the general case, we have a double limit since N and n can grow independently. Convergence is then much more difficult to prove and requires imposing strict conditions on k as we shall see in Theorem 5. On the other hand, if we consider an iterated limit where we fix N and let n tend to infinity so that

$$Q_n^N f \;\rightarrow\; \sum_{j=0}^N a_j M_j(k),$$

and then let $N\rightarrow\infty$, then we can prove that the iterated limit

$$\lim_{N \to \infty} \lim_{n \to \infty} Q_n^N f = I(kf) \tag{21}$$

for all $f \in F[a,b]$ provided that $[f,p_j]_n \to (f,p_j)$ for all $f \in F[a,b]$ and all j and provided that we impose some 'natural' conditions on k. These conditions are the same as those required in [28,44,51] for the convergence of polynomial product integration for all $f \in R[a,b]$.

Our first theorem involves any $w \in A[a,b]$ and applies also when the interval $[a,b]$ is infinite. It uses the weighted L_2-norm $\| g \|_{2,w}$ defined by

$$\| g \|_{2,w}^2 = \int_a^b w(x) g^2(x) dx ,$$

and the space $L_{2,w} = \{ g : \| g \|_{2,w} < \infty \}$. The condition on k or K is that given in [44, Theorem 1].

Theorem 3. Let $w \in A[a,b]$, $K = k/w \in L_{2,w}[a,b]$ and $f \in L_{2,w}[a,b] \cap F[a,b]$. Let $I_n(fp_j) \to (f,p_j)$ for all j and all $f \in F[a,b]$. Then (21) holds.

Proof: Let $S_N f(x) = \sum_{j=0}^N a_j f(x)$. Then, since $f \in L_{2,w}$, $\| f - S_N f \|_{2,w} \to 0$ as $N \to \infty$. Furthermore

$$|I(kf) - I(kS_N f)| \leq \| K \|_{2,w} \| f - S_N f \|_{2,w} \to 0$$

as $N \to \infty$. Fix N sufficiently large so that $|I(kf) - I(kS_N f)| < \epsilon/2$. We now show that $|I(kS_n f) - Q_n^N f| = |\sum_{j=0}^N (a_j - \hat{a}_{j,n}) M_j(k)| < \epsilon/2$ for $n > n_N$. We just choose n_N so that for all $n > n_N$

$$\max_{0 \leq j \leq N} |a_j - \hat{a}_{j,n}| < \epsilon / [2(N+1) \max_{0 \leq j \leq N} |M_j(k)|] .$$

Thus $|I(kf) - Q_n^N f| < \epsilon$ for all $n > n_N$.

In [44, Theorem 1], it is proved that if $K \in L_{2,w}$ and $I_n f$ is the n-point Gauss integration rule with respect to $w \in A[a,b]$ over the finite interval $[a,b]$, then $Q_n^{n-1} f \to I(kf)$ for all $f \in R[a,b]$, and similarly the companion rule $|Q_n^{n-1}|f \to I(|k|f)$. In Theorem 3, the conclusion (21) is much weaker but so are the hypotheses. Thus, we allow $[a,b]$ to be an infinite interval. Furthermore, we may let $F[a,b]$ include certain improperly integrable functions, for instance, when a is finite, $f \in M_d(a;k)$ as defined, for example, in [32]. In general, in the finite case, $F[a,b] = R[a,b]$ or $C[a,b]$, so that all we require is that $I_n f \to I(wf)$ for all $f \in F[a,b]$ rather than the more complicated condition in the theorem.

In the next theorem, we set $[a,b] = [-1,1]$ and $w \in GSJ$. We can then relax our condition on k to that given in [28,51] with $r = s = 0$. Here we work with the standard L_p norms and spaces.

Theorem 4. Let $w \in GSJ$ as given by (20) and let

$$v(x) = (1-x)^A (1+x)^B \prod_{k=1}^{m} |t_k - x|^{C_k}$$

where

$$A = -\max[(2\alpha+1)/4, 0]$$

$$B = -\max[(2\beta+1)/4, 0]$$

$$C_k = -\max[\gamma_k/2, 0], \quad k = 1, .., m.$$

Let $kv \in L_p[a,b]$ for some $p, 1 < p < \infty$, and let $f/v \in L_q[a,b]$ where $p^{-1} + q^{-1} = 1$. Let $f \in F[a,b]$ and assume that $I_n(fp_j) \rightarrow (f,p_j)$ for all j and all $f \in F[a,b]$. Then (21) holds.

Proof: By Badkov [5], the values of A, B and C_k, $k = 1, \ldots, m$, are chosen so that

$$\|(f - S_N f)/v\|_q \rightarrow 0 \quad \text{as} \quad N \rightarrow \infty.$$

Hence, by the Hölder inequality

$$|I(kf) - I(kS_N f)| \leq \|kv\|_p \|(f - S_N f)/v\|_q \rightarrow 0.$$

The rest of the proof proceeds as in the proof of Theorem 3.

Note that in Theorems 3 and 4 we do not have a double limit situation. That is, given $\epsilon > 0$, we cannot find N_0 and n_0 such that for all $N > N_0$ and all $n > n_0$, $|I(kf) - Q_n^N f| < \epsilon$. However, in the next theorem, this will indeed hold. This theorem which generalizes Theorem 3 in [8] restricts k considerably. Before stating it, we rewrite $Q_n^N f$ in Lagrangian form as in [8] and elsewhere, namely

$$Q_n^N f = \sum_{i=1}^{m_n} w_{in}^N f(x_{in}) \tag{22}$$

where

$$w_{in}^N = w_{in} S_N K(x_{in}), \quad K = k/w. \tag{23}$$

We are now ready to state our theorem.

Theorem 5. Assume that w and k are such that $S_N K(x)$ converges uniformly to $K(x) = k(x)/w(x)$ in $[a,b]$. If $I_n g$ and $F[a,b]$ are such that

1. $I_n(gp) \rightarrow I(wgp)$ for all $g \in F[a,b]$ and all polynomials p.

2. $\sum_{i=1}^{m_n} |w_{in}| \leq B$ for all n.

3. If $f \in F[a,b]$, then $|f|$ is bounded and $\sup|f| = M_f$,

in particular, if $F[a,b] = C[a,b]$ or $R[a,b]$, $[a,b]$ is finite and $I_n f \rightarrow I(wf)$ for all $f \in F[a,b]$, then for all $f \in F[a,b]$, the double limit converges,

$$\lim_{\substack{N \to \infty \\ n \to \infty}} Q_n^N f = I(kf). \tag{24}$$

If $F[a,b] = C[a,b]$ or $R[a,b]$ where $[a,b]$ is finite and in addition $\sum_{i=1}^{m_n} |w_{in}| \rightarrow I(w)$, in particular if all $w_{in} > 0$, then the companion rule converges to $I(|k|f)$.

Proof: Since $S_n K$ converges uniformly to K, we can find an N_0 such that $\| S_N K - S_{N_0} K \|_\infty$ is small for all $N > N_0$. We now have that

$$\left| I(wKf) - Q_n^N f \right| \leq \sum_{i=1}^{m_n} |w_{in}| \, \| S_N K(x_{in}) - S_{N_0} K(x_{in}) \| |f(x_{in})|$$

$$+ \left| I_n(S_{N_0} Kf) - I(w S_{N_0} Kf) \right| + \left| I(w(K - S_{N_0} K)f) \right|.$$

The first term is less than $B M_f \, \| S_N K - S_{N_0} K \|_\infty$ and the third term is less than $I(w) M_f \, \| S_N K - S_{N_0} K \|_\infty$, so that these terms are small for sufficiently large N_0. Finally, the second term will be small for all $n > n_0$, proving the first part of the theorem.

To prove the theorem for the companion rule, we write

$$\left| I(w|K|f) - \sum_{i=1}^{m_n} |w_{in}^N| f(x_{in}) \right| \leq \sum_{i=1}^{m_n} |w_{in}| \, \big| |S_N K(w_{in})| - |S_{N_0} K(x_{in})| \big| |f(x_{in})|$$

$$+ \left| \sum_{i=1}^{m_n} |w_{in}| \, |S_{N_0} K(x_{in})| - I_n(w|S_{N_0} K|f) \right|$$

$$+ \left| I_n(|S_{N_0} K|f) - I(w|S_{N_0} K|f) \right| + I(wf \, \big| |S_{N_0} K| - |K| \big|).$$

As before, the first and last terms are small for $N > N_0$ and N_0 sufficiently large. The third term goes to zero as $n \rightarrow \infty$ by our assumptions on $I_n f$. The additional hypothesis ensures that the second term goes to zero as $n \rightarrow \infty$ since (cf. [29])

$$\sum_{i=1}^{m_n} (|w_{in}| - w_{in}) |S_{N_0} K(x_{in})| f(x_{in}) \leq \| S_{N_0} Kf \|_\infty \sum_{i=1}^{m_n} (|w_{in}| - w_{in}) \rightarrow 0.$$

We conclude by listing some sufficient conditions on w and K which ensure that $S_N K \to K$ uniformly as $N \to \infty$. We first introduce some notation, where A is an arbitrary positive constant.

$f \in BF[-1,1]$ if f is of bounded variation on $[-1,1]$.

$f \in LD(\lambda)$, $\lambda > 0$, if the modulus of continuity of f, $\omega(f;\delta)$, satisfies $\omega(f;\delta) \leq A \log^\lambda \delta^{-1}$ as $\delta \to 0^+$.

$f \in H(\mu)$, $\mu > 0$, if $f \in C^{[\mu]}[-1,1]$ and $\omega(f^{[\mu]};\delta) \leq A \delta^{\mu-[\mu]}$ for all $\delta > 0$.

If we define $E_n(f) = \|f - q_n\|_\infty$ where q_n is the polynomial of degree n of best uniform approximation to f, then by Jackson's theorem [16, p. 18], if $f \in C^{(r)}[-1,1]$, $E_n(f) \leq An^{-r}\omega(f^{(n)};2/(n-r))$. Hence, if $f \in H(\mu)$, $E_n(f) \leq An^{-\mu}$ for sufficiently large n.

The Lebesque constant L_n is defined by

$$L_n = \sup_{a \leq t \leq b} \int_a^b w(x) \left| \sum_{k=0}^n p_k(x;w)p_k(t;w) \right| dx .$$

It is shown in [14, IV (4.16)] that

$$\|K - S_N K\|_\infty \leq (1 + L_N)E_N(K).$$

Since it has been stated in [1], that for $[a,b] = [-1,1]$ and $w(x) = (1-x)^\alpha (1+x)^\beta$, $\alpha,\beta \geq -1/2$,

$$L_n \leq A \sup_{-1 \leq x \leq 1} \left(\log n + \frac{n^{\alpha+1/2}}{(n\sqrt{1-x})^{\alpha+1/2}+1} + \frac{n^{\beta+1/2}}{(n\sqrt{1+x})^{\beta+1/2}+1} \right)$$

it follows that for any $\epsilon > 0$, if $\alpha = \beta = -1/2$ and $K \in LD(1+\epsilon)$ or if $\gamma = \max(\alpha,\beta) > -1/2$ and $K \in H(\gamma + \frac{1}{2} + \epsilon)$, then we have uniform convergence.

For $w \in GSJ$, we have the weaker result that if $K \in H(\sigma)$ where $\sigma = \max(\alpha+1,(\gamma_k+1)/2, k = 1, \ldots, m, \beta+1)$, then we have uniform convergence. This follows from the inequality [14, Proof of Th. IV.4.7]

$$L_n \leq A \max_{-1 \leq x \leq 1} (\lambda_n(w,x))^{1/2}$$

where $\lambda_n(w,x)$ is the Christoffel function and the estimate in [20]

$$\lambda_n(w,x) \sim \frac{1}{n} \left[\sqrt{1-x} + \frac{1}{n} \right]^{2\alpha+1} \prod_{k=1}^m \left[|t_k - x| + \frac{1}{n} \right]^{\gamma_k} \left[\sqrt{1+x} + \frac{1}{n} \right]^{2\beta+1} .$$

Some additional sufficient conditions for uniform convergence on the interval [-1,1] are:

$K \in H(1/2+\epsilon)$ and $w(x) \geq m(1-x^2)^{-1/2}$ [14, IV.1.4]

$K \in BV[-1,1] \cap LD(2+\epsilon)$ and $m(1-x^2)^{-1/2} \leq w(x) \leq M(1-x^2)^{-1/2}$ [14, IV.1.5]

$K \in BV[-1,1] \cap C[-1,1]$ and $w(x) \leq M(1-x^2)^{-1/2}$ [14, IV.5.2].

For infinite intervals, we have the following results from [13, p. 211]:

If $[a,b] = [0,\infty)$, $w(x) = e^{-x} x^{\alpha}$, $\alpha > -1$ and K is analytic in a parabola around the real axis with its focus at the origin or if $[a,b] = (-\infty,\infty)$, $w(x) = e^{-x^2}$ and K is analytic in a strip whose central axis is the real axis, then $S_N K \to K$ uniformly as $N \to \infty$. In both cases, K must satisfy certain growth conditions.

REFERENCES

1. S.A. Agahanov and G.I. Natanson, Approximation of functions by Fourier-Jacobi sums, *Soviet Math. Dokl.* **7** (1966) 1-4.

2. K.E. Atkinson, The numerical solution of Fredholm integral equations of the second kind, *SIAM J. Numer. Anal.* **4** (1967) 337-348.

3. K.E. Atkinson, The numerical solution of Fredholm integral equations of the second kind with singular kernels, *Numer. Math.* **19** (1972) 248-259.

4. K.E. Atkinson, *A Survey of Numerical Methods for the Solution of Fredholm Integral Equations of the Second Kind*, SIAM Publications, Philadelphia, 1976.

5. V.M. Badkov, Convergence in the mean and almost everywhere of Fourier series in polynomials orthogonal on an interval, *Math. USSR Sbornik* **24** (1974) 223-256.

6. C.T.H. Baker, *The Numerical Treatment of Integral Equations*, Clarendon Press, Oxford, 1977.

7. R.E. Beard, Some notes on approximate product integration, *J. Inst. Actu.* **73** (1947) 356-416.

8. C. Dagnino, Extended product integration rules, *BIT* **23** (1983) 488-499.

9. P.J. Davis and P. Rabinowitz, *Methods of Numerical Integration*, Second Edition, Academic Press, New York, 1984.

10. F. de Hoog and R. Weiss, Asymptotic expansions for product integration, *Math. Comp.* **27** (1973) 295-306.

11. D. Elliott and D.F. Paget, Product-integration rules and their convergence, *BIT* **16** (1976) 32-40.

12. D. Elliott and D.F. Paget, The convergence of product integration rules, *BIT* **18** (1978) 137-141.

13. A. Erdélyi, W. Magnus, F. Oberhettinger and F.G. Tricomi, *Higher Transcendental Functions*, Vol. II, McGraw Hill, New York, 1953.

14. G. Freud, *Orthogonale Polynome*, Birkhäuser, Basel, 1969.

15. L. Gatteschi, On some orthogonal polynomial integrals, *Math. Comp.* **35** (1980) 1291-1298.

16. D. Jackson, *The Theory of Approximation*, Amer. Math. Soc., New York, 1930.

17. S. Lewanowicz, Construction of a recurrence relation for modified moments, *J. Comp. Appl. Math.* **5** (1979) 193-206.

18. S. Lewanowicz, Recurrence relations for modified moments, *Rev. Téc. Ing., Univ. Zulia* **8** (1985) 49-60.

19. D.S. Lubinsky and A. Sidi, Convergence of product integration rules for functions with interior and endpoint singularities over bounded and unbounded intervals, *Math. Comp.* **46** (1986) 229-245.

20. P. Nevai, Mean convergence of Lagrange interpolation. III, *Trans. Amer. Math. Soc.* **282** (1984) 669-698.

21. D.F. Paget, Generalized Product Integration, Ph.D. Thesis, Univ. Tasmania, Hobart, 1976.

22. R. Piessens and M. Branders, The evaluation and application of some modified moments, *BIT* **13** (1973) 443-450.

23. R. Piessens and M. Branders, Computations of oscillating integrals, *J. Comp. Appl. Math.* **1** (1975) 153-164.

24. R. Piessens and M. Branders, Numerical solution of integral equations of mathematical physics, using Chebyshev polynomials, *J. Comp. Phys.* **21** (1976) 178-196.

25. R. Piessens and M. Branders, Modified Clenshaw-Curtis method for the computation of Bessel function integrals, *BIT* **23** (1983) 370-381.

26. R. Piessens, E. de Doncker-Kapenga, C.W. Überhuber and D.K. Kahaner, QUADPACK, *A Quadrature Subroutine Package*, Series in Computational Math. **1**, Springer-Verlag, Berlin, 1983.

27. P. Rabinowitz, The convergence of interpolatory product integration rules, *BIT* **26** (1986) 131-134.

28. P. Rabinowitz, On the convergence of interpolatory product integration rules based on Gauss, Radau and Lobatto points, to appear in *Israel J. Math.*

29. P. Rabinowitz, On the convergence of closed interpolatory integration rules based on the zeros of Gegenbauer polynomials, to appear in *J. Comp. Appl. Math.*

30. P. Rabinowitz, On a stable Gauss-Kronrod algorithm for Cauchy principal value integrals, to appear in *Comp. & Maths. with Appls.*

31. P. Rabinowitz, Some practical aspects in the numerical evaluation of Cauchy principal value integrals, to appear in *Int. J. Comp. Math.*

32. P. Rabinowitz and I.H. Sloan, Product integration in the presence of a singularity, *SIAM J. Numer. Anal.* **21** (1984) 149-166.

33. C. Schneider, Numerische Behandlung schwachsingulärer Fredholmschen Integralgleichungen zweiten Art, *ZAMM* **59** (1979) T77-79.

34. C. Schneider, Produktintegration mit nicht-äquidistanten Stützstellen, *Numer. Math.* **35** (1980) 35-43.

35. C. Schneider, Product integration for weakly singular integral equations, *Math. Comp.* **36** (1981) 207-213.

36. C. Schneider, Produktintegration zur Lösung schwachsingulären Integralgleichungen, *ZAMM* **6** (1981) T317-319.

37. I.H. Sloan, On the numerical evaluation of singular integrals, *BIT* **18** (1978) 91-102.

38. I.H. Sloan, On choosing the points in product integration, *J. Math. Phys.* **21** (1980) 1032-1039.

39. I.H. Sloan, The numerical solution of Fredholm equations of the second kind by polynomial interpolation, *J. Integral Eq.* **2** (1980) 265-279.

40. I.H. Sloan, A review of numerical methods for Fredholm equations of the second kind, in *The Application and Numerical Solution of Integral Equations* (R.S. Anderssen, F.R. de Hoog and M.A. Lukas, Eds.), Sijthoff and Noordhoff, Alphen aan den Rijn, 1980.

41. I.H. Sloan and B.J. Burn, Collocation with polynomials for integral equations of the second kind: a new approach to the theory, *J. Integral Eq.* **1** (1979) 77-94.

42. I.H. Sloan and W.E. Smith, Product-integration with the Clenshaw-Curtis and related points. Convergence properties, *Numer. Math.* **30** (1978) 415-428.

43. I.H. Sloan and W.E. Smith, Product-integration with the Clenshaw-Curtis and related points. Implementation and error estimates, *Numer. Math.* **34** (1980) 387-401.

44. I.H. Sloan and W.E. Smith, Properties of interpolatory product integration rules, *SIAM J. Numer. Anal.* **19** (1982) 427-442.

45. W.E. Smith and I.H. Sloan, Product-integration rules based on the zeros of Jacobi polynomials, *SIAM J. Numer. Anal.* **17** (1980) 1-13.

46. W.E. Smith, I.H. Sloan and A.H. Opie, Product integration over infinite intervals. I. Rules based on the zeros of Hermite polynomials, *Math. Comp.* **40** (1983) 519-535.

47. A. Spence, Product integration for singular integrals and singular integral equations, in *Numerische Integration* (G. Hämmerlin, Ed.) pp. 288-300, ISNM 45, Birkhäuser, Basel, 1979.

48. H.-J. Töpler and W. Volk, Die numerische Behandlung von Integral-gleichungen zweiten Art mittels Splinefunktionen, in *Numerical Treatment of Integral Equations* (J. Albrecht and L. Collatz, Eds.) pp. 228-243, ISNM 53, Birkhäuser, Basel, 1980.

49. A. Young, Approximate product-integration, *Proc. Roy. Soc. London Ser. A* **224** (1954) 552-561.

50. A. Young, The application of approximate product-integration to the numerical solution of integral equations, *Proc. Roy. Soc. London Ser. A* **224** (1954) 561-573.

51. P. Rabinowitz and W.E. Smith, Interpolatory product integration for Riemann-integrable functions (submitted for publication).

52. W.E. Smith, Erdös-Turán mean convergence for Lagrange interpolation at Lobatto points, to appear in *Bull. Austr. Math. Soc.*

SOME QUADRATURE RULES FOR FINITE TRIGONOMETRIC AND RELATED INTEGRALS[*]

J. N. Lyness
Mathematics and Computer Science Division
Argonne National Laboratory
9700 South Cass Avenue
Argonne, IL 60439-4844 (USA)

ABSTRACT. This article is a partial review of some old work and some new work. Its purpose is to connect up apparently different rules and methods. In particular, an attempt is made to put the Filon Luke rules, the Euler Expansion method, and the Fast Fourier Transform in proper relative perspective to one another.
      The emphasis here is on the analytical properties of these rules, on how to classify them, and on how to arrange the calculations. There is no discussion of the relative merits of various rules and no error expressions are given.

1. INTRODUCTORY CONCEPTS

In this article, I shall be concerned with the numerical evaluation of <u>Trigonometric Integrals</u> of the form

$$(1.1) \qquad \int_A^B f(x) e^{i\sigma x} dx \qquad\qquad \sigma \text{ real}$$

in terms of function values of $f(x)$. A familiar special case of the trigonometric integral is the <u>Fourier Coefficient</u>, where a period of the circular function coincides with the integration interval, that is

$$(1.2) \qquad \sigma = 2\pi s/(B-A) \qquad\qquad s \text{ integer .}$$

The trigonometric integrals are themselves special cases of the finite integral

$$(1.3) \qquad \int_A^B f(x) e^{\beta x} dx$$

---

[*]This work was supported by the Applied Mathematical Sciences subprogram of the Office of Energy Research, U.S. Department of Energy, under contract W-31-109-Eng-38.

P. Keast and G. Fairweather (eds.), Numerical Integration, 17–33.

where β can be real or complex; many but by no means all of the results of this article apply equally to this integral. We are primarily interested in cases where $f(x)$ is a real function of a real variable x, where $f(x)$ is well behaved on the integration interval and in quadrature rules which involve abscissas on the real axis. We shall occasionally depart from these guidelines. Our use of complex variable in formulas is generally only a notational convenience to avoid writing pairs of marginally more complicated formulas.

This article does <u>not</u> consider semifinite or doubly infinite Fourier Transforms such as

$$\int_A^\infty f(x)e^{i\sigma x}dx \quad \text{or} \quad \int_{-\infty}^\infty f(x)e^{i\sigma x}dx$$

except in cases when $f(x)$ has compact support and the problem reduces to one of the earlier ones.

From this point until Section 5, we shall use the interval $[0,1]$ as our finite integration interval.

Particularly in the theory of the calculation of Fourier Coefficients, a useful concept is the <u>periodic extension</u> $\overline{f}(x)$ of $f(x)$ with respect to the integration interval. Pro tem, we take the integration interval to be $[0,1]$. When $f(x)$ is continuous, $\overline{f}(x)$ may be defined as

$$(1.5) \qquad \begin{aligned} \overline{f}(x) &= f(x) & x \in [0,1] \\ \overline{f}(1) &= \overline{f}(0) = \frac{1}{2}\left(f(1)+f(0)\right) \\ \overline{f}(x+1) &= \overline{f}(x) & \forall\, x \end{aligned}$$

Thus $\overline{f}(x)$ generally has some sort of discontinuity at integer values of x, even when $f(x)$ is very well behaved.

Another notational convenience will be to define

$$(1.6) \qquad c^{(r)}f = \int_0^1 f(x)\cos 2\pi r x\, dx$$

and to refer to this integral as a Fourier Coefficient. The well known Fourier Series may be expressed as

$$(1.7) \qquad \begin{aligned} \overline{f}(x) &= If + 2\sum_{r=1}^\infty c^{(r)}f\cos 2\pi r x + s^{(r)}f\sin 2\pi r x \\ &= \sum_{r=-\infty}^\infty (c^{(r)}f + is^{(r)}f)e^{-2\pi i r x}dx \ . \end{aligned}$$

(Since $f(x)$ is real when x is real, this final expression has zero imaginary part.)

Before embarking on an actual method, I shall introduce without proof a few formulas which give some feeling for the magnitudes of some of the quantities involved.

The Fourier Coefficient Asymptotic Expansion is

$$(1.8) \qquad c^{(r)}f \simeq \frac{f'(1)-f'(0)}{(2\pi r)^2} - \frac{f^{(3)}(1)-f^{(3)}(0)}{(2\pi r)^4} + \ldots$$

This expression usually diverges and may converge to some result other than $c^{(r)}f$. We note here that, if $f'(1) \neq f'(0)$, an acceptable numerical result for very large r requires only good numerical approximations to $f'(1)$ and $f'(0)$. We also note that if

$$(1.9) \qquad f^{(s)}(1) = f^{(s)}(0) \qquad s = 0,1,\ldots,p-2 \ ,$$

then

$$(1.10) \qquad c^{(r)}f + is^{(r)}f = 0(r^{-p+1})$$

and the Fourier series (1.7) converges faster than is generally the case. This condition on $f(x)$ at $x = 0$ and $x = 1$ has a simple interpretation in terms of the periodic continuation function $\bar{f}(x)$ defined in (1.5). It is equivalent to

$$(1.11) \qquad \bar{f}(x) \ \varepsilon \ C^{(p-1)}(-\infty,\infty) \ .$$

One can also think of this in terms of $f(x)$ being "nearly" periodic. If $f(x)$ were periodic, the functions $f(x)$ and $f(x-1)$ would coincide and $f^{(s)}(1) = f^{(s)}(0)$ for all s. Condition (1.11) states that they do not necessarily coincide but the first p-1 terms of their Taylor series expansion agree.

In most of this article we shall be treating approximations to these integrals based on a set of m+1 function values, equally spaced over the range of integration. The m-panel endpoint trapezoidal rule $R^{[m,1]}g$ plays a major role in our results. To this end we make the following definition and use the indicated abbreviation.

$$(1.12) \qquad R^{[m,1]}f = R_x^{[m,1]}f(x) = \frac{1}{m} \sum_{j=0}^{m} {}'' f(j/m)$$

$$\equiv \frac{1}{m} \ [\frac{1}{2} \ f(0) + \sum_{j=1}^{m-1} f(j/m) + \frac{1}{2} \ f(1)] \ .$$

This approximation to

$$(1.13) \qquad If = I_x f(x) = \int_0^1 f(x)dx$$

is exact when $f(x)$ is a trigonometric polynomial of degree m-1 or less.

Further information about the quality of this approximation comes from the Poisson Summation Formula

$$(1.14) \qquad R^{[m,1]}f = If + 2 \sum_{r=1}^{\infty} c^{(rm)}f \ .$$

This is very simple to prove. What it shows is that the trape-zoidal rule error, $R^{[m,1]}f$-If is likely to be smaller if the Fourier Coefficients are smaller, i.e. if the Fourier Series converges rapidly. In view of (1.10) and (1.11) this in turn depends on the continuity properties of the periodic continuation $\bar{f}(x)$.

2.   ELEMENTARY INTERPOLATORY FORMULAS

Completely in keeping with the approach followed by Numerical Analysts of the last half century, we commence by considering an interpolation function $G(x)$ for which either

$$(2.1.a) \qquad G(j/m) = g(j/m) \qquad\qquad j = 0,1,\ldots,m$$

or

$$(2.1.p) \qquad G(j/m) = \bar{g}(j/m) \qquad\qquad j = 0,1,\ldots,m$$

and we approximate functionals involving $g(x)$ by exact evaluations of the same functional involving $G(x)$. The difference between (2.1.a) and (2.1.p) is of little moment. Its principal effect is to make correla-tion between earlier and later work in the same area marginally more difficult.

We introduce below three distinct types of interpolation functions.

(a)   Trigonometric Interpolation.
Let $G(x)$ be a trigonometric polynomial of degree $m/2 - 1$ with an additional $\cos \pi m x$ term. That is,

$$(2.2) \qquad G(x) = 2 \sum_{r=0}^{m/2}{}'' \mu_r \cos 2\pi rx + \sum_{r=1}^{m/2 - 1} \nu_r \sin 2\pi rx \ .$$

The unique function $G(x)$ of this form which satisfies (2.1.p) has coefficients

$$(2.3) \qquad \mu_r = R_x^{[m,1]}\left(g(x)\cos 2\pi rx\right) \ .$$

The Fourier Coefficients of (2.2) are therefore

$$(2.4) \qquad c^{(r)}g \simeq c^{(r)}G = \mu_r \qquad\qquad r = 0,1,\ldots,m/2 - 1$$

$$= \frac{1}{2} \mu_{m/2} \qquad\qquad r = m/2$$

$$= 0 \qquad\qquad r > m/2$$

(b) <u>Periodic Type Interpolation</u>.

If we choose the linear spline $G_{s1}(x)$ satisfying (2.1.p) we find the Fourier Coefficients to be

$$(2.5) \qquad c^{(r)}g \simeq c^{(r)}G_{s1} = \tau_1(r/m)R_x^{[m,1]}\left(g(x)\cos 2\pi r x\right)$$

with $\tau_1(r/m) = \left(\sin(\pi r/m)/(\pi r/m)\right)^2$. $\tau(z)$ is referred to as an <u>attenuation factor</u>. The results for higher order splines depend on how the spline is anchored. If it is a periodic spline, then the Fourier coefficients are of the same form as for the linear spline, but the attenuation factor is different. It has been shown that if the interpolation is of a "periodic type", the result is also of form (2.8). For a "periodic type" function $G(x)$ <u>is</u> the same function of $\bar{g}(j/m)$ $j=0,\pm1,\pm2,\ldots$ as $G(x+1/m)$ is of $\bar{g}\left((j+1)/m\right)$ $j=0,\pm1,\pm2,\ldots$.

(c) <u>Local Polynomial Interpolation</u>.

Here the interpolation function is a piecewise continuous interpolating polynomial of degree d, and d must divide m. It constitutes m/d quite independent interpolating polynomials. The resulting formulas are known as the Filon-Luke formulas. These are defined in Section 4 and again in Section 6; in general the results are of a more complicated structure than (2.5) above. However, in the case d = 1, the interpolation function is the linear spline and the Fourier Coefficient is given by (2.5). This rule is sometimes called the Filon Trapezoidal rule.

In respect of (a) and (b), let me emphasize that these somewhat elegant results apply only to Fourier Coefficients. The analogous results for general trigonometric integrals are more complicated and less elegant.

3. METHODS BASED ON THE EULER EXPANSION

A general class of methods for Fourier Coefficients may be based on the <u>Euler Expansion</u>

$$(3.1) \qquad f(x) = h_{p-1}(x) + g_p(x)$$

where

$$(3.2) \qquad h_{p-1}(x) = \sum_{q=1}^{p-1} \lambda_q B_q(x)/q!$$

$$(3.3) \qquad \lambda_q = \left(f^{(q-1)}(1) - f^{(q-1)}(0)\right)$$

and $B_q(x)$ is the Bernoulli polynomial of degree q. (See Abramowitz and Stegun 1964.) When $f(x)$ is a polynomial of degree p-1, it coincides with the polynomial If + $h_{p-1}(x)$. Otherwise, this expansion has little to recommend it. For numerical approximation, it requires both the integral of $f(x)$ and its derivatives. Moreover, it is generally divergent. However, its possible utility is presaged by the fact that

if one operates on both sides with the Trapezoidal Rule Operator, one
obtains the Euler Maclaurin asymptotic expansion, which is used as a
theoretical basis for Romberg Integration.

There is a standard integral representation for $g_p(x)$ which we
shall not use. The property we are interested in is

$$(3.4) \qquad g_p^{(s)}(1) - g_p^{(s)}(0) = 0 \qquad\qquad s = 0, 1, \ldots, p-2 \; .$$

This follows from elementary properties of the Bernoulli polynomials.
Thus, while not periodic, $g_p(x)$ is one of those functions which has a
rapidly converging Fourier Series, and so is one for which the
Trapezoidal rule is efficient when used to approximate the Fourier
Coefficients. This suggests an approach based on the formula

$$(3.5) \qquad \int_0^1 f(x)e^{\beta x}dx = \int_0^1 h_{p-1}(x)e^{\beta x}dx + \int_0^1 g_p(x)e^{\beta x}dx \; .$$

The first integrand on the right is the product of a polynomial and an
exponential and so its integral can be evaluated in closed form. And,
in view of (3.4), trigonometric interpolation of $g_p(x)$ is efficient for
evaluating the final integral. In general, the resulting expressions
are inelegant and are given in Lyness 1981. In the case of the Fourier
Coefficients, $\beta = 2\pi i s$ and the resulting formulas are very simple.
Since

$$(3.6) \qquad \int_0^1 \frac{B_q(x)}{q!} e^{2\pi i s x}dx = -\left(\frac{1}{2\pi i s}\right)^q \qquad \begin{array}{l} q > 0 \\ s > 0 \end{array}$$

we find

$$(3.7) \qquad \int_0^1 f(x)e^{2\pi i s x}dx = \sum_{q=1}^{p-1} \frac{-\lambda_q}{(2\pi i q)^s} + \int_0^1 g_p(x)e^{2\pi i s x}dx \; .$$

The reader will recognize the real part of this as the asymptotic
expansion for the cosine Fourier Coefficient given in (1.8) above.

When the final integral is evaluated using the m point trapezoidal
rule, one finds the following results

$$c^{(s)}F + is^{(s)}F = \mu_0$$

$$(3.8) \qquad \begin{aligned} &= \mu_s + i\nu_s - \sum_{q=1}^{p-1} \lambda_q/(2\pi i s)^q \qquad &1 < s < m/2 \\ &= \frac{1}{2}\mu_{m/2} - \sum_{q=1}^{p-1} \lambda_q/(\pi i m)^q \\ &= -\sum_{q=1}^{p-1} \lambda_q/(2\pi i s)^q \qquad &s > m/2 \end{aligned}$$

where, as in (3.2) above

$$(3.9) \qquad \mu_s + i\nu_s = R_x^{[m,1]}\left(g_p(x)e^{2\pi i s x}\right) .$$

The set (3.8) are the exact Fourier Coefficients of

$$(3.10) \qquad F(x) = h_{p-1}(x) + G_p^{(m)}(x)$$

where $G_p^{(m)}(x)$ is the Trigonometric interpolation function of $g_p(x)$ defined in (2.2).

This paper is about calculating these integrals in terms of real function values. It must be remarked in passing that if analytic derivatives are available, this formula should be used as it stands. And also, if f(x) can be calculated for complex values of x, then what are essentially exact derivatives at x = 0 (or 1) can be calculated by interpolation from function values on a circle surrounding x = 0 (or 1). In this case the overall configuration of abscissas resembles a dumbbell. A program which uses this formula to evaluate Fourier Coefficients is available (see Giunta and Murli (1987)).

We make three relatively minor observations about (3.8).

First, as the reader will have noticed, the first of (3.8) as written is what is known as the <u>Euler Maclaurin Quadrature Rule</u>. This is because

$$\mu_0 = R_x^{[m,1]}g_p(x) = R_x^{[m,1]}\left(f(x)-h_{p-1}(x)\right)$$

$$(3.9) \qquad = R_x^{[m,1]}f - \sum_{q=1}^{p-1}\lambda_q R_x^{[m,1]}B_q(x)/q!$$

$$= R_x^{[m,1]}f - \sum_{q=1}^{p-1}\lambda_q \overline{B}_q /m^q q!$$

where $\overline{B}_q$ is the Bernoulli number when $q \neq 1$ and $\overline{B}_1 = 0$.

Secondly, if f(x) is a polynomial of degree $\overline{p}-1$ or greater, then we know from the Euler expansion that f(x) = If + $h_{p-1}(x)$ and so $g_p(x)$ = If and is constant. In this case formulas (3.8) are exact whatever the value of m, as are the corresponding formulas for the trigonometric integrals.

Finally, if f(x) is a trigonometric polynomial of degree m/2 - 1 or less, we find $h_{p-1}(x) = 0$, $G_p^{(m)}(x) = g_p(x) = f(x)$ and approximations (3.8) are exact. When $\overline{f}(x)$ is $C^\infty[-\infty,\infty]$, $h_{p-1}(x) = 0$ and the theory becomes redundant, the method reducing to the unmodified use of the trigonometric interpolation function. This, of course, is generally accepted as the best option in this case.

4. QUADRATURE RULES DERIVED FROM THE EULER EXPANSION

In this section we look into the question of using the formula of the previous section but with inaccurate derivatives. First we take an open view in which the source of the derivative approximations is ignored, but it is simply assumed that their accuracy degrades in a familiar way. Then we look in detail at the structure of the rules obtained by approximating the derivatives using polynomial approximation. We recover some previously known rules and we suggest some new rules.

Perhaps the most important property of formulas (3.8) is that, under normal use, they are relatively insensitive to inaccuracies in the derivatives. This aspect of these formulas is analyzed in considerable detail in Lyness 1974. At first glance, it appears that an incorrect value of λ_q might affect (3.8) significantly. What one must remember is that the values of μ_s and ν_s depend on the values of λ_q. This is because $g_p(x)$ can be evaluated only after λ_q has been assigned. So, if one arbitrarily alters λ_q one must recalculate μ_s and ν_s. Bearing this in mind it turns out that λ_q for large q actually plays a relatively small role in (3.8) as its explicit role (as coefficient of $(2\pi i s)^{-q}$) is partly counterbalanced by its implicit role in μ_s and ν_s.

Another way of looking at the same effect is to recall that $h_{p-1}(x)$ is simply used as a subtraction function. In principle, we could replace $h_{p-1}(x)$ by any polynomial function, say

$$(4.1) \qquad \tilde{h}_{p-1}(x) = \sum_{q=1}^{p-1} \tilde{\lambda}_q B_q(x)/q!$$

note the identity

$$(4.2) \qquad \int_0^1 f(x)e^{\beta x}dx = \int_0^1 \tilde{h}_{p-1}(x)e^{\beta x}dx + \int_0^1 \tilde{g}_p(x)e^{\beta x}dx ,$$

carry out the second integral analytically and use trigonometric interpolation for the third. If $\tilde{h}_{p-1}(x)$ is not particularly close to $h_{p-1}(x)$, the penalty will be that the final integral is not particularly amenable to trigonometric approximation and a higher value of m would be required for comparable accuracy.

Incidentally, among its other virtues, (3.8) is a formula for which it is quite easy to gauge the accuracy. The final values of μ_s and ν_s, those with s near m/2, give a good indication of the absolute accuracy of all the approximations including those with much higher values of s. Another simple estimate is given in Lyness 1974.

The present author believes that, in practice, a user should divorce the calculation of derivatives from the rest of the calculation. However, in this section, we are interested in the construction of quadrature rules. We shall now proceed on the assumption that function values will be used to calculate approximations to derivatives, and that these should be considered as part of the overall quadrature rule.

A standard approximation to a set of derivatives $f^{(q)}(z)$ q=0,1,... may be based on the polynomial of degree d which interpolates f(x) at $x = t_0, t_1, \ldots, t_d$. We denote this polynomial by

(4.3) $\phi_d(x;f;t_0,t_1,\ldots,t_d)$ or $\phi_d(x;f;\vec{t})$

or simply by $\phi(x)$ if no confusion is likely to arise. We set

(4.4) $\vec{f} = \left(f(t_0),f(t_1),\ldots,f(t_d)\right)^T$

and

(4.5) $\vec{a} = \left(\phi(0),\phi'(0),\phi''(0)/2!,\ldots,\phi^{(d)}(0)/d!\right)$,

the Taylor coefficients of the interpolating polynomial, $\phi(x)$, and we have

(4.6) $\vec{f} = V\vec{a}$; $v_{j,k} = t_j^k$

where V is a Vandermonde matrix. To obtain a set of derivatives, one would solve these equations and set

(4.7) $\tilde{f}^{(q)}(0) = a_q q! = \phi^{(q)}(0;f;t_0,t_1,\ldots,t_d)$.

The same calculation can be carried out using (general) divided differences, or by using extrapolation employing the extended T-table.

<u>Definition 4.8.</u> Let $m \geq 1$, $p \geq 0$ be integers and $\vec{t}^{(i)}$, i=0,1, be vectors each of which has d_i+1 elements. The quadrature rule

$$Q_\beta[m;p;\vec{t}^{(0)};\vec{t}^{(1)}]$$

is the quadrature rule approximation to $\int_0^1 f(x)e^{\beta x}dx$ obtained from (4.1) and (4.2) using

(4.8) $\chi_q = \phi_{d_1}^{(q-1)}(1;f;\vec{t}^{(1)}) - \phi_{d_0}^{(q-1)}(0;f;\vec{t}^{(0)})$.

This rule with $\beta = 2\pi i$ coincides with (3.8) if we replace λ_q by $\tilde{\lambda}_q$ given by (4.8) (which induces a change in the values of μ_s,v_s^q). Clearly, this rule is a weighted linear sum of function values $f(x)$ at three sets of abscissas, namely

(4.9) $\frac{j}{m}(j=0,1,\ldots,m)$; $t_j^{(0)}(j=0,1,\ldots,d_0)$; $t_j^{(1)}(j=0,1,\ldots,d_1)$.

The abscissas in each set are distinct, but the same abscissa may occur in two or all three of these sets. This opens up the possibility of many different quadrature rules. We shall prove two simple theorems about these rules and give two sets of examples.

<u>Theorem 4.9.</u> When $d_0 \geq p-1$ and $d_1 \geq p-1$, the rule Q defined in (4.8) is of polynomial degree p-1.

Proof. When $d_j \geq p-1$, and $f(x)$ is a polynomial of degree $p-1$, $f(x)$ coincides with any interpolating polynomial $\phi_{d_j}(x)$. Thus, $\lambda_q = \tilde{\lambda}_q$ and $h_{p-1}(x) = \tilde{h}_{p-1}(x)$. For such a polynomial, in view of the Bernoulli expansion, $\tilde{g}_p(x) = g_p(x) = $ If which is constant. This constant is integrated correctly by trigonometric interpolation for all $m \geq 1$.

Theorem 4.10. When $d_0 = d_1$ and $t_j^{(1)} = 1 + t_j^{(0)}$ the rule Q defined in (4.8) is of trigonometric degree $\tilde{m}/2 - 1$.

Proof. When $f(x)$ is a trigonometric polynomial, $f(1+x) = f(x)$ and, when $t_j^{(1)} = 1 + t_j^{(0)}$ it follows that $\phi(0;f;\vec{t}^{(0)}) = \phi(1;f;\vec{t}^{(1)})$ giving $\tilde{\lambda}_q = 0$. Thus $\tilde{h}_{p-1}(x) = 0$ giving $g_p(x) = f(x)$ and application of the rule reduces to using trigonometric interpolation, based on m panels (or $m+1$ points).

Our examples are ones in which the values of t_j are chosen to be equally spaced with spacing $1/m$. Example 4.11 corresponds to using forward finite differences and backward finite differences to estimate the derivatives at $x = 0$ and at $x = 1$. Example 4.12 corresponds to using central finite differences at both ends.

Example 4.11. When we choose

$$t_j^{(1)} = j/d \qquad j = 0,1,\ldots,d$$

$$t_j^{(0)} = j/d \qquad j = 0,1,\ldots,d$$

$$m = d ; \qquad p = d+1$$

we obtain the Filon Luke Rule of degree d. This is because, in view of Theorem 4.9, the rule is of degree d and the rule uses only $d+1$ distinct function values. The Filon Luke Rule is the unique rule with this property. (m can be replaced by any factor of d, including 1 to obtain the identical rule.)

In a later section we shall suggest a way of calculating with the Filon Luke Rules. The approach there completely masks this property. Of interest here is that the Filon Luke Rule can be considered to be the closest approximation to the rule based on the Euler Expansion if one uses all the function values at one's disposal to approximate the derivatives.

However, this rule does not have positive trigonometric degree. Essentially, the approximation of the derivatives has destroyed this property.

Example 4.12. The trigonometric degree properties may be retained at extra cost. This is by using central differences instead of forward and backward differences. For example, with d even and any $m > 1$, we could choose

$$t_j^{(0)} = \left(j - \frac{d}{2}\right)/m \qquad j = 0,1,\ldots,d$$

$$t_j^{(1)} = 1 + t_j^{(0)}$$

$$p = d+1 .$$

This rule employs m+1 equally spaced points spanning the integration interval and an additional d points, d/2 at each end. Like the Filon Luke rule, it is of polynomial degree d. It is also of trigonometric degree m/2 - 1. For the purposes of discussion, let us take m = d. Then this rule differs from the Filon Luke rule in that central differences in place of forward and backward differences are used to approximate the derivatives. One would expect a better performance, both because central differences are usually more accurate and because of the additional trigonometric degree. However, the cost factor is double.

The situation with respect to cost alters completely if we consider the M-copy of these rules. That is, integrating over the interval [0,M] we apply this rule in each section [k,k+1]. The Filon Luke rule uses Mm+1 function values; rule 4.12 uses Mm+1+d function values. In fact, the M copy rule, scaled to [0,1], takes the form

$$(4.13) \qquad G_\beta[dM, d+1, \vec{u}^{(1)}, \vec{u}^{(0)}]$$

with $u_j^{(1)}-1 = u_j^{(0)} = (j - d/2)/dM$. Thus, the M copy version of rule 4.12 is of trigonometric degree dM/2 - 1, while the M copy version of rule (4.11) is of trigonometric degree 0 (unless β = 0 in which case it is of degree M-1). The trigonometric degrees given here are general. In fact, in many instances the strict trigonometric degree is higher than that stated here, particularly when β = $2\pi$is, the Fourier Coefficient case. This is closely related to the fact that the m-panel trapezoidal rule gives approximations to each Fourier Coefficient $c^{(s)}f$ of different degrees; dropping from m-1 for s = 0 to m/2 for s = m/2 - 1.

Finally, we remark that rule (4.13) with β = 0 is a variant of Gregory's rule. Gregory's rule itself is

$$(4.14) \qquad G_\beta[m, d+1, \vec{u}^{(1)}, \vec{u}^{(0)}]$$

with $m \geq d$,

$$u_j^{(0)} = 1 - u_j^{(1)} = j/m \qquad j = 0,1,\ldots,d$$

and, of course, β = 0. When m = d, this is the Filon Luke Rule (4.11) of polynomial degree d, which coincides with the Newton Cotes Rule of degree d (or d+1 if d is even).

The number of different rules one can construct and the connections between them seem to be endless. As mentioned above, the present author feels that in general a user should deal with formulas (3.8) directly.

5. AN ELEMENTARY FORMALISM FOR INTERPOLATORY RULES

If one knows the Moments, there are many ways of constructing an interpolatory quadrature rule. The most straightforward is to do the following. Let the integration interval be [A,B] and suppose we wish to construct a rule of degree d using the d+1 points $x_0, x_1, \ldots, x_d$. Let c be the point about which we have the moments. That is

(5.1) $m_k = \int_A^B (t-c)^k w(t) dt$ $k = 0, 1, \ldots, d$.

Let

(5.2) $f_j = f(x_j)$ $j = 0, 1, \ldots, d$

and define the Taylor coefficients about c as

(5.3) $a_k = f^{(k)}(c)/k!$ $k = 0, 1, \ldots, d$.

When $f(x)$ is a polynomial of degree d, its Taylor expansion terminates and we have

(5.4) $f_j = \sum_{k=0}^{d} (x_j-c)^k a_k$ $j = 0, 1, \ldots, d$

while the exact integral is

(5.5) $\int_A^B f(t)w(t)dt = \sum_{k=0}^{d} \int_A^B a_k(t-c)^k w(t)dt = \sum_{k=0}^{d} a_k m_k$.

We denote the vectors in (5.1), (5.2) and (5.3) by $\vec{m}$, $\vec{f}$ and $\vec{a}$ and the elements of the Vandermonde matrix V in (5.4) by

(5.6) $V_{j,k} = (x_j-c)^k$.

Then (5.5) and (5.7) may be written

(5.7) $\vec{f} = V\vec{a}$; $Q(f) = \vec{m}^T \vec{a}$

and, eliminating $\vec{a}$ we have

(5.8) $Qf = \vec{m} V^{-1} \vec{f}$.

In deriving this we have assumed that $f(x)$ is a polynomial of degree d. If we drop this restriction, (5.8) represents a weighted sum of function values of the form

(5.9) $Qf = \sum_{j=0}^{d} w_j f(t_j)$

which coincides with the integral when $f(x)$ is a polynomial of degree d. There is only one such sum, which is the interpolatory rule of degree d. The weights are clearly the elements of

(5.10) $\vec{w} = \vec{m}^T V^{-1}$.

One's approach to interpolatory quadrature would depend on the scope of the problem set. For a one-shot calculation, one would carry out the calculation as indicated in (5.7) solving a set of linear equations for a and then calculating the scalar product for Qf. In a situation in which many different functions were to be integrated, one might follow convention and calculate the weights first.

A non-trivial comment concerns the numerical stability. It is sometimes argued, quite correctly, that methods involving the numerical evaluation of high-order derivatives (like $f^{(k)}(c)$ above) may introduce an amplification of rounding error. The reader is quite correct to seek reassurance on this point. At a detailed level, a brief analysis indicates that $f^{(k)}(c)$ may be inaccurate; however, it appears in formula (5.8) with a small coefficient, actually precisely small enough to render this inaccuracy below the significance level. However, the larger questions of loss of significance which may occur in solving large Vandermonde systems remain. If this method is applied to obtain rules of high degree, say $d = 20$ or 30, a proper analysis of this point would be required. Our interest is in applications with $d = 4$, 6 or perhaps 8 only.

6. THE CONSTRUCTION OF THE FILON-LUKE RULES AND VARIANTS

To put the method of the previous section into effect, one needs numerical values of A, B the integration limits, d the degree, and $d+1$ abscissas $x_0, x_1, \ldots, x_d$. In addition, one has to assign the parameter c which is required for the moments (5.1) and for the Vandermonde matrix (5.6). The results are independent of c. In this section we treat in turn the Newton Cotes rule, the finite Laplace transform, and the Filon-Luke rules.

The Newton Cotes Rule: The eight-panel degree 9 rule over the interval $[0,1]$ may be obtained by setting $d = 8$, $x_j = j/8$ $j=0,1,\ldots,8$, $c = 0$, $m_k = 1/(k+1)$ and setting the elements of the Vandermonde matrix V to x_j^k. The weights if required are the elements of $\vec{w} = \vec{m}^T V^{-1}$. This produces an interpolatory rule of degree 8. However, due to symmetry considerations, this rule is actually of degree 9.

The Finite Laplace Transform: This is a finite quadrature with weight function $e^{\beta x}$. We take β to be real pro tem. In fact, β may be complex and all the results given here are valid.

We treat the integral

$$(6.1) \qquad \int_0^1 f(x)e^{\beta x}dx$$

and take the value of parameter c to be zero. Then the moments are given by

(6.2) $M_k(\beta) = \int_0^1 x^k e^{\beta x} dx = e^\beta k! \; \mu_k(\beta)$

where the normalized moment μ_k is given by the formula

(6.3) $\mu_k(-\gamma) = \frac{1}{\gamma^{k+1}} \{e^\gamma - \sum_{q=0}^{k} \gamma^q/q!\} = \sum_{s=0}^{\infty} \gamma^s/(s+k+1)!$.

It is annoying that one requires a different formula for large $|\beta|$ than
for small $|\beta|$. But apart from this, these formulas are extremely
simple. The procedure to work out interpolatory rules is just the same
as in the Newton Cotes rule case; however, one must replace $m_k = 1/(k+1)$
by $m_k = M_k(\beta)$. For subsequent convenience we write down two scaled
versions of this rule. We note first that

(6.4) $M_k(\beta;0,H) = \int_0^H x^k e^{\beta x} dx = H^{k+1} M_k(\beta H)$

where $M_k(\beta H)$ is given by (6.2) and (6.3) above. The quadrature rule, in
expanded form, is

(6.5) $\int_0^H f(t) e^{\beta t} dt = \sum_{j=0}^{d} \sum_{k=0}^{d} H^{k+1} M_k(\beta H)(V^{-1})_{kj} f(x_j)$

where as before $(V^{-1})_{kj}$ is an element of the inverse of the Vandermonde
matrix V for which $V_{j,k} = x_j^k$. A translation through a distance A of
this formula gives

(6.6) $\int_A^{A+H} f(t) \; e^{\beta t} dt = e^{\beta A} \sum_{j=0}^{d} \sum_{k=0}^{d} H^{k+1} M_k(\beta H)(V^{-1})_{k,j} f(A+x_j)$.

This is the interpolatory rule of degree d for weight function $e^{\beta t}$ over
the interval $(A,A+H)$ using abscissas $A+x_j$ $j=0,1,\ldots,d$.
    The next stage depends entirely on the circumstance that $e^{\beta x}$ in
interval $[0,H]$ is a constant multiple of $e^{\beta x}$ is interval $[A,A+H]$.
Because of this, weights are constant multiples of corresponding weights
and a rule which resembles an M copy version can be constructed by
setting

(6.7) $\int_A^{A+MH} f(t) e^{\beta t} dt = \sum_{\ell=0}^{m-1} \int_{A+\ell H}^{A+(\ell+1)H} f(t) e^{\beta t} dt$

and using (6.6) for each interval. We find

(6.8)
$$\int_A^{A+MH} f(t)e^{\beta t}dt$$
$$= \sum_{j=0}^{d} \sum_{k=0}^{d} H^{k+1} M_k(\beta H)(V^{-1})_{kj}\left[\sum_{\ell=0}^{M-1} f(A+\ell H+x_j)e^{\beta(A+\ell H)}\right].$$

When $\beta = 0$ and $x_j = jH/d$ $j=0,1,\ldots,d$, this is the M copy version of the d panel Newton Cotes rule. When $\beta = 0$, for arbitrary x_j it is the M copy of an interpolatory rule. However, for $\beta \neq 0$, it is the analogue of the M copy. It applies a differently scaled version of the same rule to each of the M intervals as appropriate.

The Filon Luke rules can be obtained from these by analytic continuation in the variable β. We replace β by $i\sigma$ where σ is real, and take $x_j = jH/d$. We note that $(V^{-1})_{kj}$ is real and the function values are real, but $M_k(i\sigma H)$ given by (2.2) and (2.3) has a real and imaginary part as does $e^{\beta A + \beta \ell H}$. If one wants to reconstruct Filon's rule (with $d = 2$) in its standard form, besides handling the complex multiplication without error, one has to assemble the contributions from x_j with $j = 0$ and $j = H$ as both contribute to internal abscissas $A+\ell H$. The author feels strongly that this exercise should not be attempted. One should use (6.6) with $\beta = i\sigma$ and let the computer do what it is good at, sorting out real and imaginary parts of products of complex numbers.

7. THE FAST FOURIER TRANSFORM

The term Fast Fourier Transform (or FFT) is used often indiscriminately to describe both a series of methods and a series of computer programs which rapidly evaluate transform approximations. I believe that much confusion will be avoided if I describe at this point what an FFT program does and how it may be used to provide approximations to Fourier Coefficients.

A typical program for cosine and sine coefficients of a real function works in the following way. One provides an even integer m, usually a power of 2, together with m items of data $d_0, d_1, \ldots, d_{m-1}$. The routine returns m items of data, these being $a_0, a_1, \ldots, a_{m/2}$, $b_1, b_2, \ldots, b_{m/2-1}$. These are defined by

(7.1)
$$a_s = \frac{2}{m}\sum_{\ell=0}^{m-1} d_\ell \cos 2\pi \ell s/m \qquad s = 0,1,\ldots,m/2$$

$$b_s = \frac{2}{m}\sum_{\ell=0}^{m-1} d_\ell \sin 2\pi \ell s/m \qquad s = 1,2,\ldots,m/2 - 1.$$

The author can hardly emphasize too strongly that this is all the program does. Anyone with an elementary knowledge of Fortran could program this himself. It is a matrix multiplication involving an $m \times m$ matrix all of whose elements are either $\cos 2\pi \ell s/m$ or $\sin 2\pi \ell s/m$. If coded in a straightforward way, it would involve m^2 multiplications and m^2 additions. However, this calculation may be coded so that it involves only $m \log_2 m$ multiplications. When m is very large (say 1024) this involves a saving in data manipulation time by a large factor (of 100). When m is smaller (say 16), this relative time saving factor is much smaller (being 4). One should note that it is only the data manipulation time which is reduced. The time spent calculating function values d_ℓ or processing the output a_s, b_s remains the same.

The relevance to Fourier Coefficients is as follows. If one sets

(7.2)
$$d_0 = \frac{1}{2} \left(f(A) + f(B) \right)$$
$$d_\ell = f\left(A + \frac{\ell}{m} (B-A) \right) = f(x_\ell) \qquad \ell = 1, 2, \ldots, m-1 \ ,$$

the quantities returned by this routine are

(7.3)
$$a_s + i b_s = \frac{2}{m} \sum_{\ell=0}^{m} {}'' \; f\left(A + \frac{\ell}{m} (B-A) \right) e^{2\pi i \ell s/m}$$

which may be expressed in the form

(7.4)
$$a_s + i b_s = \frac{2 e^{-\frac{2\pi i s A}{B-A}}}{B-A} \left[\frac{B-A}{m} \sum_{\ell=0}^{m} {}'' \; f(x_\ell) e^{2\pi i s x_\ell / B-A} \right] .$$

The quantity in square brackets is the m-panel trapezoidal rule approximation to the Fourier Coefficient integral representation

(7.5)
$$\int_A^B f(x) e^{2\pi i s x/(B-A)} dx \qquad s = 0, 1, \ldots, m/2 \ .$$

Thus, if one has available these m+1 equally spaced function values, and requires to calculate a set of m real (or m/2 complex) Fourier Coefficients using the Trapezoidal rule, this F.F.T. routine is a very useful way to do it.

It is sometimes thought that the FFT is useful only when one is using the trapezoidal rule to approximate Fourier Coefficients. This erroneous view is now less widespread than it used to be. In evaluating Filon Luke approximations to the Fourier Coefficients, it may be used to advantage. The Filon Luke rule, using M panels each having d+1 points is given by (6.8) with $\beta = 2\pi i s/MH$. One requires d+1 (possibly only d) sums of the form

$$(7.6) \qquad \sum_{\ell=0}^{M-1} f(A+\ell H+x_j)e^{\frac{2\pi i s}{MH}(A+\ell H)} \qquad\qquad \text{s integer .}$$

For the j-th sum of this form, one may set

$$(7.7) \qquad d_\ell = f(A+\ell H+x_j) \qquad\qquad \ell = 0,1,\ldots,M-1$$

in the FFT routine; this returning a_s and b_s given by (7.1) and (7.2) with m = M. The sum in (7.6) is then

$$(7.8) \qquad \frac{M}{2}(a_s+ib_s)e^{2\pi i s A/MH} .$$

Minor modification of the above remarks is necessary when $x_j = 0$ or H. Similarly, the methods mentioned in Sections 3 and 4 require sums of precisely this type, and the FFT routine can be used. After this major calculation, further computations are generally required to obtain the quantity of interest. The present author has not yet encountered a serious method for Fourier Coefficients for which the FFT cannot be effectively employed.

8. CONCLUDING REMARKS

The reader may have found some of the results presented in this paper to be new to him or her. Some may in fact be new. But the purpose of this article is to connect up several approaches to the same problem which have tended in the past to be treated in isolation. My goal is to highlight the underlying simplicity of an area which can appear to be extraordinarily complicated in detail.

REFERENCES

M. Abramowitz and I. A. Stegun, Eds., Handbook of Mathematical Functions with Formulas, Graphs, and Mathematical Tables, Nat. Bur. Standards Appl. Math. Series, 55, U.S. Government Printing Office, Washington, D.C., 1964; 3rd printing with corrections, 1965, MR 29#4914; MR 31 #1400.

G. Giunta and A. Murli, 'Algorithm XXX: Lynco: To Calculate Fourier Coefficients', ACM Trans. on Math. Soft., to appear in 1987.

J. N. Lyness, 'Computational Techniques Based on the Lanczos Representation,' Math. Comp. 28 (1974) 81-123.

J. N. Lyness, Technical Memorandum 370, Applied Mathematics Division, Argonne National Laboratory, Argonne, IL, 1981.

J. N. Lyness, 'The Calculation of Trigonometric Fourier Coefficients,' Jour. Comp. Phys. 54, pp. 57-73, 1984.

MODIFIED CLENSHAW-CURTIS INTEGRATION AND APPLICATIONS TO NUMERICAL COMPUTATION OF INTEGRAL TRANSFORMS

R. Piessens
Computer Science Department
University of Leuven
Celestijnenlaan 200 A
3030 Heverlee, Belgium

Abstract

The main difficulty in using modified Clenshaw-Curtis integration for computing singular and oscillatory integrals is the computation of the modified moments. In this paper we give recurrence formulae for computing modified moments for a number of important weight functions.

Keywords

Modified moments, Clenshaw-Curtis quadrature, Chebyshev polynomials, recurrence relations.

1. Introduction and background

When the integrand of an integral contains 'critical factors' such as singularities, nearby poles and oscillatory factors, classical numerical integration methods, such as Romberg quadrature, Gauss-Legendre quadrature and Clenshaw-Curtis quadrature are unreliable and not efficient. There is a need of a subroutine package especially designed for computing difficult integrals. A research project for constructing such a package is in progress. All integrators of this package are based on 'modified Clenshaw-Curtis quadrature' (MCCQ).

MCCQ is a straightforward extension of the classical Clenshaw-Curtis quadrature method [3]. For the computation of

$$I = \int_a^b f(x)dx \quad , \quad a,b \text{ finite or infinite} \tag{1}$$

MCCQ requires the following steps :

P. Keast and G. Fairweather (eds.), Numerical Integration, 35–51.

i. the integration interval is mapped onto [-1,1] and the integrand is written as the product of a smooth function g and a weight function w containing the 'critical factors' :

$$I = \int_{-1}^{+1} w(x)g(x)dx \qquad (2)$$

ii. the smooth function is approximated by a truncated series of Chebyshev polynomials

$$g(x) \simeq \sum_{n=0}^{N} {}' c_n T_n(x) \quad , \quad -1 \leqslant x \leqslant 1 \qquad (3)$$

where the symbol $\sum'$ indicates that the first term in the sum must be halved, and where the coefficients c_k can be computed by FFT-techniques [2,6].

iii. an approximation of the integral is obtained by

$$I \simeq \sum_{n=0}^{N} {}' c_n M_n \qquad (4)$$

where

$$M_n = \int_{-1}^{+1} w(x)T_n(x)dx \qquad (5)$$

are called 'modified moments' [5].

If $w(x) \equiv 1$, this method is the classical Clenshaw-Curtis quadrature. The main problem of MCCQ is the computation of the modified moments. It is not possible to give a general method for computing M_n for arbitrary weight function. For various useful weight functions, the modified moments can be computed using recurrence relations [1].

The practical use of recurrence relations requires an analysis of the numerical stability. It is well known that the numerical stability of forward recursion depends on the asymptotic behaviour of the modified moments M_n for $n \to \infty$, relative to the asymptotic behaviour of the solutions of the corresponding homogeneous recurrence relations.

Since

$$M_n = \int_0^{\pi} w(\cos t) \sin t \cos(nt) \, dt \qquad (6)$$

we have, if $w(x)$ is N times continuously differentiable on $(-1,1)$, the following asymptotic expansion for $n \to \infty$ (Erdélyi [4]) :

$$M_n \sim \sum_{j=0}^{[N/2]-1} (-1)^j \, [(-1)^n \, \Phi^{(2j+1)}(\pi) - \Phi^{(2j+1)}(0)] n^{-2j-2} + O(n^{-N}) \qquad (7)$$

where

$$\Phi(t) = \sin t \; w(\cos t) \; .$$

f $w(x)$ has a finite number of singularities of the type $|x-\xi|^\beta$, $|x-\xi|^\beta \mathrm{sign}(x-\xi)$, $x-\xi|^\beta \log(x-\xi)$, $|x-\xi|^\beta \log(x-\xi)\mathrm{sign}(x-\xi)$, where $\xi \in [-1,1]$, the asymptotic behaviour of M_n can be determined using asymptotic formulas for Fourier coefficients given by Lighthill [7].

Modified moments are useful for other numerical applications, such as the construction of nonclassical Gaussian quadrature formulas [5,14] and the solution of integral equations [13].

The purpose of this paper is to give recurrence relations for the computation of the modified moments and to discuss their numerical aspects for the following weight functions :

$$w_1(x) = (1-x)^\alpha (1+x)^\beta$$

$$w_2(x) = (1-x)^\alpha (1+x)^\beta \exp(-ax)$$

$$w_3(x) = (1-x)^\alpha (1+x)^\beta \ln((1+x)/2)\exp(-ax)$$

$$w_4(x) = \exp(-ax^2)$$

$$w_5(x) = (1-x)^\alpha (1+x)^\beta \exp(-a(x+1)^2)$$

$$w_6(x) = (1-x)^\alpha (1+x)^\beta \exp(-a/(x+1))$$

$$w_7(x) = (1-x)^\alpha (1+x)^\beta \exp(-a/x^2)$$

$$w_8(x) = ((1-x)^\alpha (1+x)^\beta \exp(-a/(x+1)^2)$$

$$w_9(x) = (1-x)^\alpha(1+x)^\beta \ln((1+x)/2)$$

$$w_{10}(x) = (1-x)^\alpha(1+x)^\beta \ln((1+x)/2)\ln((1-x)/2)$$

$$w_{11}(x) = |x-a|^\alpha$$

$$w_{12}(x) = |x-a|^\alpha \operatorname{sign}(x-a)$$

$$w_{13}(x) = |x-a|^\alpha \ln|x-a|$$

$$w_{14}(x) = |x-a|^\alpha \ln|x-a| \operatorname{sign}(x-a)$$

$$w_{15}(x) = (1-x)^\alpha(1+x)^\beta |x-a|^\gamma$$

$$w_{16}(x) = (1-x)^\alpha(1+x)^\beta |x-a|^\gamma \ln|x-a|$$

$$w_{17}(x) = [(x-b)^2+a^2]^\alpha$$

$$w_{18}(x) = (1+x)^\alpha J_\nu(a(x+1)/2)$$

2. Recurrence relations for modified moments

2.1 $w_1(x) = (1-x)^\alpha(1+x)^\beta$

The recurrence relation is
$$(\alpha+\beta+n+2)M_{n+1}+2(\alpha-\beta)M_n+(\alpha+\beta-n+2)M_{n-1} = 0 \ .$$

An explicit expression is given by

$$M_n = 2^{\alpha+\beta+1} \frac{\Gamma(\alpha+1)\Gamma(\beta+1)}{\Gamma(\alpha+\beta+2)} \ {}_3F_2 \left[\begin{matrix} n,-n,\alpha+1 \\[2mm] \frac{1}{2}, \alpha+\beta+2 \end{matrix} \ ; i+1 \right] \ .$$

The asymptotic expansion of M_n, $n \to \infty$ is

$$M_n \sim \sum_{k=0}^{\infty} a_k(\alpha,\beta)c(\alpha+k)n^{-2\alpha-2k-2}$$

$$+ (-1)^n \sum_{k=0}^{\infty} a_k(\beta,\alpha)c(\beta+k)n^{-2\beta-2k-2}$$

where

$$c(\alpha) = \cos[\pi(\alpha+1)]\,\Gamma(2\alpha+2)$$

and

$$a_0(\alpha,\beta) = 2^{\beta-\alpha}$$

$$a_1(\alpha,\beta) = -\frac{\alpha+2\beta+2}{3}\,2^{\beta-\alpha-2}$$

$$a_2(\alpha,\beta) = \left|\frac{4}{15} + \frac{19\alpha}{45} + \frac{\alpha^2}{9} + \frac{2\alpha\beta}{3} + \beta + \beta^2\right|2^{\beta-\alpha-5}.$$

If $\alpha < \beta$ and $\alpha \neq \frac{1}{2}$ + integer or $\alpha > \beta$ and $\beta \neq \frac{1}{2}$ + integer forward recursion is stable.

Starting values are :

$$M_0 = 2^{\alpha+\beta+1}B(\alpha+1,\beta+1)$$

$$M_1 = \frac{\beta-\alpha}{\alpha+\beta+2}\,M_0 \ .$$

where $B(\alpha,\beta)$ is the beta function.
An important application of these modified moments is the computation of integrals with algebraic end-point singularities [10,11]

$$\int_a^b (b-x)^\alpha(x-a)^\beta f(x)dx \ .$$

Special cases arise when :

i. $\alpha = \beta$
    In this case the recurrence relation is of first order and forward recursion is always stable.

ii. $\alpha = 0$
    The recurrence relation is

$$(n-1)(n+\beta+1)M_n + n(n-\beta-2)M_{n-1} = -2^{\beta+1} \ .$$

.2 $w_2(x) = (1-x)^\alpha(1+x)^\beta\exp(-ax)$

In this weight function a is real or complex, and $\alpha,\beta > -1$.

The recurrence relation is

$$\frac{a}{2} M_{n+2} - (n+\alpha+\beta+2)M_{n+1} - (2\alpha-2\beta+a)M_n$$

$$+ (n-\alpha-\beta-2)M_{n-1} + \frac{a}{2} M_{n-2} = 0 .$$

For this general recurrence relation, it is difficult to give formulas for the starting values. We consider only special cases :

i. $\alpha = 0$

Instead of a five-term homogeneous recurrence relation we can use a four-term nonhomogeneous recurrence relation

$$- a M_{n+1} + (2\beta+2n+2 - \frac{n-2}{n-1}a)M_n$$

$$+ \left[2n-\beta-1-\frac{(\beta+1-a)(n+1)}{n-1} \right]M_{n-1}$$

$$+ a \frac{n}{n-1} M_{n-2} = - \frac{2^{\beta+2}e^{-1}}{n-1} .$$

The asymptotic behaviour for $n \to \infty$ is

$$M_n \sim -2^\alpha e^{-a} n^{-2} + O(n^{-4})$$

$$-(-1)^n 2^{-\alpha}\cos(\pi\alpha)\Gamma(2\alpha+2)n^{-2\alpha-2} + O(n^{-2\alpha-4}) .$$

The starting values are

$$M_0 = f_1$$
$$M_1 = f_2-f_1$$
$$M_2 = 2f_3-4f_2+f_1$$

where

$$f_n = e^{-a} \Gamma(\alpha+n)P(\alpha+n,2a)a^{-\alpha-n}$$

where

$$P(a,x) = \frac{1}{\Gamma(a)} \int_0^x e^{-t} t^{a-1}dt$$

is the incomplete gamma function.

Forward and backward recursion are unstable, and Lozier's algorithm [8] has to be used.

ii. $\alpha = \beta = 0$

The recurrence relation is nonhomogeneous and of second order

$$\frac{a}{n+1} M_{n+1} - 2M_n - \frac{a}{n-1} M_{n-1} = \frac{2}{n^2-1} [e^{-a} + (-1)^n e^a] .$$

If $a = j\omega$ where $j = \sqrt{-1}$ and ω is real then we obtain

$$\omega^2(n-1)(n-2)M_{n+2} - 2(n^2-4)(\omega^2-2n^2+2)M_n +$$
$$\omega^2(n+1)(n-2)M_{n-2} = 24\omega\sin\omega - 8(n^2-4)\cos\omega$$

for

$$M_n = \int_{-1}^{+1} \cos(\omega x)T_n(x)dx$$

and

$$\omega^2(n-1)(n-2)M_{n+2} - 2(n^2-4)(\omega^2-2n^2+2)M_n +$$
$$\omega^2(n+1)(n-2)M_{n-2} = -24\omega\cos\omega - 8(n^2-4)\sin\omega$$

for

$$M_n = \int_{-1}^{+1} \sin(\omega x)T_n(x)dx .$$

For both recurrence relations Lozier's algorithm has to be used. Applications of these modified moments are the computation of Fourier coefficients [12]

$$\int_a^b \cos(\omega x)f(x)dx$$

and the construction of Gaussian quadrature formulas for

$$\int_0^1 \left\{ 1 + \begin{vmatrix} \cos(\omega x) \\ \sin(\omega x) \end{vmatrix} \right\} f(x)dx .$$

2.3 $w_3(x) = (1-x)^\alpha(1+x)^\beta \ln((1+x)/2)\exp(-ax)$

The parameter a may be complex and $\alpha > -2$ and $\beta > -1$.
The recurrence relation is

$$\frac{a}{2} M_{n+2} - (n+\alpha+\beta+2)M_{n+1} - (2\alpha-2\beta+a)M_n$$
$$+ (n-\alpha-\beta-2)M_{n-1} + \frac{a}{2} M_{n-2} = I_{n+1} + I_{n-1} - 2I_n$$

where

$$I_n = \int_{-1}^{+1} w_2(x)T_n dx .$$

The asymptotic expansion is

$$M_n \sim e^{-a}[-2^{\alpha-\beta-2}c(\beta+1)n^{-2\beta-4}+ \cdots]$$
$$+ (-1)^n e^a [2^{\beta-\alpha+1}c(\alpha)P_n(\alpha)n^{-2\alpha-2}+ \cdots]$$

where

$$c(\alpha) = \cos[\pi(\alpha+1)]\Gamma(2\alpha+2)$$

and

$$P_n(\alpha) = \psi(2\alpha+2)-\ln(2n) - \frac{\pi}{2} \operatorname{tg}(\pi\alpha) .$$

2.4 $w_4(x) = \exp(-ax^2)$

The recurrence relation is

$$a(n-1)M_{n+2} + 2(1-n^2-a)M_n - a(n+1)M_{n-2} = 4e^{-a}$$
$$M_n = 0 \quad \text{when} \quad n \quad \text{is odd} .$$

(see Branders [1], and Sadowski and Lozier [20]).

The asymptotic expansion for $n \rightarrow \infty$ is

$$M_n \sim 2e^{-a}n^{-2}+2(6a-1)e^{-a}n^{-4}-2(60a^2-60a+1)e^{-a}n^{-6}+ \cdots .$$

Forward recursion is not stable. Lozier's algorithm has to be used with one starting value

$$M_0 = (\pi/a)^{1/2} \operatorname{erf}(a^{1/2}) .$$

2.5 $w_5(x) = (1-x)^\alpha(1+x)^\beta\exp(-a(x+1)^2)$

In this weight function $\alpha,\beta > -1$ and a is real or complex.
The recurrence relation is

$$aM_{n+3} + 2aM_{n+2} - [a+2(\alpha+\beta+2+n)]M_{n+1} - 4(a+\alpha-\beta)M_n$$
$$- [a+2(\alpha+\beta+2-n)]M_{n-1} + 2aM_{n-2} + aM_{n-3} = 0 .$$

In the special case $\alpha = 0$, the order of the recurrence relation can be reduced.
The starting values can be expressed as generalized hypergeometric functions, for example

$$M_0 = 2^{\alpha+\beta+1} \, {}_2F_2\left[\frac{\beta+1}{2} , \frac{\beta+2}{2} ; \frac{\alpha+\beta+2}{2} , \frac{\alpha+\beta+3}{2} ; -4a \right] .$$

2.6 $w_6(x) = (1-x)^\alpha(1+x)^\beta \exp(-a/(x+1))$

The recurrence relation is

$$(\alpha+\beta+3+n)M_{n+2} + 2(a+2\alpha+2+k)M_{n+1}$$
$$+ 2(3\alpha-\beta-2a+1)M_n + 2(a+2\alpha+2-k)M_{n-1}$$
$$+ (\alpha+\beta+3-k)M_{n-2} = 0 .$$

Forward and backward recursion are numerically unstable. Lozier's algorithm with three starting values is a stable way to compute M_n. The starting values are

$$M_0(\alpha,\beta) = 2^{\alpha+\beta+1}e^{-a/2}\Gamma(\alpha+1)U(\alpha+1,-\beta,a/2)$$
$$M_1(\alpha,\beta) = - M_0(\alpha+1,\beta) + M_0(\alpha,\beta)$$
$$M_2 = ((\beta+2a-3\alpha-1)M_0(\alpha,\beta) - 2(2\alpha+a+2)M_1(\alpha,\beta)/(\alpha+\beta+3)) .$$

where $U(\alpha,\beta)$ and $M(\alpha,\beta)$ are Kummer's confluent hypergeometric functions.

2.7 $w_7(x) = (1-x)^\alpha(1+x)^\beta \exp(-a/x^2)$

The recurrence relation is

$$(\alpha+\beta+5+n)M_{n+4} + 2(\alpha-\beta)M_{n+3} + [4(\alpha+\beta+2+2a)+2n]M_{n+2}$$
$$+ 6(\alpha-\beta)M_{n+1} + [6(\alpha+\beta+1)-16a]M_n + 6(\alpha-\beta)M_{n-1}$$
$$+ [4(\alpha+\beta+2+2a)-2n]M_{n-2} + 2(\alpha-\beta)M_{n-3}$$
$$+ (\alpha+\beta+5-n)M_{n-4} = 0 .$$

The numerical aspects and the starting values of this recurrence relation require further study.

2.8 $w_8(x) = (1-x)^\alpha(1+x)^\beta \exp(-a/(x+1)^2)$

The recurrence relation is

$$(\alpha+\beta+4+n)M_{n+3} + 2(3\alpha-\beta+6+2n)M_{n+2}$$
$$+ (15\alpha-\beta+12+8a+5n)M_{n+1} + 4(5\alpha-\beta+2-4a]M_n$$
$$+ (15\alpha-\beta+12+8a-5n)M_{n-1} + 2(3\alpha+\beta+6+2n)M_{n-2}$$
$$+ (\alpha+\beta+4-n)M_{n-3} = 0 .$$

The numerical aspects and the starting values of this recurrence relation require further study.

2.9 $w_9(x) = (1-x)^\alpha(1+x)^\beta \ln((1+x)/2)$

The recurrence relation for the modified moments is

$$(\alpha+\beta+n+2)M_{n+1}+2(\alpha-\beta)M_n + (\alpha+\beta-n+2)M_{n-1}$$
$$= 2I_n - I_{n+1} - I_{n-1}$$

where

$$I_n = \int_{-1}^{+1} (1-x)^\alpha(1+x)^\beta T_n(x)dx \ .$$

Numerical aspects, starting values and applications are discussed in [10].

2.10 $w_{10} = (1-x)^\alpha(1+x)^\beta \ln((1+x)/2)\ln((1-x)/2)$

In this weight function $\alpha,\beta > -2$.
The recurrence relation is

$$(\alpha+\beta+n+2)M_{n+1} + 2(\alpha-\beta)M_n + (\alpha+\beta-n+2)M_{n-1}$$
$$= 2H_n - H_{n+1} - H_{n-1} - 2G_n - G_{n+1} - G_{n-1}$$

where

$$H_n = \int_{-1}^{+1} (1-x)^\alpha(1+x)^\beta \ln \frac{1-x}{2} T_n(x)dx$$

$$G_n = \int_{-1}^{+1} (1-x)^\alpha(1+x)^\beta \ln \frac{1+x}{2} T_n(x)dx \ .$$

The asymptotic expansion for $n \rightarrow \infty$ is

$$M_n \sim A_n(\alpha,\beta) + (-1)^n A_n(\beta,\alpha)$$

where

$$A_n(\alpha,\beta) = 2^{\beta-\alpha-1}C(\alpha+1)P_n(\alpha+1)n^{-2\alpha-4}$$
$$- 2^{\beta-\alpha-4}C(\alpha+2)[(2\alpha+6\beta+3)P_n(\alpha+2)+1]n^{-2\alpha-6}/3 + \cdots$$

where

$$C(\alpha) = \cos(\pi\alpha)\Gamma(2\alpha+2)$$

and

$$P_n(\alpha) = \psi(2\alpha+2) - \ln 2n - \frac{\pi}{2}\, tg(\pi\alpha) \ .$$

The starting values are

$$M_0 = 2^{\alpha+\beta+1}B(\alpha+1,\beta+1)[(\psi(\alpha+1)-\psi(\alpha+\beta+2))]$$

$$(\psi(\beta+1)-\psi(\alpha+\beta+2)-\psi'(\alpha+\beta+2)].$$

$$M_1 = M_0(\alpha,\beta+1) - M_0(\alpha,\beta).$$

where $\psi(x)$ is the psi function and $B(x,y)$ is the beta function.
Forward recursion is stable.

2.11 $w_{11}(x) = |x-a|^\alpha$

The conditions here are : $|a| < 1$ and $\alpha > -1$ or $|a| > 1$ and α arbitrary.
The recurrence relation for the modified moments is

$$\left[1 + \frac{\alpha+1}{n+1}\right] M_{n+1} - 2aM_n + \left[1 - \frac{\alpha+1}{n-1}\right] M_{n-1}$$

$$= \frac{2}{1-n^2} [(1-a)^{\alpha+1}-(-1)^n (1+a)^{\alpha+1}].$$

The most important case for practical use is where $|a| < 1$ and $\alpha > -1$. In that case,
the asymptotic expansion for $n \to \infty$ is

$$I_n \sim \sum_{k=0}^{\infty} (-1)^{k+1}(2k+1)! \, [r_k(a)+(-1)^n r_k(-a)]n^{-2k-2}$$

$$+ 2 \sum_{k=0}^{\infty} s_{2k} \cos(n \arccos a) \cos[\frac{\pi}{2}(\alpha+2k+1)]\Gamma(\alpha+2k+1)n^{-\alpha-2k-1}$$

$$- 2 \sum_{k=0}^{\infty} s_{2k+1} \sin(n \arccos a) \sin[\frac{\pi}{2}(\alpha+2k+2)]\Gamma(\alpha+2k+2)n^{-\alpha-2k-2}$$

where

$$r_0 = (1-a)^\alpha$$

$$r_1 = - \left|\frac{(1-a)^\alpha}{3!} + \frac{\alpha}{2}(1-a)^{\alpha-1}\right|$$

$$s_0 = (1-a^2)^{(\alpha+1)/2}$$

$$s_1 = a(1-\frac{\alpha}{2})(1-a^2)^{\alpha/2}$$

Forward recursion is stable.
Starting values for the recurrence relation are

$$M_0 = \frac{1}{\alpha+1} [(1-a)^{\alpha+1} + (1+a)^{\alpha+1}]$$

$$M_1 = \frac{1}{\alpha+2} [(1-a)^{\alpha+2} - (1+a)^{\alpha+2}] + aM_0$$

$$M_2 = \frac{2}{\alpha+2} [(1-a)^{\alpha+3} + (1+a)^{\alpha+3}] + 4aM_1 - (2a^2+1)M_0.$$

2.12 $w_{12}(x) = |x-a|^{\alpha}\text{sign}(x-a)$

The recurrence relation is

$$\left[1 + \frac{\alpha+1}{n+1}\right] M_{n+1} - 2aM_n + \left[1 - \frac{\alpha+1}{n-1}\right] M_{n-1}$$

$$= \frac{2}{1-n^2} [(1-a)^{\alpha+1}-(-1)^n (1+a)^{\alpha+1}].$$

If $|a| < 1$ and $\alpha > -1$, the asymptotic expansion is

$$M_n \sim \sum_{k=0}^{\infty} (-1)^{k+1}(2k+1)! \, [r_k(a)-(-1)^n r_k(-a)]n^{-2k-2}$$

$$+ 2 \sum_{k=0}^{\infty} s_{2k} \sin(n \arccos a) \sin[\frac{\pi}{2} (\alpha+2k+1)]\Gamma(\alpha+2k+1)n^{-\alpha-2k-1}$$

$$- 2 \sum_{k=0}^{\infty} s_{2k+1} \cos(n \arccos a) \cos[\frac{\pi}{2} (\alpha+2k+2)]\Gamma(\alpha+2k+2)n^{-\alpha-2k-2}$$

where the coefficients r_k and s_k are the same as for the weight function w_{11}.
Forward recursion is stable.
The starting values are

$$M_0 = \frac{1}{\alpha+1} [(1-a)^{\alpha+1} - (1+a)^{\alpha+1}]$$

$$M_1 = \frac{1}{\alpha+2} [(1-a)^{\alpha+2} - (1+a)^{\alpha+2}] + aM_0$$

$$M_3 = \frac{2}{\alpha+3} [(1-a)^{\alpha+3} - (1+a)^{\alpha+3}] + 4aM_1 - (2a^2+1)M_0.$$

2.13 $w_{13}(x) = |x-a|^{\alpha}\ln|x-a|$

The recurrence relation is

$$(n-1)(n+\alpha+2)M_{n+1} - 2a(n^2-1)M_n + (n+1)(n-\alpha-2)M_{n-1}$$

$$= 2(-1)^n (a+1)^{\alpha+1}\ln(a+1) - 2(1-a)^{\alpha+1}\ln(1-a)$$

$$- (n-1)I_{n+1} + (n+1)I_{n-1} .$$

where

$$I_n = \int_{-1}^{+1} |x-a|^\alpha T_n(x)dx \ .$$

f $|a| < 1$ and $\alpha > -1$, forward recursion is stable. Starting values are given by

$$M_0 = P(\alpha,a) + P(\alpha,-a) \ .$$

$$M_1 = aP(\alpha,a) + aP(\alpha,-a) + P(\alpha+1,-a) - P(\alpha+1,a)$$

$$M_2 = 2P(\alpha+2,a) + 2P(\alpha+2,-a) + (2a^2-1)[P(\alpha,a)+P(\alpha,-a)]$$
$$+ 4aP(\alpha+1,-a) - 4aP(\alpha+1,a)$$

where

$$P(\alpha,a) = \frac{(1+a)^{\alpha+1}}{\alpha+1} \ln(1+a) - \frac{(1+a)^{\alpha+1}}{(\alpha+1)^2} \ .$$

.14 $w_{14}(x) = |x-a|^\alpha \ln|x-a|\mathrm{sign}(x-a)$

The recurrence relation is

$$(n-1)(n+\alpha+2)M_{n+1} - 2a(n-1)(n+1)M_n + (n+1)(n-\alpha-2)M_{n-1}$$
$$= (n+1)I_{n-1} - (n-1)I_{n+1}$$
$$- 2(1-a)^{\alpha+1}\ln(1-a) - 2(-1)^n(a+1)^{\alpha+1}\ln(a+1)$$

where

$$I_n = \int_{-1}^{+1} |x-a|^\alpha \mathrm{sign}(x-a)T_n(x)dx.$$

Forward recursion is stable. The starting values are closely related to the starting values of 2.13.

.15 $w_{15} = (1-x)^\alpha(1+x)^\beta|x-a|^\gamma$

The recurrence relation is

$$(\alpha+\beta+\gamma+3+n)M_{n+2} + 2[(1-a)(\alpha+1)-(1+a)(\beta+1)-an]M_{n+1}$$
$$+ 2[(1-2a)(\alpha+1)+(1+2a)(\beta+1)-(\gamma+1)]M_n$$
$$+ 2[(1-a)(\alpha+1)-(1+a)(\beta+1)+an]M_{n-1}$$
$$+ (\alpha+\beta+\gamma+3-n)M_{n-2} = 0.$$

Forward recursion is stable. Starting values are

$$M_0 = J(\alpha,\beta,\gamma,a) + J(\beta,\alpha,\gamma,-a)$$

$$M_1 = aM_0 + J(\gamma+1,\beta,\alpha,-a) - J(\gamma+1,\alpha,\beta,a)$$

$$M_2 = \frac{1}{\alpha+\beta+\gamma+3} \{2[(1+a)(\beta+1)-(1-a)(\alpha+1)]M_1 +$$

$$[(\gamma+1)-(1-2a)(\alpha+1)-(1+2a)(\beta+1)]M_0\}$$

where

$$J(\alpha,\beta,\gamma,a) = 2^\alpha(1+a)^{\beta+\gamma+1}B(\beta+1,\gamma+1){}_2F_1(-\alpha,\beta+1; \gamma+\beta+2; \frac{1+a}{2})$$

2.16 $w_{16}(x) = (1-x)^\alpha(1+x)^\beta|x-a|^\gamma\ln|x-a|$

The recurrence relation is

$$(\alpha+\beta+\gamma+3+n)M_{n+2} + 2[(1-a)(\alpha+1)-(1+a)(\beta+1)-an]M_{n+1}$$

$$+ 2[(1-2a)(\alpha+1)+(1+2a)(\beta+1)-(\gamma+1)]M_n$$

$$+ 2[(1-a)(\alpha+1)-(1+a)(\beta+1)+an]M_{n-1}$$

$$+ (\gamma+\alpha+\beta+3-n)M_{n-2} = 2I_n - I_{n+2} - I_{n-2}$$

where

$$I_n = \int\limits_{-1}^{+1} (1-x)^\alpha(1+x)^\beta|x-a|^\gamma T_n(x)dx.$$

2.17 $w_{17}(x) = [(x-b)^2+a^2]^\alpha$

The recurrence relation is

$$\left[\frac{1}{4} + \frac{\alpha+1}{2(n+1)}\right]M_{n+2} - b\left[1 + \frac{\alpha+1}{n+1}\right]M_{n+1} + \left[a^2+b^2 + \frac{1}{2} + \frac{\alpha+1}{1-n^2}\right]M_n$$

$$- b\left[1 + \frac{\alpha+1}{1-n}\right]M_{n-1} + \left[\frac{1}{4} + \frac{\alpha+1}{2(1-n)}\right]M_{n-2}$$

$$= [(a^2+(1-b)^2)^{\alpha+1} + (-1)^n(a^2+(1+b)^2)^{\alpha+1}]/(1-n^2) \ .$$

Forward recursion is not stable.
If $\alpha = -1$, the starting values are

$$M_0 = \frac{1}{a}\left|\arctan\frac{1-b}{a} + \arctan\frac{1+b}{a}\right|$$

$$M_1 = bM_0 + \frac{1}{2} \ln \frac{(1-b)^2+a^2}{(1+b)^2+a^2}$$

$$M_2 = 4bM_1 - (2a^2+2b^2+1)M_0 + 4$$

f $b = 0$, the starting values are

$$M_0 = -2 \ln\frac{|a|}{(1+a^2)^{1/2}+1} \qquad \text{if } \alpha = -\frac{1}{2}$$

$$M_0 = \frac{2}{a} \arctan \frac{1}{a} \qquad \text{if } \alpha = -1$$

$$(1+2\alpha)M_0(\alpha) = 2(1+a^2)^\alpha + 2\alpha a^2 M_0(\alpha-1)$$

$$(3+2\alpha)M_2(\alpha) = 4(1+a^2)^\alpha - (3+2\alpha+2a^2)M_0(\alpha) \ .$$

$.18 \quad w_{18}(x) = (1+x)^\alpha J_\nu(a(x+1)/2)$

$J_\nu(x)$ is the Bessel function of the first kind and of order ν. The recurrence relation is

$$\frac{\omega^2}{16} M_{n+4} + [(n+3)(n+3+2\alpha)+\alpha^2-\nu^2 - \frac{\omega^2}{4}]M_{n+2}$$

$$+ [4(\nu^2-\alpha^2)-2(n+2)(2\alpha-1)]M_{n+1}$$

$$- [2(n^2-4)+6(\nu^2-\alpha^2)-2(2\alpha-1) - \frac{3\omega^2}{8}]M_n$$

$$+ [4(\nu^2-\alpha^2)+2(n-2)(2\alpha-1)]M_{n-1}$$

$$+ [(n-3)(n-3-2\alpha)+(\alpha^2-\nu^2- \frac{\omega^2}{4})]M_{n-2} + \frac{\omega^2}{16} M_{n-4} = 0.$$

The asymptotic expansion is

$$M_n \sim -\frac{1}{2} J_n(\omega)n^{-2} +$$

$$(-1)^n 2^{-3\nu-2\alpha-1} \frac{\omega^\nu}{\Gamma(\nu+1)} \cos[\pi(\alpha+1)]\Gamma(2\alpha+2)n^{-2\alpha-2\nu-2}.$$

forward recursion is stable only if $|a|$ is very large. For small $|a|$, Lozier's algorithm as to be used, with 6 initial values and 2 end values. starting values are

$$M_0 = G(\omega,\nu,\alpha)$$

$$M_1 = 2G(\omega,\nu,\alpha+2) - G(\omega,\nu,\alpha)$$

$$M_2 = 8G(\omega,\nu,\alpha+2) - 8G(\omega,\nu,\alpha+1) + G(\omega,\nu,\alpha)$$

$$M_3 = 32G(\omega,\nu,\alpha+3) - 48G(\omega,\nu,\alpha+2) + 18G(\omega,\nu,\alpha+1) - G(\omega,\nu,\alpha)$$

where

$$G(\omega,\nu,\alpha) = \int_0^1 x^\alpha J_\nu(\omega x)dx.$$

The computation of these integrals is described by Luke [9]. See also [19].

Conclusion

The efficiency of MCCQ depends on the computation method of the modified moments. For the computation of modified moments related to weight functions with end point singularities and/or internal singularities of algebraic and/or logarithmic type and with oscillatory factors of trigonometric and Bessel function type, recurrence relations exist and are practically useful. A further step in this research project is the construction of a software package for the computation of difficult integrals by MCCQ.

Acknowledgment

This research was supported by the 'Onderzoeksfonds' of the Catholic University of Leuven.

REFERENCES

1. M. Branders : *Toepassingen van Chebyshev-veeltermen in de numerieke integratie.* Doctoral thesis, Catholic University of Leuven, 1976.

2. M. Branders and R. Piessens : 'Algorithm 1, An extension of Clenshaw-Curtis quadrature', *J. Comp. Appl. Math.* 1, pp. 55-65, 1965.

3. C.W. Clenshaw and A.R. Curtis : 'A method for numerical integration on an automatic computer', *Num. Math.* 2, pp. 197-205, 1960.

4. A. Erdélyi : 'Asymptotic representations of Fourier integrals and the method of stationary phase, *J. Soc. Indust. Appl. Math.* 3, pp. 17-27, 1955.

5. W. Gautschi : 'On the construction of Gaussian rules from modified moments', *Math. Comp.* 24, pp. 245-260, 1970.

6. W.M. Gentleman : 'Algorithm 424, Clenshaw-Curtis quadrature', *Comm. ACM* 15, pp. 353-355, 1972.

7. M.J. Lighthill : *Fourier analysis and generalized functions*, Cambridge University Press, Cambridge 1959.

8. D.W. Lozier : 'Numerical solution of linear difference equations', report NBSIR 80-1976, Nat. Bureau of Standards, Washington, 1980.

9. Y.L. Luke : *Integrals of Bessel functions*, McGraw-Hill Book Co., New York, 1962.

0. R. Piessens and M. Branders, 'The evaluation and application of some modified moments', *BIT* 13, pp. 443-450, 1973.

1. R. Piessens, I. Mertens and M. Branders : 'Automatic integration of functions having algebraic end points singularities', *Angewandte Informatik* 2, pp. 65-68, 1974.

2. R. Piessens and M. Branders : 'Computation of oscillating integrals', *J. Comp. Appl. Math.* 1, p. 153, 1975.

3. R. Piessens and M. Branders : 'Numerical solution of integral equations of mathematical physics using Chebyshev polynomials', *J. Comp. Phys.* 21, pp. 178-196, 1976.

4. R. Piessens, M. Chawla and N. Jayarajan : 'Gaussian quadrature formulas for the numerical calculation of integrals with logarithmic singularity', *J. Comp. Phys.* 21, pp. 356-360, 1976.

5. R. Piessens, M. Branders and I. Mertens : 'The automatic evaluation of Cauchy principal values', *Angewandte Informatik*, pp. 31-35, 1976.

6. R. Piessens and M. Branders : 'Modified Clenshaw-Curtis method for the computation of Bessel integrals', *BIT* 23, pp. 370-382,1983.

7. R. Piessens and M. Branders : 'Computation of Fourier transform integrals using Chebyshev series expansions', *Computing* 32, pp. 177-186, 1984.

8. R. Piessens and M. Branders : 'Algorithm for the computation of Bessel function integrals', *Journal Comp. Appl. Math.* 11, pp. 119-137, 1984.

9. R. Piessens and M. Branders : 'A survey of numerical methods for the computation of Bessel function integrals', *Rend. Sem. Mat. Univers Politecn. Torino, Fascicolo speciale*, pp. 250-265, 1985.

0. W.L. Sadowski and D.W. Lozier : 'Use of Olver's algorithm to evaluate certain definite integrals of plasma physics involving Chebyshev polynomials', *J. Comp. Phys.* 10, pp. 607-613, 1972.

FAST GENERATION OF QUADRATURE RULES WITH SOME SPECIAL PROPERTIES

Chr. T. H. Baker & M. S. Derakhshan
Department of Mathematics
The University
Manchester M13 9PL
England

ABSTRACT. We consider fast methods, based on FFT techniques, for obtaining families of quadrature rules approximating
$$\int_0^{nh} (nh-s)^{-\frac{1}{2}} \varphi(s)ds \qquad (n = 1, 2, 3, \ldots,N; \quad N= 2^R; \quad R \epsilon Z_+).$$
The rules are fractional quadrature rules derived from approximations to $\int_0^{nh} \varphi(s)ds$ ($n \epsilon Z_+$) generated by implicit linear multistep formulae. Suitable "starting" weights, computed using the Björck−Pereyra algorithm and FFT techniques, produce formulae with good order accuracy for functions $\varphi(s)$ of the form $\varphi(s) = \varphi_0 +\varphi_1 s^{1/2} + \varphi_2 s + \varphi_3 s^{3/2} + \varphi_4 s^2 + \ldots$ as $s \to 0$. The discussion is associated with FORTRAN 77 code given in [2].

1. INTRODUCTION TO REDUCIBLE AND FRACTIONAL QUADRATURE

A k−step linear multistep method (LMM) for the initial−value problem in ordinary differential equations is defined by its first and second characteristic polynomials

(1.1)
$$\rho(\mu) = \alpha_0 \mu^k + \alpha_1 \mu^{k-1} +\ldots+ \alpha_k$$

(1.2)
$$\sigma(\mu) = \beta_0 \mu^k + \beta_1 \mu^{k-1} +\ldots+ \beta_k$$

and is here assumed to be implicit ($\beta_0 \neq 0$). The formula is of interest to us only if it is strongly−stable and consistent, which we assume. The order of the LMM is p if $\rho(e^z)/\sigma(e^z) = z + O(z^{p+1})$ as $z \to 0$. The Adams−Moulton formulae correspond to $\rho(\mu) = \mu^k-\mu^{k-1}$ and a choice of $\sigma(\mu)$ which yields order $k+1$, whilst the backward differentiation formulae (BDF$_k$ with $k \leqslant 6$) correspond to $\sigma(\mu) = \mu^k$ and a choice of $\rho(\mu)$ which yields order k.

The initial−value problem $y'(t) = \varphi(t)$ ($t \geqslant 0$, $y(0) = 0$) has solution $\int_0^t \varphi(s)ds$; when the LMM is applied to this problem the approximations

(1.3)
$$\int_0^{nh}\varphi(s)ds \simeq h \sum_{j=0}^n w_{nj} \varphi(jh) \qquad (n \geqslant \kappa)$$

arise where w_{nj} ($0 \leqslant j \leqslant \kappa-1$) are determined by the starting approximations for $\int_0^{nh} \varphi(s)ds$, $n = 0,1,\ldots,\kappa-1$ (with $\kappa \geqslant k$, in general) and
(1.4)
$$w_{nj} = \omega_{n-j} \qquad (\kappa \leqslant j \leqslant n).$$

53

P. Keast and G. Fairweather (eds.), Numerical Integration, 53–60.
© 1987 by D. Reidel Publishing Company.

The weights $\omega_0, \omega_1, \omega_2, \ldots$ are the coefficients of the expansion

(1.5) $\dot{\omega}(\mu) := \sigma(\mu^{-1})/\rho(\mu^{-1}) = \omega_0 + \omega_1 \mu + \omega_2 \mu^2 + \ldots$.

Rules (1.3) obtained in this manner are termed $\{\rho, \sigma\}$−reducible; the Gregory rules of a fixed order provide an example since they correspond to the Adams−Moulton formulas of corresponding order. We assume (1.1) and (1.2) to be supplied and introduce the notation

(1.6) $\alpha(\mu) = \alpha_0 + \alpha_1 \mu + \alpha_2 \mu^2 + \ldots + \alpha_k \mu^k$

(1.7) $\beta(\mu) = \beta_0 + \beta_1 \mu + \beta_2 \mu^2 + \ldots + \beta_k \mu^k$

so that $\sigma(\mu^{-1})/\rho(\mu^{-1}) = \beta(\mu)/\alpha(\mu)$. The coefficients $\Omega_\ell^{(1/2)}$ of the expansion

(1.8) $\Omega^{(1/2)}(\mu) := \{\omega(\mu)\}^{1/2} = \Omega_0^{(1/2)} + \Omega_1^{(1/2)} \mu + \Omega_2^{(1/2)} \mu^2 + \ldots$,

provide approximations for weakly−singular convolutions of the form

(1.9) $\{1/\sqrt{\pi}\} \int_0^{nh} (nh-s)^{-\frac{1}{2}} \varphi(s)ds \simeq h^{\frac{1}{2}} \sum_{j=0}^n W_{nj}^{(1/2)} \varphi(jh)$

where

(1.10) $W_{nj}^{(1/2)} = \Omega_{n-j}^{(1/2)}$ $(\lambda \leqslant j \leqslant n; \lambda = \lambda(p) \ \epsilon Z_+)$

and a suitable choice of "starting" weights $W_{nj}^{(1/2)}$ $(0 \leqslant j \leqslant \lambda-1)$ ensures that the approximations (1.9) are of order p for the class of functions $\varphi(s) = \varphi_0 + \varphi_1 s^{1/2} + \varphi_2 s + \varphi_3 s^{3/2} + \varphi_4 s^2 + \ldots + \Phi(s)$, where $\Phi(s)$ is smooth.

Strong stability implies that $\rho(\mu)/(\mu-1)$ is a Schur polynomial, and that $\omega(\mu)$ +c is a positive definite function where $1/c$ is the diameter of the stability disk of the LMM. (In particular, for an A−stable method ω is positive−definite, and Nevanlinna [9] calls the associated quadrature rules (1.1) (A−) "positive".) Related properties make formulae (1.3) and (1.9) of practical value in the numerical solution of Volterra and Abel equations, where smoothness and stability are relevant considerations. We refer to Lubich [7,8] and his references for theoretical details.

In [1], the authors considered the use of the FFT for the more general problem of approximating $\int_0^{nh} (nh-s)^{\nu-1} \varphi(s)ds$ for $0 < \nu < 1$. There, as here, we permit arbitrary ρ, σ satisfying the given conditions. The present contribution differs from [1] by devoting consideration to the case $\nu = 1/2$, when a special choice of method allows savings to be made; also, we here consider the generation of the "starting" weights (which is not described, but is already presaged, in [1]). The quadrature weights for the general case $\nu \ \epsilon \ (0,1)$ are associated with the coefficients of

(1.11) $\Omega^{(\nu)}(\mu) := \{\omega(\mu)\}^\nu = \Omega_0^{(\nu)} + \Omega_1^{(\nu)} \mu + \Omega_2^{(\nu)} \mu^2 + \ldots$,

and this observation explains the use of the notation in (1.8) (which could otherwise be simplified).

We now address the question of calculating the weights of the rules (1.9), with the aim of assessing the savings obtainable over the efficient algorithm, given in [1], for treating the general case.

2. ALGORITHMIC CONSIDERATIONS

2.1. Introduction

Our first task is to compute $\omega_0, \omega_1, \omega_2, \ldots$, the coefficients of $\omega(\mu)$, and the coefficients $\Omega_{\ell}^{(1/2)}$ of $\Omega^{(1/2)}(\mu) = [\omega(\mu)]^{1/2}$. The technique which we will describe employs FFT routines to generate (1.9) given (1.1) and (1.2), or (equivalently) (1.6) and (1.7).

The determination of the coefficients in the general series (1.11) requires (see [5]) the construction of $\exp\{\nu \, \ell n\{\omega(\mu)\}\}$, whilst the coefficients in the special case $\nu = \frac{1}{2}$ in (1.8) can be obtained more directly. (The case $\nu = \frac{1}{2}$ appears to be the commonest to arise in practice, for example in discretizing classical Abel equations of the first and second kinds. In the numerical solution of (simple) classical Abel equations by FFT techniques, approximately half of the time is consumed by the calculation of the weights of the quadrature formulae.)

We shall compare the cost of executing the coded method for the general case (with arbitrary $\nu \, \epsilon \, (0,1)$), with the execution time of code available in the case $\nu = 1/2$. We also mention the effect of restricting the choice of ρ and σ. Our objective is to allow the reader to assess the relative merits of providing generality in contrast to a special case. Whilst the situation is in some ways analogous to the comparison of determining $b^{\frac{1}{2}}$ for $b \, \epsilon \, R_+$ by Heron's method with setting $\nu = \frac{1}{2}$ in $b^{\nu} = \exp\{\nu \ell n(b)\}$, the operations employed here are substantially different from those used for $b \, \epsilon \, R_+$, and value judgements have to be adjusted accordingly.

2.2. The role of FFT

As is well known (cf [5]), the FFT can be employed to determine the coefficients of a product of two polynomials. Thus, if $u(\mu) = u_0 + u_1\mu + \ldots + u_{n-1}\mu^{n-1}$ and $v(\mu) = v_0 + v_1\mu + \ldots + v_{n-1}\mu^{n-1}$ then the product $p(\mu) := u(\mu) \times v(\mu) = p_0 + p_1\mu + \ldots + p_{2n-2}\mu^{2n-2}$ has coefficients

$$(2.1) \qquad p_{\ell} = \sum_{j=0}^{\ell} u_j \, v_{\ell-j} \qquad (\ell = 0,1,\ldots,2n-2).$$

The computation of the set of coefficients p_{ℓ} involves the determination of a family of discrete convolutions. Later, we encounter another context in which a large family of discrete convolutions of this type is required; for this task the FFT provides a fast computational tool. To be specific, suppose $\{u_0, u_1, \ldots, u_{n-1}, 0, \ldots, 0\}$ and $\{v_0, v_1, \ldots, v_{n-1}, 0, \ldots, 0\}$ to be sequences each of 2^m elements, where $n = 2^{m-1}$, $m \, \epsilon \, Z$. Let $\{U_0, U_1, \ldots, U_{2n-1}\}$ and $\{V_0, V_1, \ldots, V_{2n-1}\}$ be the sequences obtained by computing the discrete Fourier transforms (DFTs) of these sequences:

$$(2.2) \qquad U_k = \{1/\sqrt{(2n)}\} \sum_{j=0}^{2n-1} u_j \, \epsilon^{jk}, \qquad \epsilon = \exp(-i\pi/n)$$

etc. Then

$$p_k = \{1/(2n)\} \sum_{j=0}^{2n-1} V_j U_j \, \epsilon^{-jk}$$

$(k = 0,1,\ldots, 2n-2)$ are the required coefficients of the product polynomial. The transforms from $\{u_j\}$ to $\{U_k\}$, $\{v_j\}$ to $\{V_k\}$ and from $\{V_k U_k\}$ to $\{p_j\}$ can be accomplished by fast routines such as those presently in the NAG library [10] under the names C06FAF, C06GBF and C06FBF, to find a transform, the conjugate of a Hermitian sequence, and an inverse transform (respectively).

In view of (1.10), our task is partially completed by finding the coefficients in the formal power series (fps)

$$(2.3)) \qquad \Omega^{(1/2)}(\mu) = \Omega_0^{(1/2)} + \Omega_1^{(1/2)}\mu + \Omega_2^{(1/2)}\mu^2 + \dots .$$

With a finite algorithm it is possible to compute only a finite number of these coefficients. For any fps $\gamma(\mu) = \gamma_0 + \gamma_1\mu + \gamma_2\mu^2 + \dots$ we introduce the notation

$$(2.4) \qquad [\gamma(\mu)]_N = \gamma_0 + \gamma_1\mu + \gamma_2\mu^2 + \dots + \gamma_{N-1}\mu^{N-1}$$

for the *truncated* power series; we shall be able to determine the sequence $\{\Omega_0^{(1/2)}, \Omega_1^{(1/2)}, \Omega_2^{(1/2)}, \dots \Omega_{N-1}^{(1/2)}\}$ defining $[\Omega^{(1/2)}(\mu)]_N$, $N = 2^R$, given R. Once this sequence is found, it is possible to solve systems of linear algebraic equations in order to determine the "starting" weights.

The ability to determine the product of two polynomials has the following consequences: (1) it allows the implementation of Newton's method to find the truncated reciprocal of a polynomial; such a facility is clearly needed, for example to determine $[\omega(\mu)]_m = [\beta(\mu)/\alpha(\mu)]_m$ for $m = 2,4,8,16,\dots$; (2) it allows the implementation of Heron's iteration to find $[\{\omega(\mu)\}^{1/2}]_m$.

2.3. Construction of $\Omega_{\varrho}^{(\frac{1}{2})}$

In view of the preceding remarks, we address the question   (cf [1], and [5]) of finding reciprocals and square roots of certain fps.

We first consider, briefly, the problem of finding a reciprocal of a fps. Recall that Newton's iteration for finding a^{-1} where $0 \neq a \epsilon \mathbb{R}$ reads $r_{n+1} = (2 - ar_n)r_n$; this suggests that, given a fps $a(\mu) = a_0 + a_1\mu + a_2\mu^2 + \dots$, where $a_0 \neq 0$, an approximation $r_*(\mu)$ to the reciprocal $1/a(\mu) := r(\mu) = r_0 + r_1\mu + r_2\mu^2 + \dots$ can be improved by constructing

$$(2.5) \qquad r^*(\mu) = \{2 - a(\mu) \times r_*(\mu)\} \times r_*(\mu).$$

Indeed, suppose m is such that

$$(2.6) \qquad r_*(\mu) = [1/a(\mu)]_m$$

and let

$$(2.7) \qquad r^*(\mu) = [\{2r_*(\mu) - r_*(\mu) \times ([a(\mu)]_{2m} \times r_*(\mu))\}];$$

then we find "quadratic improvement", in the sense that

$$(2.8) \qquad [r^*(\mu)]_{2m} = [1/a(\mu)]_{2m}.$$

The order of operations in (2.7) differs from that in (2.5) to reflect, in part, computational practice. The polynomial multiplications ($\times$) in (2.7) can be achieved via FFT techniques, and proper organization entails   obtaining one fast transform of the sequence $\{r_0, r_1, \dots, r_{m-1}, 0, \dots 0\}$, which is stored for later use, obtaining the FFT associated with $[a(\mu)]_{2m}$, and (in view of the linearity of the FFT) obtaining only one inverse transform to determine the relevant coefficients of r^*. Thus, finding

(2.8) involves 3 FFT evaluations and $O(m)$ additional arithmetic operations. If one commences *ab initio* with the task of finding a reciprocal, one takes $1/a_0$ as starting approximation and corrects repeatedly using (2.7) in the form $r_{n+1}(\mu) = [\{2r_n(\mu) - r_n(\mu) \times ([a(\mu)]_{2m} \times r_n(\mu))\}]_{2m}$, $m = 2^n$. By (2.8), successive iterates have 2, 4, 8, 16, ... correct coefficients, showing quadratic convergence.

We now consider the question of finding a square root via the analogue of Heron's algorithm. If $b \in R_+$, Heron's algorithm for determining $\sqrt{b}$ reads $s_{n+1} = \frac{1}{2}\{s_n + b/s_n\}$. If $b(\mu) = b_0 + b_1\mu + b_2\mu^2 + ...$ where $b_0 > 0$, a sequence of approximations $\hat{s}_n(\mu)$ to the square root $\{b(\mu)\}^{1/2}$ can be found by the iteration $\hat{s}_{n+1}(\mu) = \frac{1}{2}\{\hat{s}_n(\mu) + b(\mu)/\hat{s}_n(\mu)\}$. In fact, we employ the quadratically−convergent iteration

$$(2.9) \quad s_{n+1}(\mu) = \frac{1}{2}\left[s_n(\mu) + [b(\mu)]_{2m} \times [1/s_n(\mu)]_{2m}\right]_{2m} \quad (m = 2^n; \; n \in Z).$$

This requires the determination of $[1/s_n(\mu)]_{2m}$. For this determination, we do not commence *ab initio* in Newton's iteration (unless $n = 1$), but employ the correction (2.7) twice, starting with the "short" approximation $[1/s_{n-1}(\mu)]_{\frac{1}{2}m}$ which is $[1/s_n(\mu)]_{\frac{1}{2}m}$. *This observation achieves significant improvement in efficiency*, and should not be overlooked by the reader. We note that the subscripts on $[s_n(\mu)]$ allow us to determine the lengths of sequences transformed by the FFT at any stage.

We are now in a position to exploit our ability to construct reciprocals and square roots in order to compute $[\Omega^{(1/2)}(\mu)]_N$, $N = 2^R$, given α, β determined by ρ and σ. We double the number of coefficients found at any stage. Given the weights $\{\omega_\ell \mid \ell = 0,1,...,\frac{1}{2}m-1\}$ $(m = 2^n)$ then $\omega_0, \omega_1, ... , \omega_{m-1}$ can be obtained by computing $[1/\alpha(\mu)]_m$ and its product with $[\beta(\mu)]_m$, using 6 FFT calls[*]. A further 9 FFT calls are required to determine $[\Omega^{(1/2)}(\mu)]_m$ from $[\Omega^{(1/2)}(\mu)]_{\frac{1}{2}m}$. At various stages, the sequences transformed double in length; we note (see [1]) that 15 FFT calls are required, in lieu of the 9 FFT calls, if the algorithm for general ν is employed.

2.4. Construction of starting weights

The "starting" weights $W_{nj}^{(1/2)}$ for $0 \leqslant j \leqslant \lambda-1$, with the choice $\lambda = (2p-1)$ are to be determined so that, with the choice (1.10),

$$\{1/\sqrt{\pi}\}\int_0^{nh} (nh-s)^{-\frac{1}{2}} \varphi(s)ds - h^{\frac{1}{2}} \sum_{j=0}^n W_{nj}^{(1/2)}\varphi(jh) = O(h^p) \text{ as } h \to 0$$

with $nh = T$ fixed, for any $\varphi(s) = \varphi_0 + \varphi_1 s^{1/2} + \varphi_2 s + \varphi_3 s^{3/2} + \varphi_4 s^2 + ... + \varphi_{2p-3}s^{p-3/2} + \varphi_{2(p-1)}s^{p-1} + \Phi(s)$, with $\Phi \in CP[0,T]$. To achieve this, we solve the equations

$$(2.10) \quad h^{\frac{1}{2}} \sum_{j=0}^{\lambda-1} W_{nj}^{(1/2)} (jh)^{\ell/2} = \{1/\sqrt{\pi}\} \int_0^{nh}(nh-s)^{-\frac{1}{2}} s^{\ell/2}ds$$
$$- h^{\frac{1}{2}} \sum_{j=\lambda}^n \Omega_{n-j}^{(1/2)} (jh)^{\ell/2}$$

---

[*] For the Gregory rules, these operations can be eliminated, with a substantial saving, since the weights ω_ℓ are easily written down. In the case of rules generated by a 4−th order BDF formula, Hairer, Lubich & Schlichte [4] avoid the generation of ω and cast the problem in terms of finding $\{\alpha(\mu)/\beta(\mu)\}^{-\frac{1}{2}}$.

for $\ell = 0,1,2,...,\lambda-1$ (with $\lambda=2p-1$, $n \in \mathbb{Z}$).

3. IMPLEMENTATION AS CODE

Equations (2.10) constitute a system of linear equations for $W_{n j}^{(\frac{1}{2})}$ ($j = 0$, $1,2,...,2p-2$) given n; in these equations the coefficient matrix is a Vandermonde matrix V, with (i,j)-th element of the form $(\bar{\jmath}i)^j$ in which the sequence $\{\bar{\jmath}i\}$ is ordered. Although such a matrix may have a relatively large condition number, a result of Higham [6] shows that the Björck–Pereyra algorithm does not lead to unduly amplified rounding errors. However, the right–hand side involves a cancellation of terms which can be a source of mild rounding.

Regarding the right–hand side, an analytical expression for the integrals is available in the form $h^{\frac{1}{2}\ell} \Gamma(\frac{1}{2}\ell+1)n^{\frac{1}{2}(\ell+1)}/\Gamma(\frac{1}{2}\ell+\frac{3}{2})$ (so the terms in h can be cancelled in (2.10)).

We observe the need to calculate a family of discrete convolutions arising from the choice $n = 1,2,3,...,$. Inspecting (2.10), we see that, since $\{\Omega_h^{(1/2)}\}$ is independent of $\ell \in \{0,1,2,...,\lambda-1\}$ we are required to evaluate its DFT only once as ℓ varies. Because the coefficient matrix is fixed in (2.10) and the right–hand sides vary, the question which arises is whether it is preferable to find the inverse of V; since the Björck–Pereyra algorithm requires $O(\lambda^2)$ operations the computation of the inverse has not been judged worthwhile.

3.1 Example

The fourth–order BDF formula (BDF$_4$) yields $\beta(\mu) = 1$, $\alpha(\mu) = (25/12) - 4\mu + 3\mu^2 - (4/3)\mu^3 + (1/4)\mu^4$. The first few starting weights $W_{n\ell}^{(1/2)}$ read, to four figures, and with $\lambda = 7$:

n⁼	ℓ⁼0	1	2	3	4	5	6
1	0.0565	2.8928	-6.7497	11.6491	-11.1355	5.5374	-1.1223
2	0.0371	1.7401	-2.8628	6.5207	-6.4058	3.2249	-0.6583
3	0.0300	1.3207	-2.4642	6.3612	-5.4478	2.7025	-0.5481
4	0.0258	1.1217	-2.2620	5.3683	-3.7553	2.2132	-0.4549
5	0.0230	0.9862	-2.0034	4.5005	-3.2772	2.7262	-0.4320
6	0.0208	0.9001	-1.8989	4.2847	-3.5881	2.8201	0.2253
7	0.0190	0.8506	-1.9250	4.4166	-4.0183	2.7933	0.1564

The first few coefficients ω_ℓ and $\Omega_\ell^{(\frac{1}{2})}$ (for $\ell = 0,1,2,...,7$) are:

ω_ℓ	$\Omega_\ell^{(\frac{1}{2})}$
0.4800	0.6928
0.9216	0.6651
1.0783	0.4589
1.0504	0.3175
0.9962	0.2622
0.9797	0.2451
0.9894	0.2323
1.0001	0.2164

3.2 Timings and comparisons

We have produced a code using the NAG Mark 11 FFT routines CO6FAF and CO6FBF. These routines were employed in an algorithm for the more general problem of approximating $\int_0^{nh} (nh-s)^{\nu-1} \varphi(s)ds$ for $0 < \nu < 1$. NAG routines CO6EAF and CO6EBF are employed by Hairer Lubich & Schlichte [4] in a code which treats the generation of weights from the fourth−order BDF formula, with $\nu = 1/2$.

We raise the question whether savings are to be obtained by developing codes for special cases rather than for the general case ($0 < \nu < 1$, ρ,σ arbitrary), and we give in Table 1 comparative timings when the code produced by the authors (1a) for the case $\nu = 1/2$, ρ,σ arbitrary, and (1b) for the case $\nu \in (0,1)$, ρ,σ arbitrary, is contrasted with the code (2) by Hairer et al [4] for the fourth−order BDF formula with $\nu = 1/2$. The timings are those obtained on the CDC 7600 of UMRCC when undertaking the task solved by Hairer et al.

$N = 2^R$	Code 1a ($\nu = 1/2$)	Code 1b ($0 < \nu < 1$)	Code 2 ($\nu = 1/2$, BDF_4 rule)
64	0.084	0.111	0.064
128	0.147	0.202	0.112
256	0.281	0.385	0.215
512	0.597	0.814	0.480
1024	1.171	1.603	0.916

Table 1: CDC7600 times in seconds

4. CONCLUDING OBSERVATIONS

The values in the preceding table show the savings in computer time available when various restrictions are placed upon the degree of generality. Additional considerations may concern the reader; for example, the code 1(a) requires approximately 0.7 of the storage required by the code 1(b), and (although there are sections of code in common, for example the code to generate ω) there are fewer lines of code in the method 1(a) than in the method 1(b). (The length of the code reflects, in part, the fact that Heron's algorithm for a fps requires about 25 lines of code whilst the general technique requires about 60 lines of code.)

Returning to the timings, the code in [4], being purpose−built, proves least expensive (as might be expected) but in our view the savings do not appear to merit opting for a lack of generality. Where space and time are not at a premium, we favour as much generality as possible. In contrast, Hairer, Lubich, and Schlichte [4] opted for a particular choice of formula as well as the special case $\nu = \frac{1}{2}$. Our own view would be open to change if new FFT codes were used or different computers employed; moreover, the case $\nu = \frac{1}{2}$ is so much more common than the general case that some readers are likely to be persuaded that the particular case $\nu = \frac{1}{2}$ merits special provision.

Acknowlegement. We are grateful for the opportunity provided by the comments of a referee to amplify some areas of our discussion.

REFERENCES

[1] BAKER, C. T. H & DERAKHSHAN, M.S. 'The use of NAG FFT routines in the construction of functions of power series used in fractional quadrature rules.' Numer Anal Tech Rept. 115, University of Manchester (April 1986).

[2] BAKER, C. T. H & DERAKHSHAN, M.S. 'A code for fast generation of quadrature rules with special properties' *Appendix* Numer Anal Tech Rept, 121 University of Manchester (August 1986).

[3] GOLUB, G. & VAN LOAN, C.F. 'Matrix Computations' North Oxford Academic, Oxford 1983.

[4] HAIRER, E. LUBICH, C. & SCHLICHTE, M. 'Fast numerical solution of weakly singular Volterra integral equations'. Tech Rept Dept Math., University of Geneva, May 1986.

[5] HENRICI, P. 'Fast Fourier methods in computational complex analysis'. SIAM Review 21 (1979) pp 481−529.

[6] HIGHAM, N. 'Error analysis of the Björck −Pereyra algorithm for solving Vandermonde systems' Numer Anal Tech Rept 108, University of. Manchester, Dec.1985.

[7] LUBICH, C. 'Discretized fractional calculus'. SIAM J. Math. Anal. (to appear; preprinted 1985).

[8] LUBICH, C. 'Fractional linear multistep methods for Abel−Volterra integral equations of the second kind'. Math Comp 45 (1985) pp 463−469.

[9] NEVANLINNA, O. 'Positive quadratures for ˉVolterra equations' Computing 16 (1976) pp 349−357.

[10] NUMERICAL ALGORITHMS GROUP. The NAG manual (Mark 11) NAG Central Office, Banbury Rd. Oxford.

QUADRATURE METHODS FOR THE DETERMINATION OF ZEROS OF TRANSCENDENTAL FUNCTIONS - A REVIEW

Nikolaos I. Ioakimidis
Division of Applied Mathematics and Mechanics
School of Engineering
University of Patras
P.O. Box 1120
GR-261.10 Patras
Greece

ABSTRACT. A review of quadrature methods for the numerical determination of zeros of algebraic or transcendental functions is presented. Most of these methods are based on the classical theory of analytic functions, but, recently, relevant methods based on the elementary theory of real functions were also developed. On the other hand, purely numerical methods were also recently proposed. The common point of these methods is the use of numerical integration rules for the determination of the aforementioned zeros. This makes these methods completely different from the classical (usually iterative) competitive methods. Emphasis is placed on the description of the fundamental principles of each method and, further, on making appropriate reference to the existing relevant literature. Some generalizations of these methods (e.g. to the cases of poles or systems of two equations in two unknowns) are also presented.

1. INTRODUCTION

The appearance of nonlinear algebraic or transcendental equations in problems of science (e.g. physics and chemistry), engineering and applied mathematics is very frequent. The determination of the roots of these equations, that is, of the zeros of the corresponding functions, is a classical topic in numerical analysis (see, e.g., [48, pp. 567-583; 50; 91; 116]). The Newton-Raphson method is one of the most popular of these methods. Systems of nonlinear equations also appear frequently in practical applications and their numerical solution was also considered in detail (see, e.g., [48, pp. 583-588; 87]). In general, iterative methods are used both for single nonlinear equations and for systems of such equations. The convergence of these methods is frequently not assured in advance.

P. Keast and G. Fairweather (eds.), Numerical Integration, 61–82.
© *1987 by D. Reidel Publishing Company.*

Here we will review a completely different class of methods for the solution of nonlinear algebraic or transcendental equations. This class of methods is always based on the numerical evaluation of appropriate integrals by using one or more quadrature rules of the general (and classical) form (for a real interval $[c,d]$)

$$\int_c^d w(x)F(x)dx = \sum_{i=1}^n \mu_{in} F(x_{in}) + E_n , \qquad (1.1)$$

where $w(x)$ is a nonnegative weight function, $F(x)$ is the integrand, x_{in} are the nodes, μ_{in} are the weights and E_n is the error term. The topic of numerical integration (both in one and in several dimensions) is also classical (see, e.g., [2, pp. 885-924; 36; 38; 79, pp. 347-439; 90]. Gaussian quadrature rules are generally preferred (see, e.g., [44; 48, pp. 379-445; 111], since their accuracy is great and the conditions for convergence are mild (see, e.g., [31, 32, 36]). In the case of numerical integration over closed contours, appropriate quadrature rules are also available (see, e.g., [36, pp. 168-172; 39; 41; 82]). In the case of complicated singularities, appropriate algorithms were also proposed (see, e.g., [40, 45]) exactly as in the case of Cauchy-type principal value integrals (see, e.g., [33, 39, 51, 78]).

In this paper we will present a short review of quadrature methods for the solution of nonlinear algebraic or transcendental equations. Beyond the classical papers by Delves and Lyness [37], Abd-Elall, Delves and Reid [1] and Burniston and Siewert [25], several more relevant and completely new methods were recently proposed by Anastasselou [10], the author and Papadakis. We will consider all of these methods in brief in the next three sections. Finally, the convergence of these methods is generally always assured.

2. COMPLEX-VARIABLE METHODS

These methods are the only quadrature methods for the determination of zeros of transcendental functions which are somewhat known and were used in practice up to now by several researchers. They are based on the determination of these zeros as zeros of simple polynomials (usually of degree up to four) after the numerical evaluation of appropriate integrals. The basis of these methods is the classical theory of analytic functions (see, e.g., [3, 34, 86, 110]). A strict classification of these methods and their various modifications is difficult. Here we classify these into three categories although the first one may be considered as a special case of the third. It is hoped that several misunderstandings about the origin and correlation of these methods will be clarified below.

2.1. The Delves - Lyness Method and Relevant Methods

We consider an analytic function $F(z)$ of the complex variable $z = x + iy$ on a simple smooth closed contour C in the complex plane and the finite domain S^+ surrounded by C. The number m of the zeros a_i of $F(z)$ in S^+ (under the assumption that $F(z)$ has no zeros on C) is given by the classical formula [34, p. 119]

$$m = \frac{1}{2\pi i} \int_C \frac{F'(t)}{F(t)} dt = \frac{1}{2\pi}[\arg F(t)]_C, \; t = x + iy, \; t \in C, \qquad (2.1.1)$$

a direct consequence of the Cauchy theorem of residues in complex analysis. In addition, if $g(z)$ is an analogous analytic function, then [34, p. 119]

$$\frac{1}{2\pi i} \int_C g(t) \frac{F'(t)}{F(t)} dt = \sum_{i=1}^{m} g(a_i) \qquad (2.1.2)$$

(because of the same theorem) whence for $m = 1$ (with $g(t) = t$)

$$a_1 = s_1 \equiv \frac{1}{2\pi i} \int_C t \frac{F'(t)}{F(t)} dt . \qquad (2.1.3)$$

More generally, for $m > 1$ (with $g(t) = t^k$)

$$\sum_{i=1}^{m} a_i^k = s_k \equiv \frac{1}{2\pi i} \int_C t^k \frac{F'(t)}{F(t)} dt . \qquad (2.1.4)$$

In this way, for $m = 1$ the numerical evaluation of the integral s_1 gives us directly the sought zero a_1: $a_1 = s_1$ (whereas $m = s_0$). On the other hand, for $m > 1$, the sought zeros a_i $(i = 1(1)m)$ are determined as the zeros of the polynomial

$$P_m(z) = \sum_{l=0}^{m} b_l z^l = \prod_{i=1}^{m} (z - a_i), \; b_m = 1 , \qquad (2.1.5)$$

with its coefficients b_l obtained from the classical Newton identities [50, pp. 36-37; 29; 81]

$$s_1 + b_{m-1} = 0 ,$$

$$s_2 + s_1 b_{m-1} + 2 b_{m-2} = 0 ,$$

$$s_m + s_{m-1} b_{m-1} + s_{m-2} b_{m-2} + \cdots + s_1 b_1 + m b_0 = 0 \qquad (2.1.6)$$

on the basis of the integrals $s_k (k = 1(1)m)$. These integrals (2.1.4) have to be evaluated by quadrature methods. Frequently, if $m > 4$, we divide the region S^+ into appropriate subregions with $m \leq 4$.

This is the principle of the more or less classical Delves-Lyness method. It is not known where this classical method appeared and was used for the first time at least implicitly. For $m = 1$, (2.1.3) can be considered as an example in the classical book by Copson [34, p. 119]. In the same book [34, pp. 123-124] this approach was explicitly used for the determination of the inverse function $G(z)$ of an analytic function $F(z)$ (obviously, with $m = 1$). McCune [84] considered also the same method for $m = 1$ and used it in a physical problem. Equation (2.1.4) is also obvious for $m = 1$ [110, p. 91]. In the case when $m > 1$, the above approach seems due in its explicit form to Delves and Lyness [37, 82], who studied this method in great detail from a numerical point of view (that is, from the point of the efficient evaluation of s_k from (2.1.4) with consideration of the appropriate subdivision of the domain S^+ into subdomains). Therefore, the method is generally called the Delves-Lyness method. This method is also reported by Householder [50, pp. 191-192], Abd-Elall, Delves and Reid [1] (who studied it further from the numerical point of view), Trulsen [113] (who gave the explicit formulae for b_l in (2.1.5) for $m \leq 9$ with a different notation) and Li [81] (who revised this method, by using the homotopy continuation method, in such a way that no division of the region S^+ is required).

Because of the importance of the Delves-Lyness method in practice, at least two computer programs were prepared for this method by Lampariello and Sorrentino [80] and by Botten, Craig and McPhedran [24]. Moreover, this method was used extensively in a long series of paper for the solution of transcendental equations appearing mainly in physical applications (see, e.g., [18, 19, 21, 23, 83, 93, 113]), whereas reference to the same method is made in several more papers (see, e.g., [43, 92, 94]).

One disadvantage of the Delves-Lyness method consists in the fact that the poles of the integrand in (2.1.4) (due to the zeros of $F(z)$ in S^+) contribute signficantly to the error term of the quadrature rule used for the evaluation of the integrals s_k. This causes a significant loss of accuracy for the sought zeros a_i when a quadrature rule of the form (1.1) is used. We can easily get rid of this disadvantage of the Delves-Lyness method by modifying it through the use of the Cauchy integral theorem in complex analysis [34, p. 60]. Therefore, in the case of one zero a_1 of $F(z)$ in S^+, we have

$$\frac{1}{2\pi i} \int_C \left[t \frac{F'(t)}{F(t)} - \frac{a_1}{t - a_1} \right] dt = 0 \qquad (2.1.7)$$

instead of (2.1.3). The integrand in (2.1.7) does not possess a pole at $z = a_1$ and, consequently, the accuracy of the numerical results for a_1 by applying (1.1) directly to (2.1.7) is dramatically increased. This approach was proposed and illustrated by numerical results by Ioakimidis [57]. The case of several zeros of $F(z)$ in S^+ was also considered in the same reference by using integrals generalizing (2.1.7). We will not enter into further details of this method.

Finally, another disadvantage of the Delves-Lyness method is the need to evaluate the derivative $F'(t)$ of $F(t)$ as is clear from (2.1.4). An attempt to avoid this inconvenience by Delves and Lyness [37] was rather unsuccessful. Carpentier and Dos Santos [29] proposed a rather complicated method avoiding the evaluation of $F'(t)$, valid for circular regions only. In the general case of the Delves-Lyness method, Ioakimidis and Anastasselou proved recently [69] that the integrals s_k in (2.1.4) can be evaluated by the equivalent formula

$$s_k = \frac{-m}{2\pi i} \int_C t^{k-1} \log[(t-c)^{-m} F(t)] dt + mc^k, \tag{2.1.8}$$

where c is an arbitrary point of S^+. This was achieved simply by an integration by parts in (2.1.4). Obviously, no multivaluedness problems arise in the integrand of (2.1.8). These have been the most important modifications of the Delves-Lyness method known to the author.

2.2. The Abd-Elall - Delves - Reid Method and Related Methods.

In their fundamental paper [1], Abd-Elall, Delves and Reid studied not only the Delves-Lyness method for the location of zeros of analytic functions in S^+, but also a competitive and, probably, much more convenient method, quite different from that of Delves and Lyness [37]. The major advantage of this method is that the derivative $F'(z)$ of $F(z)$ does not appear at all (exactly as in the method decribed in the last paragraph of the previous subsection). We report below the simple derivation of the aforementioned method proposed by Ioakimidis and Anastasselou [74]. We consider the meromorphic function

$$M(z) = 1/F(z) \tag{2.2.1}$$

(with poles a_i coinciding with the zeros of $F(z)$) as well as the polynomial $P_m(z)$ defined by (2.1.5). Then we obtain directly from the Cauchy integral theorem [34, p. 60]

$$\int_C t^k P_m(t) M(t) dt = 0, \quad k = 0(1)(m-1), \tag{2.2.2}$$

and, equivalently,

$$\sum_{l=0}^{m} d_{l+k} b_l = 0, \quad k = 0(1)(m-1), \tag{2.2.3}$$

with

$$d_k = \frac{1}{2\pi i} \int_C t^k M(t) dt, \quad k = 0(1)(2m-1). \tag{2.2.4}$$

Having evaluated these integrals, d_k, we can determine the coefficients b_l of $P_m(z)$ by solving the system of linear equations (2.2.3). (Of course, in the case of one simple zero a_1 of $F(z)$, $m = 1$, we have simply that $a_1 = d_1/d_0$.) The integrals d_k in (2.2.4) play the role of the integrals s_k in (2.1.4) in the present method and they have to be evaluated by numerical integration rules. Unfortunately, their number is equal to $2m$ and not m as was the case in the Delves-Lyness method. Some modifications of the Abd-Elall-Delves-Reid method (which, unfortunately, were ignored up to now) are also described in brief by Ioakimidis and Anastasselou [74].

As a generalization of the method of Abd-Elall, Delves and Reid [1], we can consider the identities (generalizations of the Cauchy integral formula)

$$\frac{1}{2\pi i}\int_C \frac{M(t)}{t-z}dt = M(z)-\sum_{i=1}^m \frac{A_i}{z-a_i}, \quad z \in S^+, \tag{2.2.5}$$

$$\frac{1}{2\pi i}\int_C \frac{M(t)}{t-z}dt = -\sum_{i=1}^m \frac{A_i}{z-a_i}, \quad z \in S^-, \tag{2.2.6}$$

(where S^- denotes the infinite region outside C). Since $M(z)$ is a known function, we can use the above formulae (and the relevant formulae resulting by differentiation) in a variety of ways (since z is arbitrary) in order to determine the residues A_i at the poles a_i of $M(z)$ (coinciding with the zeros of $F(z)$) and these poles. Several such algorithms are described by Ioakimidis and Anastasselou [74]. One possibility is to consider the very special case of the identity (2.2.6) when $z \rightarrow \infty$. Then, since (as is seen directly)

$$\frac{1}{z-c} = \sum_{k=0}^\infty \frac{c^k}{z^{k+1}}, \tag{2.2.7}$$

(2.2.6) reduces to

$$\sum_{i=1}^m A_i a_i^k = d_k, \quad k = 0(1)(2m-1), \tag{2.2.8}$$

(where the d_k were defined by (2.2.4)). These equations (and their solutions) are completely analogous to the corresponding equations resulting during the construction of Gaussian quadrature rules by the algebraic approach [36, pp. 109-112] and their solution is similar. It follows directly that the sought zeros a_i of $F(z)$ are the zeros of the polynomials $P_m(z)$ defined by (2.1.5), (2.2.3) and (2.2.4), and the method of Abd-Elall, Delves and Reid [1] is rederived in this way. Yet, it should be emphasized that (2.2.5) and (2.2.6) are not a simple generalization of the method of Abd-Elall, Delves and Reid, since the aforementioned equations are identities in the complex plane and not simple algebraic equations as is the case in the Abd-Elall - Delves - Reid method ((2.2.3) and (2.2.4)).

Furthermore, in the case where we seek the m zeros a_i of $F(z)$ in the infinite region S^- outside C, we have simply to use the identities [74]

$$\frac{1}{2\pi i}\int_C \frac{M(t)}{t-z}dt = -M(z)+G_l(z)+\sum_{i=1}^m \frac{A_i}{z-a_i}, \quad z \in S^-, \tag{2.2.9}$$

$$\frac{1}{2\pi i}\int_C \frac{M(t)}{t-z}dt = G_l(z)+\sum_{i=1}^m \frac{A_i}{z-a_i}, \quad z\in S^+, \tag{2.2.10}$$

(where $G_l(z)$ denotes the principal part of $M(z)$ at infinity, which is simply a polynomial) and proceed in an analogous way to derive the relevant algorithms [74].

A special case is that when the closed contour C shrinks into an open arc L in the complex plane. Then (2.2.10) reduces to [74]

$$-\frac{1}{2\pi i}\int_L \frac{M^+(t)-M^-(t)}{t-z}dt = -M(z)+G_l(z)+\sum_{i=1}^m \frac{A_i}{z-a_i}, \quad z\notin L, \tag{2.2.11}$$

where M^+ and M^- are the restrictions of M to S^+ and S^-, respectively. This is also an identity in the complex plane useful for the determination of the zeros a_i of the sectionally analytic function $F(z)$ in the cut complex plane. This approach and its practical implementation are described in detail by Anastasselou and Ioakimidis [11]. Practical applications to physics, mechanics, algebra and numerical analysis are presented in [11, 6], [8, 16], [17] and [7], respectively. The peculiarity in [17], where the solution of the quintic equation is considered, is that Cauchy-type principal value integrals or even finite-part integrals appear.

Another interesting possibility for sectionally analytic functions is to consider the function $P_m(z)M(z)$ (without zeros and poles) instead of $M(z)$. Then (2.2.11) reduces to

$$P_m(z)M(z) = \frac{1}{2\pi i}\int_L P_m(t)\frac{M^+(t)-M^-(t)}{t-z}dt + Q_{m+l}(z), \quad z\notin L, \tag{2.2.12}$$

where $M(z)$ has at infinity an expansion of the form

$$M(z) = \sum_{j=-\infty}^l B_j z^j \tag{2.2.13}$$

and $Q_{m+l}(z)$ denotes the principal part of $P_m(z)M(z)$ at infinity (with $Q_{m+l}(z)\equiv 0$ if $l<-m$). By taking into account (2.2.7) and letting $z\to\infty$ in (2.2.12), we obtain for the coefficients b_j of $P_m(z)$ in (2.1.5)

$$\sum_{j=0}^m b_j(B_{-k-j-1}+I_{k+j}) = 0, \quad k = 0(1)(m-1), \tag{2.2.14}$$

where now

$$I_j = \frac{1}{2\pi i}\int_L t^j[M^+(t)-M^-(t)]dt, \quad j=0(1)(2m-1). \tag{2.2.15}$$

This algorithm was proposed by Anastasselou and Ioakimidis [15] and applied to the classical transcendental equation

$$w = \tanh(pw+q) \qquad (2.2.16)$$

appearing in the theory of ferromagnetism.

The special case (2.2.14) of the general identity (2.2.12) was also derived directly by using the Cauchy integral theorem in complex analysis [34, p. 60] by Anastasselou and Ioakimidis [13]. In that reference the complex contour integrals

$$K_k = \int_L z^k P_m(z)M(z)dz \qquad (2.2.17)$$

were considered and (2.2.14) and (2.2.15) were easily rederived without using (2.2.12). This approach is analogous to the original Abd-Elall - Delves - Reid method [1] for a finite domain S^+ as opposed to the methods based on (2.2.5) and (2.2.6) and should be considered as the most elementary method for locating zeros of sectionally analytic functions (although not general at all). Equation (2.2.16) was also considered in [13] with a relevant discussion on the Gaussian quadrature rule used there and the proposal of competitive quadrature rules in [47]. Applications of the method of [13] to mechanics and elementary calculus can be found in [14] and [73], respectively. All the methods of this subsection are also applicable to the location of a whole contour in a problem of optics and mechanics [54].

2.3. The Siewert-Burniston Method and Related Methods

Another possibility for the location of the zeros a_i of an analytic function $F(z)$ inside a closed contour C consists in "transforming" this function into a polynomial. By introducing a new function $\Phi(z)$ defined by

$$\Phi(z) = F(z), \ z \in S^+, \ \Phi(z)=1, \ z \in S^-, \qquad (2.3.1)$$

we have the following homogeneous Riemann-Hilbert boundary value problem [42, p. 90-96) along C

$$\Phi^+(t)/\Phi^-(t) = F(t), \ t \in C . \qquad (2.3.2)$$

The solution of this problem is [42, pp. 92-96]

$$\Phi(z)=F(z)=P_m(z)X(z), \ z \in S^+, \ \Phi(z)=1=P_m(z)X(z), \ z \in S^-, \qquad (2.3.3)$$

where (2.3.1) was taken into account, $P_m(z)$ is a polynomial with zeros coinciding with those of $F(z)$ in S^+ and $X(z)$ is the canonical function of the Riemann-Hilbert boundary value problem under consideration determined by

$$X(z)=\exp[\Gamma(z)], \ z \in S^+, \ X(z)=z^{-m}\exp[\Gamma(z)], \ z \in S^-, \qquad (2.3.4)$$

with

$$\Gamma(z) = \frac{1}{2\pi i}\int_C \frac{\log[t^{-m}F(t)]}{t-z}dt . \qquad (2.3.5)$$

Therefore, because of (2.3.3), the sought zeros of $F(z)$ are the zeros of the polynomial

$$P_m(z) = F(z)/X(z), \ z \in S^+, \ \text{or} \ P_m(z) = 1/X(z), \ z \in S^-, \tag{2.3.6}$$

where $X(z)$ is completely determined from the values $F(t)$ of $F(z)$ on C. The identities (2.3.6) (for the whole complex plane) are analogous to the identities (2.2.5) and (2.2.6) of the previous subsection. The identities (2.3.3) (coinciding with (2.3.6)) "transform" $F(z)$ into a polynomial, $P_m(z)$.

This is the basic principle of the generalized Siewert-Burniston method for the location of zeros of analytic functions $F(z)$ in the region S^+ surrounded by C. Anastasselou and Ioakimidis [12] proposed a generalization of the method in the case when the zeros of $F(z)$ in the infinite region S^- outside C are sought. In practice, it is advisable to use Laurent series in the second of (2.3.6) to obtain a system of linear equations for the coefficients b_l of $P_m(z)$. This point was made by Ioakimidis [55]. It was also shown by Anastasselou [9] that the Delves-Lyness method described in brief in Subsection 2.1 is simply a special case of the generalized Siewert-Burniston method described above (with $z \to \infty$ in the second of (2.3.6) in order that the Delves-Lyness method be obtained). This result is analogous to the corresponding result for the Abd-Elall - Delves - Reid method already reported in Subsection 2.2.

Originally, the Siewert-Burniston method was proposed by these authors [25] in the case of sectionally analytic functions with an open arc of discontinuity L, probably based on results from linear transport theory [30]. Because of the rather complicated presentation of the results in reference [30], the method did not receive the attention it deserved although it was used in a very long series of papers for physical, mechanical and mathematical applications (transcendental functions) by Siewert, Burniston and their collaborators (see, e.g., [26-28, 95-109, 112]. The relevant results were also used in [114, 115] and just referenced in [22, 35] by other researchers. Finally, recently Anastasselou [7] and Ioakimidis [53, 63] also used the Siewert-Burniston method in its original form. The disadvantage of this method is that it is only applicable to sectionally analytic functions $F(z)$ with an open arc of discontinuinty L contrary to its generalization for closed contours C described in [12] (and above) and [55].

2.4. Other Complex-Variable Methods

In addition to the methods described above for the location of zeros of analytic functions, we wish to report here a few more related results: In [65] Ioakimidis proposed a method combining the results of Subsections 2.1 and 2.3 and based on the determination of the reciprocals of the sought zeros a_i of $F(z)$. In [66] the same author proposed a very general method based on the solution of a nonhomogeneous Riemann-Hilbert boundary value problem on C [42, pp. 406-436]. All the methods of the previous subsections for sectionally analytic functions turn out to be special cases of this general method. Another method, equivalent to the Siewert-Burniston method, but based simply on the Cauchy integral theorem·in complex analysis, was proposed by Ioakimidis and Anastasselou [70]. Moreover, the approach of Subsection 2.3 was used by Ioakimidis and Anastasselou [71] for the simultaneous determination of zeros of analytic or sectionally analytic functions (this result generalizing previous analogous results for polynomials [85]).

Finally, the case of two analytic functions in two complex variables (that is, a system of two nonlinear equations in two unknowns) was considered in brief by Haydock and Temple [46]. Much more detailed results in this area are reported in the book by Aizenberg and Yuzhakov [4] (systems of nonlinear equations in several complex variables), where further relevant references are reported.

2.5. Poles of Meromorphic Functions

The methods for the determination of the poles of a meromorphic function $M(z)$ in S^+ (or S^-) are analogous to those described above for the zeros of an analytic function $F(z)$ since we can always assume that $M(z)=1/F(z)$. The simultaneous determination of zeros and poles of a meromorphic function $M(z)$ inside a simple closed contour C is described in detail by Abd-Elall, Delves and Reid [1], and by Ioakimidis [55]. The case of poles of sectionally meromorphic functions satisfying a Riemann-Hilbert boundary value problem on the contours of their discontinuity was also studied by Ioakimídis [60].

2.6. Other Applications

Beyond the simple determination of zeros and poles of analytic and meromorphic functions, it is possible to apply the results of the previous subsections, appropriately generalized, to the determination of crack tips in mechanics [52, 56], as well as complete curves of interest in mechanical-optical [5, 53, 54, 63] and physical [63] problems. For example, in the last reference equipotential lines and relevant curves defined by harmonic functions are determined by the Siewert-Burniston method.

3. COMPLETELY NUMERICAL METHODS

This class of methods is based exclusively on the application of numerical integration rules to the evaluation of appropriate integrals. We will describe in brief three of these methods just for one simple zero a of $F(x)$ in a real interval (c,d). Several generalizations and modifications are completely possible. The present class of methods is due to Ioakimidis and Anastasselou and it is decribed in sufficient detail in several papers [59, 62, 68, 72], where several possibilities of using completely numerical methods for the solution of transcendental equations are presented and illustrated through appropriate numerical results for the simple transcendental equations

$$F_1(x) = xe^z - b = 0 \quad \text{and} \quad F_2(x) = x\tan x - b = 0, \tag{3.1}$$

appearing in problems of physics and engineering. The most convenient of these methods seems to be the following one [72].

We consider the integral

$$J = \int_c^d [(d-x)(x-c)]^{-\frac{1}{2}} \frac{x-a}{F(x)} dx . \tag{3.2}$$

In this way, the zero $x = a$ of $F(x)$ does not give rise to a pole in the integrand because of the term $x - a$ in the numerator. For the numerical evaluation of J, we use both the Gauss- and the Lobatto-Chebyshev quadrature rules [79, pp. 383-384, 396], but for the interval $[c, d]$ in our case:

$$\int_c^d [(d-x)(x-c)]^{-\frac{1}{2}} g(x) dx = \frac{\pi}{n} \sum_{i=1}^n g(x_{in}) + E_n , \tag{3.3}$$

$$\int_c^d [(d-x)(x-c)]^{-\frac{1}{2}} g(x) dx = \frac{\pi}{n} \sum_{i=0}^n {}'' g(x_{in}^*) + E_n^* , \tag{3.4}$$

respectively (where the double prime denotes that the first and the last terms of the sum should be halved). The nodes in these quadrature formulae are given by

$$x_{in} = \{c + d - (c-d)\cos[(2i-1)\pi/(2n)]\}/2, \quad i = 1(1)n , \tag{3.5}$$

$$x_{in}^* = \{c + d - (c-d)\cos[i\pi/n]\}/2, \quad i = 0(1)n . \tag{3.6}$$

Applying both quadrature rules (3.3) and (3.4) to (3.2) and ignoring the error terms E_n and E_n^*, we obtain the following approximation a_n to the sought zero a of $F(x)$:

$$a_n = \sum_{j=0}^{2n} {}'' \frac{(-1)^j y_{jn}}{F(y_{jn})} \bigg/ \sum_{j=0}^{2n} {}'' \frac{(-1)^j}{F(y_{jn})} , \tag{3.7}$$

where now (because of (3.5) and (3.6))

$$y_{jn} = \{c + d - (c-d)\cos[j\pi/(2n)]\}, \quad j = 0(1)2n . \tag{3.8}$$

(Of course, if $F(y_{jn}) = 0$, then $a = y_{jn}$.) The unrestricted convergence of the method (as $n \to \infty$) was also proved in [72] under the assumptions that $F(x)$ is continuous in $[\dot{c}, d]$ and possesses a continuous second derivative in a neighborhood of the zero a. In this proof of convergence, the results of [31, 33, 78] have been taken into account.

Another possibility, somewhat more complicated from a practical point of view, is to consider the Cauchy-type principal value integral

$$J^* = \int_c^d \left[(d-x)(x-c)\right]^{-\frac{1}{2}} \frac{1}{F(x)} dx \tag{3.9}$$

instead of J in (3.2) and to work in a way analogous to the above one. This approach is described in detail in [68] and its convergence was also proved. We will not give the details nor the respective formulae, but we wish to mention that both methods can also be used when the sought zero a of $F(x)$ lies outside (but near) the integration interval $[c,d]$. This fact is simply stated in [72] (with appropriate numerical results) for the first method and considered in further detail in [62] (with a long series of displayed numerical results) for the second method.

We can mention at this point that the first of the above methods was also generalized to the case of two transendental functions in two unknowns $\left(F(x,y)=0, \ G(x,y)=0\right)$ by Ioakimidis [61], where the following integral was used

$$I = \int \int_D w \frac{(F_x+iG_x)(x-a)+(F_y+iG_y)(y-b)}{F+iG} dD, dD = dx dy, \tag{3.10}$$

where D is the two-dimensional region under consideration, $w(x,y)$ is an appropriate weight function, F_x, F_y, G_x, G_y are the first partial derivatives of F and G and (a,b) is the sought root of $F(x,y)=0$, $G(x,y)=0$. Several numerical results were also presented and generalizations of the method to systems of three or four simultaneous transcendental equations are also possible. We will give no further details here.

Returning to the case of one equation $F(x)=0$, we describe one more possibility for the derivation of a purely numerical method for the determination of the zero $x=a$ of $F(x)$ in (c,d). We consider the function $(x-a)/F(x)$ and its derivative

$$\frac{d}{dx}\left[\frac{x-a}{F(x)}\right] = -\frac{(x-a)F'(x)}{F^2(x)} + \frac{1}{F(x)}. \tag{3.11}$$

Then, by integrating (3.11), we have

$$\int_c^d \left[\frac{(x-a)F'(x)}{F^2(x)} - \frac{1}{F(x)}\right] dx = \left[\frac{a-x}{F(x)}\right]_c^d \tag{3.12}$$

By applying a quadrature rule of the form (1.1) to the left-hand side of (3.12) (e.g., the classical Gauss quadrature rule) and ignoring the error term E_n, we find for the approximation a_n to a

$$a_n = \frac{\sum_{i=1}^{n} \mu_{in} \left[\dfrac{x_{in} F'(x_{in})}{F^2(x_{in})} - \dfrac{1}{F(x_{in})} \right] + \dfrac{d}{F(d)} - \dfrac{c}{F(c)}}{\sum_{i=1}^{n} \mu_{in} \dfrac{F'(x_{in})}{F^2(x_{in})} + \dfrac{1}{F(d)} - \dfrac{1}{F(c)}} \tag{3.13}$$

Another (heuristic) approach to the derivation of this formula was proposed by Ioakimidis [59]. The convergence of this method (as $n \to \infty$) was proved in the same reference under the assumption that $F(x)$ possesses a continuous third derivative in $[c,d]$. Numerical results for the transcendental equations (3.1) were also presented in [59].

Comparing the first method of this section with the last one, we observe that the advantage of the first method is that it avoids the use of the derivative $F'(x)$ of $F(x)$. On the other hand, the advantage of the last method is that it uses just one numerical integration rule (usually Gaussian) instead of two such rules (as is the case in the first method). Of course, the principles of both methods are quite elementary and their unrestricted convergence should not be forgotten.

4. REAL-VARIABLE METHODS

This class of quadrature methods for the determination of zeros of transcendental functions $F(x)$ in a real interval (c,d) seems to be a very recent one in spite of its simplicity compared to the complex-variable methods described in Section 2. The method was inspired by the results of Picard [88, 89], who found an integral formula for the number of roots of a system of nonlinear equations and is based on elementary calculus [20]. The results of Picard were recently reconsidered (from the numerical point of view) by Hoenders and Slump [49]. In this section we will generalize this approach to the real determination of zeros of transcendental functions $F(x)$ (in contrast to Picard, who evaluated just the number of these zeros). Moreover, we adopt a very simple approach to the derivation of the relevant fundamental formulae. This approach is described by Ioakimidis and Papadakis [77].

We consider a continuous function $F(x)$ on an interval $[c,d]$ of the real axis with one simple zero a in $[c,d]$. Then we have the obvious formula

$$\int_c^d \operatorname{sgn} F \, dx = \pm[(d-a)-(a-c)] = \mp[2a-(c+d)], \tag{4.1}$$

where the upper signs hold true for $f(c)<0$ (and $f(d)>0$) and the lower signs for $f(c)>0$ (and $f(d)<0$). Therefore, we have for the sought zero a of $F(x)$

$$a = \tfrac{1}{2}(c+d\mp I), \tag{4.2}$$

where

$$I = \int_c^d \operatorname{sgn} F \, dx = \int_c^d \frac{|F|}{F} \, dx \, .$$

From the numerical point of view, the evaluation of I by using a quadra-ture rule is not very convenient because of the jump discontinuity of the integrand at $x = a$. An integration by parts in (4.3) permits us to get a formula with a continuous integrand in I, which is much more appropriate for the appli-cation of numerical integration rules. The result is [77]

$$I = \int_c^d \frac{F''}{(F')^2} |F| \, dx + \left[\frac{|F|}{F'} \right]_c^d . \tag{4.4}$$

A second integration by parts yields [77]

$$I = \frac{1}{2} \int_c^d \left(-\frac{F'''}{(F')^3} + \frac{3(F'')^2}{(F')^4} \right) F \, |F| \, dx + \left[\frac{|F|}{F'} + \frac{F''}{2(F')^3} F \, |F| \right]_c^d , \tag{4.5}$$

where the integrand has now a continuous first derivative. Additional formulae of higher order (under the assumption of existence of the appropriate deriva-tives of $F(x)$) are also reported in [77] and can be directly obtained by the method of integration by parts. The method was applied to the solution of Kepler's transcendental equation in celestial mechanics:

$$f(x) = x - e \sin(x + M) = 0$$

(where e and M are given constants) by application of the classical Gauss and Lobatto quadrature rules and excellent results were obtained especially by using higher-order formulae for I. In a subsequent paper [76], Ioakimidis and Papa-dakis generalized the above method and applied it to the solution of the Lagrange quintic equation in the three-body problem in celestial mechanics also with very good numerical results.

Instead of the discontinuous function $\operatorname{sgn} F$ (at the zero a of $F(x)$), one can also use the discontinuous function $\tan^{-1}(\epsilon/F)$ (where ϵ is a nonzero constant) or other analogous discontinuous functions. (The function $\tan^{-1}$ was originally used by Picard [88] in his investigation of the number of zeros of $F(x)$.) Results analogous to the previous ones for the function $\tan^{-1}(\epsilon/F)$ were derived by Ioakimidis [58] and Ioakimidis and Papadakis [75] accompanied by particularly accurate numerical results in the last paper. To save space, we will not give the relevant formulae here.

Furthermore, we should mention that it is also possible to derive the funda-mental formulae (of order 0) by using the classical Green's theorem in elemen-tary two-variable calculus [20] following the approach of Picard [88, 89]. This was done by Ioakimidis [64] and Ioakimidis and Papadakis [75] for the derivation of the fundamental formulae of the method. In this case, the function $\tan^{-1}$ results in a natural way. An illustration of the formula is also contained in [64] (beyond the aforementioned extensive numerical results of [75]). Finally, the above results were recently generalized by Ioakimidis [67] to the case of a system of two real transcendental equations in two unknowns.

5. REFERENCES

[1]   ABD-ELALL, L. F., DELVES, L. M. and REID, J. K.: 'A numerical
      method for locating the zeros and poles of a meromorphic func-
      tion', in: *Numerical Methods for Nonlinear Algebraic Equations*
      (P. Rabinowitz, ed.), Gordon and Breach, London, 1970, Chap. 3,
      pp. 47-59.
[2]   ABRAMOWITZ, M. and STEGUN, I. A. (eds.): *Handbook of Mathematical
      Functions with Formulas, Graphs, and Mathematical Tables*, Dover,
      New York, 1965.
[3]   AHLFORS, L. V.: *Complex Analysis*, 2nd ed., McGraw-Hill, New York,
      1966.
[4]   AÏZENBERG, I. A. and YUZHAKOV, A. P.: *Integral Representations
      and Residues in Multidimensional Complex Analysis* (Translations
      of Mathematical Monographs, Vol. **58**), American Mathematical So-
      ciety, Providence, Rhode Island, 1983.
[5]   ANASTASSELOU, E. G.: 'A new method for the theoretical determina-
      tion of isochromatic fringes in plane elasticity crack problems',
      International Journal of Fracture, Vol. **24**, pp. R103-R106, 1984.
[6]   ANASTASSELOU, E. G., 'New formulae for the discrete eigenvalues
      of the homogeneous one-speed transport equation', *Physica Scripta*,
      Vol. **30**, pp. 97-98, 1984.
[7]   ANASTASSELOU, E. G., 'Computation of the roots of Cauchy type
      principal value integrals', *International Journal of Computer
      Mathematics*, Vol. **16**, pp. 85-96, 1984.
[8]   ANASTASSELOU, E. G.: 'A new quadrature method for the computation
      of critical loads in problems of elastic buckling', *Acta Mechani-
      ca*, Vol. **57**, pp. 85-98, 1985.
[9]   ANASTASSELOU, E. G.: 'A formal comparison of the Delves-Lyness and
      Burniston-Siewert methods for locating the zeros of analytic fun-
      ctions', *IMA Journal of Numerical Analysis*, 1986 (in press).
[10] ANASTASSELOU, E. G.: *Methods for the Determination of Zeros, Poles
      and Discontinuity Intervals of Analytic Functions with Applica-
      tions to Engineering Problems*, Doctoral Thesis at the National
      Technical University of Athens, 1986 (in preparation).
[11] ANASTASSELOU, E. G. and IOAKIMIDIS, N. I.: 'A new method for ob-
      taining exact analytical formulae for the roots of transcendental
      functions', *Letters in Mathematical Physics*, Vol. **8**, pp. 135-143,
      1984.
[12] ANASTASSELOU, E. G. and IOAKIMIDIS, N. I.: 'A generalization of
      the Siewert-Burniston method for the determination of zeros of
      analytic functions', *Journal of Mathematical Physics*, Vol. **25**,
      pp. 2422-2425, 1984.
[13] ANASTASSELOU, E. G. and IOAKIMIDIS, N. I.: 'Application of the
      Cauchy theorem to the location of zeros of sectionally analytic
      functions', *Zeitschrift für Angewandte Mathematik und Physik*,
      Vol. **35**, pp. 705-711, 1984.
[14] ANASTASSELOU, E. G. and IOAKIMIDIS, N. I.: 'A closed-form formula
      for the critical buckling load of a bar with one end fixed and
      the other pinned', *Strojnícky Časopis*, Vol. **35**, pp. 621-628, 1984.
[15] ANASTASSELOU, E. G. and IOAKIMIDIS, N. I.: 'A new approach to the

derivation of exact analytical formulae for the zeros of sectionally analytic functions', *Journal of Mathematical Analysis and
Applications*, Vol. **112**, pp. 104–109, 1985.

[16]   ANASTASSELOU, E. G. and PANAYOTOUNAKOS, D. E.: 'A method for the
closed-form evaluation of critical buckling loads for planar
strongly curved bars', *Journal of the Franklin Institute*, Vol.
319, pp. 397–403, 1985.

[17]   ANASTASSELOU, E. G. and PANAYOTOYNAKOS, D. E.: 'Analytical solution of polynomial equations with an application to the quintic
equation', *Applied Mathematics and Computation*, 1986 (to appear).

[18]   ANTAR, B. N.: 'On the solution of two-point linear differential
eigenvalue problems', *Journal of Computational Physics*, Vol. **20**,
pp. 208–219, 1976.

[19]   ANTIA, H. M.: 'Finite-difference method for generalized eigenvalue
problem in ordinary differential equations', *Journal of Computational Physics*, Vol. **30**, pp. 283–295, 1979.

[20]   APOSTOL, T. M.: *Calculus*, Vols. I and II, 2nd ed., Wiley, New
York, 1967, 1969.

[21]   BARAKAT, R.: 'Moment estimator approach to the retrieval problem
in coherence theory', *Journal of the Optical Society of America*,
Vol. **70**, pp. 688–694, 1980.

[22]   BERKSHIRE, F. H.: 'Two-dimensional linear lee wave modes for models
including a stratosphere', *Quarterly Journal of the Royal Meteoro-
logical Society*, Vol. **101**, pp. 259–266, 1975.

[23]   BOTTEN, L. C., CRAIG, M. S. and McPHEDRAN, R. C.: 'Highly conduct-
ing lamellar diffraction gratings', *Optica Acta*, Vol. **28**, pp. 1103–
1106, 1981.

[24]   BOTTEN, L. C., CRAIG, M. S. and McPHEDRAN, R. C.: 'Complex zeros
of analytic functions', *Computer Physics Communications*, Vol. **29**,
pp. 245–259, 1983.

[25]   BURNISTON, E. E. and SIEWERT, C. E.: 'The use of Riemann problems
in solving a class of transcendental equations', *Proceedings of
the Cambridge Philosophical Society*, Vol. **73**, pp. 111–118, 1973.

[26]   BURNISTON, E. E. and SIEWERT, C. E.: Exact analytical solutions of
the transcendental equation $\alpha \sin \zeta = \zeta$', *SIAM Journal of Applied Mathematics*, Vol. **24**, pp. 460–466, 1973.

[27]   BURNISTON, E. E. and SIEWERT, C. E.: 'Exact analytical solutions
basic to a class of two-body orbits', *Celestial Mechanics*, Vol. **7**,
pp. 225–235, 1973.

[28]   BURNISTON, E. E. and SIEWERT, C. E.: 'Further results concerning
exact analytical solutions basic of two-body orbits', *Celestial
Mechanics*, Vol. **10**, pp. 5–15, 1974.

[29]   CARPENTIER, M. P. and DOS SANTOS, A. F.: 'Solution of equations
involving analytic functions', *Journal of Computational Physics*,
Vol. **45**, pp. 210–220, 1982.

[30]   CASE, K. M. and ZWEIFEL, P. F.: *Linear Transport Theory*, Addison-
Wesley, Reading, Massachusetts, 1967.

[31]   CHAWLA, M. M.: 'Error bounds for the Gauss-Chebyshev quadrature
formula of the closed type', *Mathematics of Computation*, Vol. **22**,
pp. 889–891, 1968.

[32]   CHAWLA, M. M. and JAIN, M. K.: 'Error estimates for Gauss quadra-

ture formulas for analytic functions', *Mathematics of Computation*,
Vol. **22**, pp. 82-90, 1968.

[33] CHAWLA, M. M. and RAMAKRISHNAN, T. R.: 'Modified Gauss-Jacobi quad-
      rature formulas for the numerical evaluation of Cauchy type singu-
      lar integrals', *BIT*, Vol. **14**, pp. 14-21, 1974.

[34] COPSON, E. T.: *An Introduction to the Theory of Functions of a
      Complex Variable*, Oxford University Press, Oxford, 1935.

[35] DANBY, J. M. A. and BURKARDT, T. M.: 'The solution of Kepler's
      equation, I', *Celestial Mechanics*, Vol. **31**, pp. 95-107, 1983.

[36] DAVIS, P. J. and RABINOWITZ, P.: *Methods of Numerical Integration*,
      2nd ed., Academic Press, New York, 1984.

[37] DELVES, L. M. and LYNESS, J. N.: 'A numerical method for locating
      the zeros of an analytic function', *Mathematics of Computation*,
      Vol. **21**, pp. 543-560, 1967.

[38] DONALDSON, J. D. and ELLIOTT, D.: 'A unified approach to quadra-
      ture rules with asymptotic estimates of their remainders', *SIAM
      Journal on Numerical Analysis*, Vol. **9**, pp. 573-602, 1972.

[39] ELLIOTT, D. and DONALDSON, J. D.: 'On quadrature rules for ordi-
      nary and Cauchy principal value integrals over contours', *SIAM
      Journal on Numerical Analysis*, Vol. **14**, pp. 1078-1087, 1977.

[40] EVANS, G. A., FORBES, R. C. and HYSLOP, J.: 'Polynomial trans-
      formations for singular integrals', *International Journal of Com-
      puter Mathematics*, Vol. **14**, pp. 157-170, 1983.

[41] FORNARO, R. J.: 'Numerical evaluation of integrals around simple
      closed curves', *SIAM Journal on Numerical Analysis*, Vol. **10**, pp.
      623-634, 1973.

[42] GAKHOV, F. D.: *Boundary Value Problems*, Pergamon Press and Addi-
      son-Wesley, Oxford, 1966.

[43] GARG, V. K. and ROULEAU, W. T.: 'Linear spatial stability of pipe
      Poiseuille flow, *Journal of Fluid Mechanics*, Vol. **54**, pp. 113-127,
      1972.

[44] GAUTSCHI, W.: 'A survey of Gauss-Christoffel quadrature formulae';
      in: *E. B. Christoffel, The Influence of His Work on Mathematics
      and the Physical Sciences* (P. L. Butzer and F. Fehér, eds.), Birk-
      häuser Verlag, Basel, 1981, pp. 72-147.

[45] HARRIS, C. G. and EVANS, W. A. B.: 'Extension of numerical quadra-
      ture formulae to cater for end point singular behaviours over fin-
      ite intervals', *International Journal of Computer Mathematics*,
      Vol. **6**, 1977, pp. 219-227.

[46] HAYDOCK, R. and TEMPLE, W. B.: 'Self-avoiding paths on a general-
      ized Cayley tree', *Journal of Physics A: Mathematical and General*,
      Vol. **8**, pp. 697-709, 1975.

[47] HENRICI, P.: 'Remark on numerical integration', *Zeitschrift für
      Angewandte Mathematik und Physik*, Vol. **35**, pp. 712-714, 1984.

[48] HILDEBRAND, F. B.: *Introduction to Numerical Analysis*, 2nd ed.,
      Tata McGraw-Hill, New Delhi, 1974.

[49] HOENDERS, B. J. and SLUMP, C. H.: 'On the calculation of the exact
      number of zeroes of a set of equations, *Computing*, Vol. **30**, pp.
      137-147, 1983.

[50] HOUSEHOLDER, A. S.: *The Numerical Treatment of a Single Nonlinear
      Equation*, McGraw-Hill, New York, 1970.

[51] HUNTER, D. B.: 'Some Gauss-type formulae for the evaluation of Cauchy principal values of integrals', *Numerische Mathematik*, Vol. **19**, pp. 419–424, 1972.

[52] IOAKIMIDIS, N. I.: 'Application of the optical method of pseudo-caustics to locating crack tips in plane elasticity problems', *International Journal of Fracture*, Vol. **23**, pp. R117–R120, 1983.

[53] IOAKIMIDIS, N. I.: 'Closed-form solution of the equations of caustics about cracks in fracture mechanics', *Journal of the Franklin Institute*, Vol. **317**, pp. 27–33, 1984.

[54] IOAKIMIDIS, N. I.: 'On the inversion of the first equation of caustics', *International Journal of Fracture*, Vol. **29**, pp. R11–R12, 1985.

[55] IOAKIMIDIS, N. I.: 'Application of the generalized Siewert-Burniston method to locating zeros and poles of meromorphic functions', *Zeitschrift für Angewandte Mathematik und Physik*, Vol. **36**, pp. 733–742, 1985.

[56] IOAKIMIDIS, N. I.: 'Locating a straight crack in an infinite elastic medium by using complex path-independent integrals', *Acta Mechanica*, Vol. **57**, pp. 241–246, 1985.

[57] IOAKIMIDIS, N. I.: 'A modification of the classical quadrature method for locating zeros of analytic functions', *BIT*, Vol. **25**, pp. 681–686, 1985.

[58] IOAKIMIDIS, N. I.: 'Two elementary analytical formulae for roots of nonlinear equations', *Applicable Analysis*, Vol. **20**, pp. 73–77, 1985.

[59] IOAKIMIDIS, N. I.: 'Application of the Gauss quadrature rule to the numerical solution of nonlinear equations,' *International Journal of Computer Mathematics*, Vol. **18**, pp. 311–322, 1986.

[60] IOAKIMIDIS, N. I.: 'Determination of poles of sectionally meromorphic functions', *Journal of Computational and Applied Mathematics*, 1986 (in press).

[61] IOAKIMIDIS, N. I.: 'A quadrature method for the numerical solution of two real nonlinear equations in two unknowns', *Journal of Computational Physics* (under revision).

[62] IOAKIMIDIS, N. I.: 'A new quadrature method for locating the zeros of analytic functions with applications to engineering problems', *Applied Numerical Mathematics* (under revision).

[63] IOAKIMIDIS, N. I.: 'Application of quadrature rules to the determination of plane equipotential lines and other curves defined by harmonic functions', *Journal of Computational and Applied Mathematics* (submitted).

[64] IOAKIMIDIS, N. I.: 'A real closed-form integral formula for roots of nonlinear equations', *Serdica* (submitted).

[65] IOAKIMIDIS, N. I.: 'A new approach to the derivation of exact integral formulae for zeros of analytic functions (under submission).

[66] IOAKIMIDIS, N. I.: 'A unified Riemann-Hilbert approach to the analytical determination of zeros of sectionally analytic functions', *Journal of Mathematical Analysis and Applications* (to appear).

[67] IOAKIMIDIS, N. I.: 'A real analytical integral formula for a simple root of a system of two nonlinear equations', *Facta Universitatis, University of Niš, Series: Mathematics and Informatics* (to appear).

[68] IOAKIMIDIS, N. I. and ANASTASSELOU, E. G.: 'A simple quadrature-
      type method for the computation of real zeros of analytic functions
      in finite intervals', *BIT*, Vol. **25**, pp. 242-249, 1985.
[69] IOAKIMIDIS, N. I. and ANASTASSELOU, E. G.: 'A modification of the
      Delves-Lyness method for locating the zeros of analytic functions',
      Journal of Computational Physics, Vol. **59**, pp. 490-492, 1985.
[70] IOAKIMIDIS, N. I. and ANASTASSELOU, E. G.: 'A new, simple approach
      to the derivation of exact analytical formulae for the zeros of
      analytic functions', *Applied Mathematics and Computation*, Vol. **17**,
      pp. 123-127, 1985.
[71] IOAKIMIDIS, N. I. and ANASTASSELOU, E. G.: 'On the simultaneous
      determination of zeros of analytic or sectionally analytic func-
      tions', *Computing*, Vol. **36**, pp. 239-247, 1986.
[72] IOAKIMIDIS, N. I. and ANASTASSELOU, E. G.: 'An elementary noniter-
      ative quadrature-type method for the numerical solution of nonlin-
      ear equations', *Computing*, 1986 (in press).
[73] IOAKIMIDIS, N. I. and ANASTASSELOU, E. G.: 'Formulae for the expon-
      ential, the hyperbolic and the trigonometric functions in terms of
      the logarithmic function', *Zeitschrift für Analysis und ihre Anwen-
      dungen* (under revision).
[74] IOAKIMIDIS, N. I. and ANASTASSELOU, E. G.: 'A new method for the
      computation of the zeros of analytic functions', *Acta Applicandae
      Mathematicae* (under revision).
[75] IOAKIMIDIS, N. I. and PAPADAKIS, K. E.: 'A new simple method for
      the analytical solution of Kepler's equation', *Celestial Mechanics*,
      Vol. **35**, pp. 305-316, 1985.
[76] IOAKIMIDIS, N. I. and PAPADAKIS, K. E.: 'Analytical solution of
      the Lagrange quintic equation in the three-body problem in celes-
      tial mechanics', *Acta Mechanica*, Vol. **55**, pp. 267-272, 1985.
[77] IOAKIMIDIS, N. I. and PAPADAKIS, K. E.: 'A new class of quite ele-
      mentary closed-form integral formulae for roots of nonlinear equa-
      tions', *Journal of Mathematical Analysis and Applications* (under
      revision).
[78] IOAKIMIDIS, N. I. and THEOCARIS, P. S.: 'On the numerical evalua-
      tion of Cauchy principal value integrals', *Revue Roumaine des
      Sciences Techniques, Série de Mécanique Appliquée*, Vol. **22**, pp.
      803-818, 1977.
[79] KOPAL, Z.: *Numerical Analysis*, 2nd ed., Chapman & Hall, London,
      1961.
[80] LAMPARIELLO, P. and SORRENTINO, R.: 'The ZEPLS program for solving
      characteristic equations of electromagnetic structures, *IEEE Tran-
      sactions on Microwave Theory and Techniques*, Vol. MTT-**23**, pp. 457-
      458, 1975.
[81] LI, T.-Y.: 'On locating all zeros of an analytic function within
      a bounded domain by a revised Delves/Lyness method', *SIAM Journal
      on Numerical Analysis*, Vol. **20**, pp. 865-871, 1983.
[82] LYNESS, J. N. and DELVES, L. M.: 'On numerical contour integration
      round a closed contour', *Mathematics of Computation*, Vol. **21**, pp.
      561-577, 1967.
[83] MARTYNOVA, T. A. and SHEVCHENKO, V. V.: 'Waves in an asymmetrical
      gas-dielectric optical waveguide with resonances in the walls',

Radio Engineering and Electronic Physics, No. 7, pp. 17-25, 1976.
[84] McCUNE, J. E.: 'Exact inversion of dispersion relations', *The Physics of Fluids*, Vol. **9**, pp. 2082-2084, 1966.
[85] MILOVANOVIĆ, G. V. and PETKOVIĆ, M. S.: 'On the convergence order of a modified method for simultaneous finding polynomial zeros, *Computing*, Vol. **30**, pp. 171-178, 1983.
[86] NEVANLINNA, R. and PAATERO, V.: *Introduction to Complex Analysis*, Addison-Wesley, Reading, Massachusetts, 1969.
[87] ORTEGA, J. M. and RHEINBOLDT, W. C.: *Iterative Solution of Nonlinear Equations in Several Variables*, Academic Press, New York, 1970.
[88] PICARD, É.: 'Sur le nombre des racines communes à plusieurs équations simultanées', *Journal de Mathématiques Pures et Appliquées*, 4ème série, Vol. **8**, pp. 5-24, 1892.
[89] PICARD, É.: *Traité d'Analyse*, 3rd ed., Vol. I, Gauthier-Villars, Paris, 1922, pp. 150-156.
[90] PIESSENS, R., DE DONCKER, E., ÜBERHUBER, C. and KAHANER, D.: *QUADPACK, A Quadrature Subroutine Package*, Springer-Verlag, Berlin, 1983.
[91] RABINOWITZ, P. (ed.): *Numerical Methods for Nonlinear Algebraic Equations*, Gordon and Breach, London, 1970.
[92] REIS, M. F.: 'Numerical investigation of the Friedrichs model dispersion relation', *Journal of Computational Physics*, Vol. **15**, pp. 375-384, 1974.
[93] ROME, J. A. and BRIGGS, R. J.: 'Stability of sheared electron flow', *The Physics of Fluids*, Vol. **15**, pp. 796-804, 1972.
[94] SCARTON, H. A.: 'The method of eigenvalleys', *Journal of Computational Physics*, Vol. **11**, pp. 1-14, 1973.
[95] SIEWERT, C. E.: 'An exact analytical solution of an elementary critical condition', *Nuclear Science and Engineering*, Vol. **51**, p. 78, 1973.
[96] SIEWERT, C. E.: 'Explicit results for the quantum-mechanical energy states basic to a finite square-well potential', *Journal of Mathematical Physics*, Vol. **19**, pp. 434-435, 1978.
[97] SIEWERT, C. E.: 'On computing eigenvalues in radiative transfer', *Journal of Mathematical Physics*, Vol. **21**, pp. 2468-2470, 1980.
[98] SIEWERT, C. E.: 'An exact expression for the Wien displacement constant', *Journal of Quantitative Spectroscopy & Radiative Transfer*, Vol. **26**, p. 467, 1981.
[99] SIEWERT, C. E. and BURKARDT, A. R.: 'On double zeros of x-tanh(ax +b)', *Zeitschrift für Angewandte Mathematik und Physik*, Vol. **24**, pp. 435-438, 1973.
[100] SIEWERT, C. E. and BURKART, A. R.: 'An asymptotic solution in the theory of neutron moderation', *Nuclear Science and Engineering*, Vol. **54**, pp. 455-456, 1974.
[101] SIEWERT, C. E. and BURNISTON, E. E.: 'An explicit closed-form result for the discrete eigenvalue in studies of polarized light', *The Astrophysical Journal*, Vol. **173**, pp. 405-406, 1972.
[102] SIEWERT, C. E. and BURNISTON, E. E.: 'An exact analytical solution of Kepler's equations', *Celestial Mechanics*, Vol. **6**, pp. 294-304, 1972.
[103] SIEWERT, C. E. and BURNISTON, E. E.: 'Exact analytical solutions

of $ze^z = a'$, *Journal of Mathematical Analysis and Applications*, Vol. **43**, pp. 626–632, 1973.

[104] SIEWERT, C. E. and BURNISTON, E. E.: 'Solutions of the equation $ze^z = a(z+b)'$, *Journal of Mathematical Analysis and Applications*, Vol. **46**, pp. 329–337, 1974.

[105] SIEWERT, C. E., BURNISTON, E. E. and THOMAS, J. R., Jr.: 'Discrete spectrum basic to kinetic theory', *The Physics of Fluids*, Vol. **16**, pp. 1532–1533, 1973.

[106] SIEWERT, C. E. and DOGGETT, W. O.: 'An exact analytical solution for the position-time relationship for an inverse-distance-squared force', *International Journal of Engineering Science*, Vol. **12**, pp. 861–863, 1974.

[107] SIEWERT, C. E. and ESSIG, C. J.: 'An exact solution of a molecular field equation in the theory of ferromagnetism', *Zeitschrift für Angewandte Mathematik und Physik*, Vol. **24**, pp. 281–286, 1973.

[108] SIEWERT, C. E. and KRIESE, J. T.: 'An exact solution of a transcendental equation basic to the theory of intermediate resonance absorption of neutrons', *Journal of Nuclear Energy*, Vol. **27**, pp. 831–837, 1973.

[109] SIEWERT, C. E. and PHELPS, J. S., III: 'On solutions of a transcendental equation basic to the theory of vibrating plates', *SIAM Journal on Mathematical Analysis*, Vol. **10**, pp. 105–111, 1979.

[110] SMIRNOV, V. I.: *A Course of Higher Mathematics*, Vol. III, Part 2: *Complex Variables - Special Functions*, Pergamon Press and Addison-Wesley, Oxford, 1964.

[111] STROUD, A. H. and SECREST, DON: *Gaussian Quadrature Formulas*, Prentice-Hall, Englewood Cliffs, New Jersey, 1966.

[112] THOMAS, J. R., Jr.: 'Eigenvalues of a Vlasov plasma: exact analytical solutions', *Plasma Physics*, Vol. **18**, pp. 715–717, 1976.

[113] TRULSEN, J.: 'Cyclotron radiation in hot magnetoplasmas', *Journal of Plasma Physics*, Vol. **6**, pp. 367–400, 1971.

[114] WILLIAMS, M. M. R.: 'The spatial dependence of the energy spectrum of slowing down particles - I. Applications to reactor physics and atomic sputtering', *Annals of Nuclear Energy*, Vol. **6**, pp. 145–173, 1979.

[115] WILLIAMS, M. M. R.: 'An exact solution of the extinction problem in supercritical multiplying systems', *Annals of Nuclear Energy*, Vol. **6**, pp. 463–472, 1979.

[116] ZAGUSKIN, V. L.: *Handbook of Numerical Methods for the Solution of Algebraic and Transcendental Equations*, Pergamon Press, Oxford, 1961.

Acknowledgements: The present results belong to a research project supported by the Greek Ministry of Research and Technology. The author gratefully acknowledges the financial support of this project. Similarly, the author acknowledges with many thanks the financial support by NATO for the presentation of this paper to the NATO Advanced Research Workshop in Halifax, Canada, August 11-15, 1986, on Numerical Integration: Recent Developments, Software and Applications, co-directed by Professors Patrick Keast and Graeme Fairweather.

COMPUTATION OF THE INDEX OF AN ANALYTIC FUNCTION

M. P. Carpentier
Departamento de Matemática
Instituto Superior Técnico
Av. Rovisco Pais
1096 Lisboa Codex
Portugal
or
Centro de Análise e Processamento de Sinais
Instituto Superior Técnico
Av. Rovisco Pais
1096 Lisboa Codex
Portugal

ABSTRACT. The evaluation of the index n of an analytic function f relative to a contour Γ defined by $n = (2\pi i)^{-1} \oint_\Gamma f'(z)/f(z)dz$ is important in several applications e.g. solution of Wiener–Hopf equations and calculation of zeros of analytic functions (Delves-–Lyness method). The most popular method for computing the index n is based on a direct integration in the above formula and corresponds to the well-known principle of the argument $n = \sum_{k=1}^{N} \phi_k(f)/2\pi$ where $\phi_k(f) = arg[f(z_k)/f(z_{k-1})]$ where $z_k, k = 0,1,\ldots,N$ are points on Γ. Because of the multivalued character of arg the computation of $\phi_k(f)$ may give erroneous values. The method proposed is aimed at overcoming this difficulty. It involves approximating f on Γ between z_{k-1} and z_k by a suitable function g_k for which $\phi_k(g_k)$ is known exactly. Numerical results are presented for the case where the g_k are polynomials of the 3rd degree.

1. INTRODUCTION

The evaluation of the index of an analytic function f relative to a contour Γ is important in several applications e.g. solution of Wiener–Hopf equations [3] and calculation of zeros of analytic functions [1,2]. This problem has been dealt with in [1] where a numerical test is presented in order to avoid errors due to the fact that the argument of complex number is necessarily restricted to its principal value i.e. to the interval $(-\pi,\pi]$. However numerical experiments have shown that in many examples the test is too pessimistic leading to computation of the index with an unnecessarily high number of points.

P. Keast and G. Fairweather (eds.), Numerical Integration, 83–90.

In this communication we propose a method to overcome this difficulty. It involves approximating f on Γ between pairs of evaluations points by polynomials of the 3rd degree. Several numerical experiments have shown that the method is practically foolproof for a number of evaluation points on Γ less or equal to that corresponding to the method introduced in [1].

It may be seen that the technique proposed for the computation of the index of the function f can be used in the evaluation of integrals of the form $I = \oint_{\Gamma} h(z) \ln[f(z)] dz$ where h is single valued in a simply-connected region containing the contour Γ.

2. PROPOSED METHOD

We consider a function f with n zeros in a bounded simply-connected region X of the complex plane and analytic in X + Γ, where Γ is the boundary of X. As is well known the index n of f relative to Γ is given by

$$n = \frac{1}{2\pi i} \oint_{\Gamma} f'(z)/f(z) \, dz \qquad (1)$$

Let $\Gamma = \bigcup_{k=1}^{N} \Gamma_k$ where Γ_k is an arc with end points z_{k-1} and $z_k (z_N = z_0)$. Then by a direct integration in (1) we obtain

$$n = \frac{1}{2\pi} \sum_{k=1}^{N} \phi_k \, (f) \qquad (2)$$

where

$$\phi_k(f) = \text{Im}\left\{\ln[f(z_k)/f(z_{k-1})]\right\} = \arg[f(z)/f(z_{k-1})]\Big|_{z=z_k} \qquad (3)$$

where $\arg [f(z)/f(z_{k-1})]$ is a continuous function on Γ_k which is zero at z_{k-1}. The formula (2) is a representation of the principle of the argument.

The values of $\alpha_k = \phi_k(f)$ calculated on the computer, henceforth denoted by $\tilde{\alpha}_k$, are necessarily restricted to the range $(-\pi, \pi]$. Then if $\alpha_k \notin (-\pi, \pi]$ we obtain an erroneous value for $\tilde{\alpha}_k$. The aim of this communication is to propose a method that calculates exactly the value of α_k for α_k in an interval wider than $(-\pi, \pi]$.

For simplicity we restrict ourselves to circular contours, and assume the contour Γ to be a circle centred at the origin. The radius of the circle will be denoted by ρ and the evaluation points by $z_k = \rho \exp(i \theta_k)$, k = 0,1,...,N, where $\theta_k = 2\pi k/N$.

Before we introduce the new method we review the previous work published by the author together with A.F. dos Santos [1]. This consists, basically, in computing the index n directly using (2) together with the following test:
    if

$$\alpha_M = \frac{1}{\pi} \max_{1 \leq k \leq N} |\tilde{\alpha}_k| < 0.75 \qquad (4)$$

and

$$M_f = \max_{1 \leq k \leq N} \left\{ \left| f(z_k)/f(z_{k-1}) \right|, \left| f(z_{k-1})/f(z_k) \right| \right\} < 6.1 \quad (5)$$

the value obtained on the computer is accepted as correct.

Conditions (4) and (5) together ensure that if a zero is so close to the contour Γ that $|\alpha_k| \geq \pi$ this situation is detected and the value of the index is rejected. The test appears to be safe for zeros of order less or equal 3. However this test is sometimes too restrictive and may give erroneous values for such simple functions as z^n for $n \geq N/2$.

Now, we consider a function g that interpolates f on $z_0, z_1, \ldots, z_N$ and for which $\phi_k(g)$ can be obtained exactly. Then for every $k = 1, \ldots, N$ we have

$$\alpha_k = \phi_k(g) + 2\pi s_k, \quad s_k \in \mathbb{Z} \quad (6)$$

If g is a sufficiently good approximation of f on the arc Γ_k between z_{k-1} and z_k then $s_k = 0$. In this communication we shall assume the function g to be a polynomial p_k of the third degree for each arc Γ_k. We shall consider two possibilities:

version 1 for p_k interpolating f on $z_{k-2}, z_{k-1}, z_k, z_{k+1}$;

version 2 for p_k interpolating f on $z_{k-1}, z_{k-1} \exp(i 2\pi\epsilon)$, $z_k, z_k \exp(i2\pi\epsilon)$, where ϵ is a small real number.

On the computer we choose $\epsilon = 1/1024$.

To determine $\phi_k(p_k)$ we need to know the zeros t_{k1}, t_{k2}, t_{k3} of p_k. We can write

$$p_k(z) = a_k \prod_{r=1}^{3} m_{kr}(z) = a_k \prod_{r=1}^{3} (z - t_{kr}) \quad (7)$$

where a_k is assumed to be nonzero. Then

$$\phi_k(p_k) = \sum_{r=1}^{3} \phi_k(m_{kr}) = \sum_{r=1}^{3} \alpha_{kr} \quad (8)$$

where

$$\alpha_{kr} = \arg\left(\frac{z_k - t_{kr}}{z_{k-1} - t_{kr}} \right) \quad (9)$$

are the angles indicated in Fig. 1. The values of α_{kr} satisfy the inequality

$$|\alpha_{kr} - \pi/N| < \pi \quad (10)$$

as can be seen by taking into account that the contour Γ is a circle.

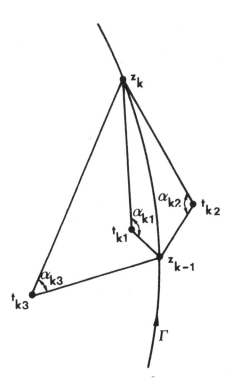

Figure 1. Geometrical relation between the zeros t_{kr} of p_k and the angles α_{kr}.

From (6) and (8) we obtain

$$\alpha_k = \sum_{r=1}^{3} \alpha_{kr} + 2\pi s_k, \quad s_k \in \mathbf{Z} \tag{11}$$

3. EVALUATION TEST

We wish to find a test that detects whether $s_k \neq 0$ in (11). To this end we begin to noting that from (10) and (11) $s_k \neq 0$ if $|\alpha_k - 3\pi/N| \geqslant 3\pi$. If $|\alpha_k - 3\pi/N| < 3\pi$ but $|\alpha_k - 3\pi/N|$ near 3π we also expect that $s_k \neq 0$. In order to prevent this possility we calculate

$$\alpha_{kS} = \sum_{r=1}^{3} |\alpha_{kr}| \tag{12}$$

and impose the condition

$$\alpha_{kS} < \beta \tag{13}$$

where $0 < \beta < 3\pi$. To obtain an additional safety margin we take $\beta = 3\pi/2$. It should be noted that with the condition (13) α_k may be exactly calculated even if $|\alpha_k| > \pi$ as long as $|\alpha_k| < \beta$. This should lead to a more efficient method for computing the index.

However we may still have $s_k \neq 0$ if a zero z_c of f is too close to the contour. Indeed, if this is the case, then at least one of the zeros of p_k, say t_{k1}, is close to the contour; then, if z_c and t_{k1} are on different sides of the contour we can have $s_k \neq 0$. To prevent this we impose

$$\max_{r=1,2,3} |\alpha_{kr} - \pi/N| < d\pi \tag{14}$$

where $0 < d < 1$ (if $|\alpha_{kr} - \pi/N| = \pi$ then t_{kr} is on Γ_k). A number of numerical experiments have convinced us that a good choice is $d = d_1 = 0.75$ and $d = d_2 = 0.98$ respectively for version 1 and version 2 for any $N \geqslant 16$. If t_{k1} is near z_{k-1} or z_k then $|\alpha_{kr} - \pi/N|$ may be much smaller than π and in this case (14) will not detect the proximity of z_c to Γ_k. To prevent this we impose

$$M^{-1} < |f(z_k)/f(z_{k-1})| < M \tag{15}$$

where M is an adequate real number. We choose $M = M_1 = 25$ and $M = M_2 = 10^4$ respectively for version 1 and version 2.

From the above considerations we can now state the following test. Putting

$$\alpha_S = \frac{1}{\pi} \max_{1 \leqslant k \leqslant N} \sum_{r=1}^{3} |\alpha_{kr}| \tag{16}$$

$$\alpha_N = \frac{1}{\pi} \max_{1 \leqslant k \leqslant N} \max_{r=1,2,3} |\alpha_{kr} - \pi/N| \qquad (17)$$

$$M_f = \max_{1 \leqslant k \leqslant N} \{|f(z_k)/f(z_{k-1})|, |f(z_{k-1})/f(z_k)|\} \qquad (18)$$

We have:

Test. If for any $N \geqslant 16$ and

(i) version 1: $\alpha_S < 1.5$, $\alpha_N < 0.75$, $M_f < 25$

(ii) version 2: $\alpha_S < 1.5$, $\alpha_N < 0.98$, $M_f < 10^4$

the computed value of the index is accepted.

4. NUMERICAL RESULTS

To compare the present method with the previous one [1], we illustrate in Tables I, II and III the application of the above tests to functions with a single or a double zero. In these tables, the subscripts 0, 1 and 2 of the variables correspond, respectively, to the method of [1] and to versions 1 and 2 of the present method. The computed values $\tilde{n}$ of the index n are marked with an asterisk if $\tilde{n} \neq n$. The values of $\tilde{n}$ that satisfy the test of the preceding section are indicated inside a polygonal line.

Examination of the Tables I and II shows that:

(i) If there are no zeros too close ($\rho_0/\rho = .95$) to the contour version 1 may be more efficient (note that for the same value of N the number of evaluation points for version 2 is double that of version 1).

(ii) If there is a zero close to the contour ($\rho_0/\rho = .99$) version 2 has a clear advantage over version 1 and, a fortiori, over the method dealt with in [1].

As regards Table III it is seen that only version 2 of the present method ensure that the computed value of the index is correct.

5. ACKNOWLEDGEMENT

The author wishes to thank his colleague A.F. dos Santos for reading the manuscript and making a few useful comments.

6. REFERENCES

1. M.P. Carpentier and A.F. dos Santos, *J. Comput. Phys.*, 45 (1982), 210-220.

2. L.M. Delves and J.N. Lyness, *Math. Comp.*, 21 (1967), 543-560.

3. M.G. Krein, *Amer. Math. Soc. Transl.*, 22 (1962), 162-288.

TABLE I

Simple Zeros: $f(z) = \sin(\pi z - \pi/4)$

ρ_o/ρ	N	$\tilde{n}_o$	$\tilde{n}_1$	$\tilde{n}_2$	M_f	α_M	α_{N1}	α_{S1}	α_{N2}	α_{S2}
	16	0*	0*	8	99.0	0.86	0.81	0.87	0.90	1.89
0.95	32	8	8	8	10.3	0.77	0.87	1.56	0.48	1.14
	64	8	8	8	3.4	0.39	0.37	0.73	0.35	0.68
	16	0*	0*	8	400.1	0.82	0.88	0.90	0.86	2.00
0.99	32	8	8	8	42.5	0.74	0.75	1.47	0.47	1.21
	64	8	8	8	12.2	0.48	0.47	0.82	0.46	0.80
	128	8	8	8	5.3	0.44	0.44	0.60	0.44	0.60

Note. $\rho_o = |z_c|$, where z_c is the zero closest to the contour Γ (ρ_o=3.75).
ρ=radius of contour Γ. * erroneous value of n. ☐ accepted value of n.

TABLE II

Double Zeros: $f(z) = \sin^2(\pi z - \pi/4)$

ρ_o/ρ	N	$\tilde{n}_o$	$\tilde{n}_1$	$\tilde{n}_2$	M_f	α_M	α_{N1}	α_{S1}	α_{N2}	α_{S2}
	16	-1*	0*	8	251.7	0.90	0.93	1.10	0.83	1.85
0.95	32	8	8	8	23.9	0.86	0.71	1.20	0.55	1.24
	64	8	8	8	5.2	0.69	0.47	0.81	0.39	0.84
	16	-1*	0*	8	5571.5	0.85	0.99	1.24	0.84	2.23
	32	6*	8	8	556.3	0.96	0.79	1.09	0.58	1.17
0.99	64	8	8	8	106.3	0.96	0.73	0.97	0.51	0.98
	128	8	8	8	25.5	0.88	0.53	0.89	0.45	0.90
	256	8	8	8	7.0	0.76	0.40	0.77	0.39	0.77
	512	8	8	8	2.8	0.57	0.29	0.57	0.29	0.57

Note. Same notation as in Table I (ρ_o=1.75).

TABLE III

Double Zeros: $f(z) = [z\ e^{-i\pi/16})^4 - 1]^2$

ρ_o/ρ	N	$\tilde{n}_o$	$\tilde{n}_1$	$\tilde{n}_2$	M_f	α_M	α_{N1}	α_{S1}	α_{N2}	α_{S2}
	16	4*	4*	4*	5.5	0.63	0.41	0.86	0.89	1.86
0.95	32	0*	8	8	14.9	0.90	0.77	1.10	0.55	1.10
	64	8	8	8	4.6	0.82	0.47	0.82	0.41	0.82
	128	8	8	8	2.4	0.55	0.44	0.55	0.28	0.55
	16	4*	4*	4*	5.8	0.53	0.31	0.94	0.77	1.86
	32	0*	0*	8	363.4	0.78	0.95	1.11	0.63	1.22
0.99	64	0*	8	8	95.2	0.94	0.70	1.06	0.54	1.06
	128	8	8	8	24.8	0.93	0.52	0.93	0.47	0.93
	256	8	8	8	7.0	0.78	0.41	0.78	0.39	0.78
	512	8	8	8	2.8	0.58	0.29	0.58	0.29	0.58

Note. Same notation as in Table I ($\rho_o = 1$).

APPROXIMATE COMPUTATION OF STRONGLY OSCILLATORY INTEGRALS WITH COMPOUND PRECISION

L.C. Hsu
Department of Mathematics
Texas A&M University
College Station, Texas 77843
U.S.A.

Y.S. Zhou
Department of Mathematics
Jilin University
Changchun
China

ABSTRACT   This paper is concerned with a constructive process for the numerical integration of rapidly oscillating functions of the form $F(x,\lambda x)$, where $F(\cdot,y)$ is periodic of period unity in y, and λ is a large parameter. It is shown that there can be found a class of quadrature formulas which possess any preassigned compound precision $[p,q,\lambda^{-m}]$. Moreover, the results announced in our abstract [6] are refined and improved.

1. INTRODUCTION

It is known that various numerical integration methods have been developed for the Fourier type of oscillatory integrals and the like, and a comprehensive exposition can be found in Davis and Rabinowitz's book [1]. In particular, Havie [3] has already shown that an Euler-Maclaurin type expansion formula could be used as an effective tool to treat oscillatory integrals (cf. also [4]). The object of this paper is to construct a class of quadrature formulas of any given compound precision for integrals of the general form

$$(1.1) \qquad I(\lambda) \equiv I(F;\lambda) := \int_0^1 F(x,\lambda x)dx ,$$

where $F(\cdot,y)$ is a periodic function of y with period 1, and λ is a real parameter that may take very large values. (e.g. $\lambda = 100$, 500, 10^6, 10^{10}, etc.)

Throughout, we will assume that

$$(1.2) \qquad F_x^{(m)}(x,y) \equiv \partial^m F/\partial x^m \in C(R) ,$$

where $m \geq 2$, $R \equiv [0,1] \times [0,1]$, and $C(R)$ denotes the set of continuous functions defined on R.

P. Keast and G. Fairweather (eds.), Numerical Integration, 91–101.

For given positive integers p,q and m, we shall say that a numerical integration formula $I(F;\lambda) \cong S(F;\lambda)$ possesses compound precision $[p,q,\lambda^{-m}]$, if for any algebraic-trigonometric polynomial in x and y of degrees not exceeding p and q, respectively, of the form

$$(1.3) \qquad P(x,y) = \sum_{\substack{0 \le \nu \le p \\ 0 \le \mu \le q}} \left(c_{\nu\mu} x^{\nu} \cos(2\pi\mu y) + d_{\nu\mu} x^{\nu} \sin(2\pi\mu y) \right),$$

$c_{\nu\mu}$, $d_{\nu\mu}$ being arbitrary constants, there exists a constant $\mathbb{K} > 0$, independent of λ, such that

$$(1.4) \qquad |S(P;\lambda) - I(P;\lambda)| < \mathbb{K} \cdot \lambda^{-m}$$

holds for all sufficiently large λ (say, $\lambda \ge 50$, 100, etc.).

Generally, for the same polynomial of the form (1.3) the constant $\mathbb{K}$ depends only upon m and the norm $\|P\| = \max_{R} |P(x,y)|$, or the norms of derivatives $P_x^{(k)}(x,y)$, $(k = m, m-1)$. An explicit estimate of the constant will be given near the end of the paper.

2. AN APPROXIMATE REPRESENTATION OF $I(\lambda)$

We will show that $I(\lambda)$ defined by (1.1) may be represented approximately by some easily computed integrals with an error term $O(\lambda^{-m})$, $(\lambda \uparrow \infty)$. For large λ, the constant implied by $O(\cdot)$ may be evaluated explicitly.

LEMMA 1.   For λ large we have the following

$$(2.1) \qquad I(\lambda) = \int\int_R F(x,y)\,dx\,dy + \sum_{k=1}^{m-1} \frac{\lambda^{-k}}{k!} \int_0^1 I(y,k,\lambda)\,dy + \rho_m ,$$

$$(2.2) \qquad I(y,k,\lambda) = \left[F_x^{(k-1)}(t,y) \bar{B}_k(y - t\lambda) \right]_{t=0}^{t=1} ,$$

$$(2.3) \qquad \rho_m = \frac{1}{m!} \lambda^{-m} \left\{ \int_0^1 I(y,m,\lambda)\,dy - \int_0^1 dx \int_0^1 \bar{B}_m(y - \lambda x) F_x^{(m)}(x,y)\,dy \right\},$$

where $\bar{B}_k(t)$ is the Bernoulli function of period 1 that coincides with the k-th degree Bernoulli polynomial $B_k(t)$ on [0,1], and the notation $\left[\psi(t) \right]_{t=a}^{t=b}$ means the difference $\psi(b) - \psi(a)$.

For λ an integer parameter, the expressions (2.1)-(2.3) have already been proved in [4] (cf. [5]). The general case for any real $\lambda > 0$ may be verified by using the Euler-Maclaurin summation formula of the form

$$(2.4) \quad \sum_{a \leq k < b} f(k) = \int_a^b f(x)dx + \sum_{\nu=1}^m \frac{1}{\nu!} \left[\bar{B}_\nu(-t)f^{(\nu-1)}(t) \right]_a^b$$

$$- \frac{1}{m!} \int_a^b \bar{B}_m(-x)df^{(m-1)}(x) \ ,$$

which is valid for any $f(x)$ such that $f^{(m-1)}(x)$ is of bounded variation on $[a,b]$. Here we need only the case where $f^{(m)}(x)$ is continuous.

By substituting $y = \lambda x$ we have

$$(2.5) \quad \int_0^1 F(x, \lambda x)dx = \frac{1}{\lambda} \int_0^\lambda F(\frac{y}{\lambda}, \{y\})dy = \int_0^1 \frac{1}{\lambda} \sum_{-y \leq k < \lambda - y} F(\frac{k+y}{\lambda}, y)dy \ ,$$

where $\{y\} = y - [y]$ denotes the decimal (fractional) part of y. Thus an application of (2.4) to the inner summation of (2.5) (with $a = -y$, $b = \lambda - y$) yields the following

$$\frac{1}{\lambda} \sum_{-y \leq k < \lambda - y} F(\frac{k+y}{\lambda}, y) = \frac{1}{\lambda} \int_{-y}^{\lambda-y} F(\frac{u+y}{\lambda}, y)du$$

$$+ \sum_{\nu=1}^m \frac{1}{\nu! \lambda^\nu} \left[\bar{B}_\nu(y-\lambda)F_x^{(\nu-1)}(1,y) - \bar{B}_\nu(y)F_x^{(\nu-1)}(0,y) \right]$$

$$- \frac{1}{m! \lambda^m} \int_{-y}^{\lambda-y} \bar{B}_m(-u)d_u F_x^{(m-1)}(\frac{u+y}{\lambda}, y)$$

$$= \int_0^1 F(x,y)dx + \sum_{\nu=1}^m \frac{1}{\nu! \lambda^\nu} I(y,\nu,\lambda) - \frac{1}{m! \lambda^m} \int_0^1 \bar{B}_m(y-\lambda x)F_x^{(m)}(x,y)dx \ .$$

Hence Lemma 1 is justified by means of (2.5); cf. also Hsu and Wang [7], §4 of chap. 4.

Let us denote

$$\|B_m\| = \max_{0 \leq x \leq 1} |B_n(x)|, \quad \|F_x^{(k)}\| = \max_R |F_x^{(k)}(x,y)| \ .$$

It is easy to get an estimate of $|\rho_m|$ as follows

$$(2.6) \quad |\rho_m| \leq \frac{1}{m! \lambda^m} \left\{ 2\|F_x^{(m-1)}\| + \|F_x^{(m)}\| \right\} \|B_m\| \ .$$

In order to obtain quadrature formulas using only integrand function values, it is necessary to approximate the partial derivatives contained in

(2.2). This may be accomplished by using a pair of numerical differentiation formulas of the form

$$(2.7) \qquad h^k f^{(k)}(0) = \sum_{\mu=0}^{m-1} A_\mu^{(k)} \cdot f(\mu h) + E_1 \cdot h^m \; ,$$

$$(2.8) \qquad h^k f^{(k)}(1) = \sum_{\mu=0}^{m-1} B_\mu^{(k)} \cdot f(1-\mu h) + E_2 \cdot h^m \; .$$

Assume that $f^{(m)}(t) \in C[0,1]$, $(m-1)h \leq 1$, $0 \leq k \leq m-1$, and let $(t)_m := t(t+1)...(t+m-1)$. Then for convenience we may simply make use of Markov's formula

$$h^k f^{(k)}(0) = \sum_{\nu=k}^{m-1} (-1)^{\nu-k} \frac{1}{\nu!} \Delta_h^\nu f(0) \cdot \left[\left(\frac{d}{dt}\right)^k (t)_\nu \right]_{t=0}$$

$$+ (-1)^{m-k} h^m \frac{1}{m!} f^{(m)}(\xi) \left[\left(\frac{d}{dt}\right)^k (t)_m \right]_{t=0} \; , \quad \xi \in [0,(m-1)h] \; ,$$

which may be rewritten in the form (2.7) with the remainder factor

$$E_1 = \frac{(-1)^{m-k}}{m!} f^{(m)}(\xi) \left[\left(\frac{d}{dt}\right)^k (t)_m \right]_{t=0}$$

$$= \frac{(-1)^{m-k}}{m!} f^{(m)}(\xi) \left| S_1(m,k) \right| \cdot k! \; ,$$

where $S_1(m,k)$ is the Stirling number of the first kind defined by

$$t(t-1) \cdots (t-m+1) = \sum_{k=1}^m S_1(m,k) t^k \; , \quad (m \geq 1) \; .$$

Consequently we have an estimate for $\left| E_1 \right|$:

$$|E_1| \leq \frac{k!}{m!} \|f^{(m)}\| \cdot |S_1(m,k)| \; , \quad \left(\|f^{(m)}\| = \max_{0 \leq t \leq 1} |f^{(m)}(t)| \right) .$$

Similarly, with h being replaced by $-h$, and assuming $(m-1)h \leq 1$, we get another Markov's formula (2.8) in which $|E_2|$ has the same estimate

$$|E_2| \leq \frac{k!}{m!} \|f^{(m)}\| \cdot |S_1(m,k)| \; .$$

Now for each fixed $y \in [0,1]$, applying (2.7) and (2.8) with $h = \lambda^{-\alpha} (0 < \alpha \le 1)$ to $F_x^{(k)}(t,y)$ at $t=0$ and $t=1$ respectively, one may obtain an approximation to $I(y,k+1,\lambda)$, $1 \le k \le m-2$, viz.

$$(2.9) \qquad \tilde{I}(y,k+1,\lambda) = \lambda^{\alpha k} \bar{B}_{k+1}(y-\lambda) \sum_{\mu=0}^{m-1} B_\mu^{(k)} \cdot F(1-\mu\lambda^{-\alpha}, y)$$

$$- \lambda^{\alpha k} \bar{B}_{k+1}(y) \sum_{\mu=0}^{m-1} A_\mu^{(k)} \cdot F(\mu\lambda^{-\alpha}, y) .$$

In particular, we define

$$(2.10) \qquad \tilde{I}(y,1,\lambda) \equiv I(y,1,\lambda) = \left[F(t,y) \bar{B}_1(y-t\lambda) \right]_{t=0}^{t=1} .$$

Then, replacing $I(y,k,\lambda)$ of (2.1) by (2.9)-(2.10), we get an approximation for $I(\lambda)$, namely $\tilde{I}(\lambda)$, where $\tilde{I}(\lambda)$ takes the form

$$(2.11) \qquad \tilde{I}(\lambda) \equiv \int\int_R F \, dx \, dy + \sum_{k=0}^{m-2} \frac{\lambda^{-k-1}}{(k+1)!} \int_0^1 \tilde{I}(y,k+1,\lambda) \, dy .$$

LEMMA 2. For a given integer $m \ge 3$ and a real number α with $(m-2)/(m-1) \le \alpha \le 1$, the error term involved in $\tilde{I}(\lambda)$ is of order $0(\lambda^{-m})(\lambda\uparrow\infty)$, provided that $F_x^{(m)}(x,y) \in C(R)$. More precisely, we have

$$(2.12) \qquad |I(\lambda) - \tilde{I}(\lambda)| \le \frac{\lambda^{-m}}{m!} \left(\sum_{k=2}^{m} \gamma_k \|B_k\| \right) ,$$

where γ_k is given by the following

$$(2.13) \qquad \gamma_k = \begin{cases} (2/k)|S_1(m,k-1)| \cdot \|F_x^{(m)}\|, & \text{when } 2 \le k \le m-1 , \\ 2\|F_m^{(m-1)}\| = \|F_x^{(m)}\|, & \text{when } k = m . \end{cases}$$

In fact, by (2.1), (2.2), (2.7), (2.8), (2.9) and (2.10) we have

$$|I(\lambda) - \tilde{I}(\lambda)| \le |\rho_m| + \sum_{k=1}^{m-2} \frac{\lambda^{-k-1}}{(k+1)!} \int_0^1 |\tilde{I}(y,k+1,\lambda) - I(y,k+1,\lambda)| \, dy$$

$$\le |\rho_m| + \sum_{k=1}^{m-2} \frac{\lambda^{-k-1}}{(k+1)!} \|B_{k+1}\| \cdot \left(|E_1| + |E_2| \right) h^{m-k} .$$

Here both E_1 and E_2 involve a parameter $y \in [0,1]$. However, they can be estimated by the following

$$\max\left\{|E_1|, |E_2|\right\} \leq \frac{k!}{m!} \, \|F_x^{(m)}(x,y)\| \cdot |S_1(m,k)| \, .$$

Recall that $h = (1/\lambda)^\alpha$, $((m-2)/(m-1) \leq \alpha \leq 1)$, so that for λ large

$$\lambda^{-k-1} h^{m-k} = (1/\lambda)^{\alpha(m-k)+k+1}$$

$$\leq (1/\lambda)^{\left[\frac{m-k-1}{m-k}\right](m-k)+k+1} = \lambda^{-m} \, .$$

Consequently

$$|I(\lambda) - \tilde{I}(\lambda)| \leq |\rho_m| + \sum_{k=1}^{m-2} \frac{2}{k+1} \, \|B_{k+1}\| \cdot \|F_x^{(m)}\| \cdot |S_1(m,k)| \frac{\lambda^{-m}}{m!} \, .$$

Hence (2.12) is obtained with the aid of (2.6).

3. CONSTRUCTION OF FORMULAS WITH COMPOUND PRECISION

Take an interpolatory quadrature formula (or Gaussian formula) of algebraic precision p for an integral on $[0,1]$, viz.

$$\int_0^1 \psi(x)dx = \sum_\mu a_\mu \psi(x_\mu), \quad (x_\mu \in [0,1])$$

and also a formula of trigonometric precision (degree) q with equidistant knots and equal coefficients for the integral of a periodic function of period unity, viz.

$$\int_0^1 \phi(y)dy = \sum_{\nu=1}^{q+1} b_\nu \phi(y_\nu), \quad \left(b_\nu = \frac{1}{q+1}, \, y_\nu = \frac{\nu}{q+1}\right) \, .$$

Then the product formula

$$(3.1) \qquad \int\int_R F(x,y)dxdy = \sum_\mu \sum_\nu a_\mu b_\nu F(x_\mu, y_\nu)$$

has the algebraic-trigonometric precision (degrees) (p,q) (cf. [2], etc.).

Notice that $\tilde{I}(y,\cdot,\cdot)$ defined by (2.9)-(2.10) are periodic functions of period 1 in y. We have to treat integrals of the form

$$(3.2) \qquad J_k = \int_0^1 \overline{B}_k(y-\theta)F(\varsigma,y)dy \ ,$$

which may be rewritten as follows

$$(3.3) \qquad J_k = \int_{-\theta}^{1-\theta} \overline{B}_k(t)F(\varsigma,t+\theta)dt = \int_0^1 B_k(t)F(\xi,t+\theta)dt \ ,$$

where $\theta=0$ or λ, and $\xi=0$ or 1 for $k=1$; and $\xi=\mu\lambda^{-\alpha}$ or $1-\mu\lambda^{-\alpha}$ for $k = 2, ..., m$-1.

For a given weight function $B_k(x)(1\le k\le m-1)$, let us now construct a formula, with trigonometric precision q of the form

$$(3.4) \qquad \int_0^1 B_k(x)\phi(x)dx \cong \sum_{\nu-1}^N C_\nu\cdot\phi(x_\nu) \ , \quad (N=2q+1) \ ,$$

where $\phi(x)$ is periodic of period 1, and $0\le x_1< x_2< \cdots < x_N\le 1$ is a given partition of $[0,1]$. The precision q requires (3.4) to be exact for all trigonometric polynomials $\phi(x):=T(x)$ of the form

$$T(x) = c_o + \sum_{j-1}^q \left(c_j\cos(2\pi jx)+d_j\sin(2\pi jx)\right).$$

Consequently we get a system of linear equations as follows

$$(3.5) \qquad \sum_{\nu=1}^N C_\nu = \int_0^1 B_k(x)dx = 0 \ ,$$

$$\sum_{\nu=1}^N C_\nu\cos(2\pi jx_\nu) = \xi_{kj} \ ,$$

$$\sum_{\nu=1}^N C_\nu\sin(2\pi jx_\nu) = \eta_{kj} \ ,$$

where $j=1,2,...,q \ (=(N-1)/2)$, and ξ_{kj} and η_{kj} are given by

$$(3.6) \qquad \xi_{kj} = \int_0^1 B_k(x)\cos(2\pi jx)dx = -(-1)^{k/2}k!/(2\pi j)^k \ , \quad (k:\text{even})$$

$$\eta_{kj} = \int_0^1 B_k(x)\sin(2\pi jx)dx = -(-1)^{(k-1)/2}k!/(2\pi j)^k \ , \quad (k:\text{odd})$$

and $\xi_{kj} = 0$ for k odd, $\eta_{kj} = 0$ for k even.

Let $M \equiv M(x_1, \cdots x_N)$ denote the coefficient matrix of the linear system (3.5). In what follows we always take

$$x_\nu = \nu/N = \nu/(2q+1) \ , \quad (\nu=1, \cdots, N) \ .$$

Certainly $M(1/N, 2/N, ..., N/N)$ is non-singular. In fact, its corresponding determinant $\det M$ has the value

$$(3.7) \qquad \det M = 2^{-\nu} \cdot (-1)^{\frac{1}{2}q(q+2)} \prod_{0 \le \nu < \mu \le 2q} (\varsigma^\nu - \varsigma^\mu) \ ,$$

where $\varsigma = e^{2\pi i/N}$, $(i = \sqrt{-1})$, is the N-th root of unity with $N = 2q+1$, and the product is taken over all the integer pairs (ν, μ) within the interval $[0, 2q]$.

Thus it is clear that a quadrature formula of any given trigonometric precision q of the form

$$(3.8) \qquad \int_0^1 B_k(t)\phi(t)dt \cong \sum_{\nu=1}^N C_\nu^{(k)} \phi(\nu/N)$$

can always be constructed explicitly, where $N = 2q+1$.

Let us now denote

$$(3.9) \qquad \sigma(F,\lambda,0) = \sum_{\nu=1}^N C_\nu^{(1)} \left[F\left(1, \frac{\nu}{N}+\lambda\right) - F\left(0, \frac{\nu}{N}\right) \right],$$

and for $k \ge 1$,

$$(3.10) \qquad \sigma(F,\lambda,k) = \sum_{\mu=0}^{m-1} B_\mu^{(k)} \sum_{\nu=1}^N C_\nu^{(k+1)} F\left(1-\mu\lambda^{-\alpha}, \frac{\nu}{N}+\lambda\right)$$

$$- \sum_{\mu=0}^{m-1} A_\mu^{(k)} \sum_{\nu=1}^N C_\nu^{(k+1)} F\left(\mu\lambda^{-\alpha}, \frac{\nu}{N}\right) \ .$$

Thus, according to (2.9)-(2.11) and (3.2)-(3.3), and making use of (3.8) we obtain

$$(3.11) \qquad \int_0^1 \tilde{I}(y,k+1,\lambda)dy \cong \lambda^{\alpha k} \sigma(F,\lambda,k) \ ,$$

where $k = 0,1, \cdots, (m-2)$.

Finally, substituting (3.11) into (2.11) and using (3.1), we arrive at the desired result $I(F;\lambda) \cong S(F;\lambda)$ with

$$(3.12) \qquad S(F;\lambda) = \sum_\mu \sum_\nu a_\mu b_\nu F(x_\mu, y_\nu) + \sum_{k=0}^{m-2} \frac{\lambda^{(\alpha-1)k-1}}{(k+1)!} \sigma(F,\lambda,k) ,$$

where $b_\nu = 1/(q+1), y_\nu = \nu/(q+1), (\nu=1, \cdots, q+1), (m-2)/(m-1) \le \alpha \le 1$, and $N = 2q+1$.

THEOREM. For any given positive integers p, q, and $m \ge 3$, and any fixed parameter α with $(m-2)/(m-1) \le \alpha \le 1$, the numerical integration formula $I(F;\lambda) \cong S(F;\lambda)$ does always possess the compound precision $[p,q,\lambda^{-m}]$. More precisely, for all polynomials $P(x,y)$ of the form (1.3), the following inequality

$$(3.13) \qquad |I(P;\lambda) - S(P;\lambda)| \le \frac{\lambda^{-m}}{m!} \left(\sum_{k=2}^{m} \gamma_k \, \|B_k\| \right)$$

holds for all sufficiently large λ, where γ_k's are given by (2.13) with $F := P$.

PROOF. In view of (2.12) given by Lemma 2, it suffices to show that for $F(x,y) := P(x,y)$ we have

$$\tilde{I}(\lambda) \equiv \tilde{I}(P;\lambda) = S(P;\lambda) .$$

Let us recall (2.11) and notice that

$$\int\int_R P(x,y)\,dx\,dy = \sum_\mu \sum_\nu a_\mu b_\nu P(x_\mu, y_\nu) .$$

Thus in accordance with (3.12) it suffices to verify that (3.11) is exact for $F := P(x,y)$. More precisely, let us denote the espressions contained in (2.9)-(2.10) by $\tilde{I}(y,k+1,\lambda;F)$ and $\tilde{I}(y,1,\lambda;F)$ respectively. Observe that all those $P(\cdot,y)$ as well as $P(\cdot,y+\theta)(\theta=0$ or $\lambda)$ are trigonometrical polynomials in y with period unity, and of degrees not exceeding q. Thus from (2.9)-(2.10), (3.2)-(3.3) and (3.8) it follows that

$$\int_0^1 \tilde{I}(y,k+1,\lambda;P)\,dy = \lambda^{\alpha k} \sigma(P,\lambda,k) ,$$

where $k = 0,1,...,(m-2)$. This shows that (3.11) is exact for $F := P$, and consequently (3.13) is proved.

Notice that for $m \ge 2$ we have

$$\|B_m\| = \max_{0 \le t \le 1} |B_m(t)| \le 2 \cdot \frac{m!}{(2\pi)^m} \varsigma(m) \, ,$$

where $\varsigma(m) = \sum_1^\infty (1/k)^m$, and

$$\sum_{k=1}^m |S_1(m,k)|(2\pi)^k = \prod_{j=0}^{m-1} (1/2\pi + j) < \frac{m!}{2\pi} \, .$$

Utilizing these facts one may deduce from (3.13) the following

COROLLARY. There is an explicit estimation for the left-hand side of (3.13), namely

$$(3.14) \qquad |I(P;\lambda) - S(P;\lambda)| < \mathbb{K} \cdot \lambda^{-m} \, ,$$

where $\mathbb{K}$ is given by

$$(3.15) \qquad \mathbb{K} = \left((m-2)!/6 + 2\varsigma(m)(2\pi)^{-m}\right) \|P_x^{(m)}\| + 4\varsigma(m)(2\pi)^{-m} \|P_x^{(m-1)}\| \, .$$

REMARKS. (i) Clearly (3.12) involves a class of quadrature formulas, in which the parameter α may be chosen with some freedom, and all the constructive coefficients $a_\mu, A_\mu^{(k)}, B_\mu^{(k)}$ and $C_\nu^{(k+1)}$ may be found by solving linear systems of equations with the aid of computers. Of course, one may also make use of other kind of numerical differentiation formulas instead of Markov's.

(ii) A specialization of (3.12) to the case $F(x,y) = f(x)g(y)$ may be used to compute transformation integrals containing a large parameter λ. For such a case the constructive process we have employed can be simplified.

ACKNOWLEDGMENT. The authors wish to thank the referee for suggestions that lead to an improved version of the paper.

REFERENCES

[1]   P.J. Davis and P. Rabinowitz, Methods of Numerical Integration, Academic Press, New York, 1975.

[2]   P.C. Hammer, A.W. Wymore and A.H. Stroud, Numerical evaluation of multiple integrals, I, II, *Math. Tables Aids Comput.* **11** (1957), 59-67; **12** (1958), 272-280.

[3]   T. Havie, Remarks on an expansion for integrals of rapidly oscillating functions, *BIT* **13** (1973), 16-29.

[4]   L.C. Hsu, Approximate integration of rapidly oscillating functions and of periodic functions, *Proc. Cambridge Philos. Soc.* **59** (1963), 81-88.

[5]   L.C. Hsu and Y.S. Chou, An asymptotic formula for a type of singular oscillatory integrals, *Math. Comp.* **37** (1981), 403-407.

[6]   L.C. Hsu and Y.S. Zhou, Approximate integration of rapidly oscillating functions with compound precision (Abstract), to appear in *Approximation Theory V* (Proceedings of Fifth Conference on Approximation Theory, College Station, Texas 1986).

[7]   L.S. Hsu and X.H. Wang, Methods of Mathematical Analysis with Selective Examples, Press of High Education, Beijing (Peking, China), 1983.

BOUNDS AND APPROXIMATIONS FOR THE ZEROS OF CLASSICAL ORTHOGONAL POLYNOMIALS. THEIR USE IN GENERATING QUADRATURE RULES.

Luigi Gatteschi
Dipartimento di Matematica
University di Torino
Via Carlo Alberto 10
I-10123 Torino, Italy

ABSTRACT

We have recently shown [2] that certain asymptotic approximations for the zeros $\vartheta_{n,k}$, $k=1,2,...,n$, of Jacobi polynomials $P_n^{(\alpha,\beta)}(\cos\theta)$ are in fact inequalities. In particular a lower bound for $\vartheta_{n,k}$ is obtained by omitting the $O(n^{-3})$ term in Frenzen-Wong formula [1]; that is the following inequality holds, under the hypotheses $-\frac{1}{2}\leq\alpha\leq\frac{1}{2}$ and $-\frac{1}{2}\leq\beta\leq\frac{1}{2}$,

$$\vartheta_{n,k} \geq t - \left[\left(\tfrac{1}{4}-\alpha^2\right)\left(\tfrac{2}{t}-\cot\tfrac{t}{2}\right) + \left(\tfrac{1}{4}-\beta^2\right)\tan\tfrac{t}{2}\right]/(4N^2),$$

with

$$N = n + \frac{\alpha+\beta+1}{2}, \quad t \equiv t_{n,k}(\alpha,\beta) = \frac{j_{\alpha,k}}{N}$$

for $k=1,2,...,n$, and where $j_{\alpha,k}$ is the k-th zero of the Bessel function $J_\alpha(x)$. Other bounds can be derived from the more elementary asymptotic formula [3]

$$\vartheta_{n,k} = v + \left[\left(\tfrac{1}{4}-\alpha^2\right)\cot\tfrac{v}{2} - \left(\tfrac{1}{4}-\beta^2\right)\tan\tfrac{v}{2}\right]/(4N^2) + O(n^{-4}),$$

with

$$N = n + \frac{\alpha+\beta+1}{2}, \quad v \equiv v_{n,k}(\alpha,\beta) = \frac{2k+\alpha-\frac{1}{2}}{2n}\pi,$$

for all the zeros $\vartheta_{n,k}$ belonging to the interval $a<\vartheta<b$, with $0<a<b<\pi$, and under the previous hypotheses for the parameters α and β. The main tool that we use in deriving the bounds is the well-known Sturm comparison theorem.

In the first part of this lecture we shall discuss the above formulas, the case $\alpha=0$, $\beta=1$, and we shall give a survey of the results we have obtained for the zeros $\lambda_{n,k}$ of the Laguerre polynomial $L_n^{(\alpha)}(x)$. In particular we shall

P. Keast and G. Fairweather (eds.), Numerical Integration, 103–104.

present two new uniform asymptotic approximations of $\lambda_{n,k}$ involving the zeros of the Bessel function $J_\alpha(x)$ or the zeros of the Airy function $Ai(-x)$.

The second part of the lecture will be devoted to the study of the accuracy of the previous approximations and to the comparison of some higher-order iteration methods for solving $P_n^{(\alpha,\beta)}(x) = 0$ or $L_n^{(\alpha)}(x) = 0$, that is, for generating classical Gaussian quadrature rules. We shall describe a number of numerical experiments which use a fifth-order iteration method, analogous to the one used by Lether [4] in the construction of Gauss-Legendre rules, or the Piessens method [5] for solving algebraic equations.

REFERENCES

1.   C.L. Frenzen and R. Wong, A uniform asymptotic expansion of the Jacobi polynomials with error bounds, Can. J. Math. 37 (1985), pp. 979-1007.

2.   L. Gatteschi, New inequalities for the zeros of Jacobi polynomials, SIAM J. Math. Anal., to appear.

3.   L. Gatteschi and G. Pittaluga, An asymptotic expansion for the zeros of Jacobi polynomials, in J.M. Rassias (Ed.), Mathematical Analysis, Teubner-Texte zür Math., Bd. 79 (1985), pp. 70-86.

4.   F.G. Lether, On the construction of Gauss-Legendre quadrature rules, J. Comp. and Appl. Math., 4 (1978), pp. 47-51.

5.   R. Piessens en J. Denef, Hoge-orde iteratieformules voor de berekening van nulpunten van bijzondere funkties, Simon Stevin, Wis-en Nat. Tijdschrift, 46 (1972-73), pp. 53-61.

INDEFINITE INTEGRATION FORMULAS BASED ON THE SINC EXPANSION

S. Haber
National Bureau of Standards
Mathematical Analysis Division
Gaithersburg, MD 20899
U.S.A.

ABSTRACT

The "Tanh rule" of C. Schwartz approximates $\int_{-1}^{1} g(u)\,du$ by

$$I_N = h \sum_{j=-N}^{N} \frac{g\left(tanh\,\dfrac{jh}{2}\right)}{2\cosh^2 \dfrac{jh}{2}},$$

where $h = h(N)$ is given by a certain formula; $h \sim cN^{-\frac{1}{2}}$. This integration rule has been shown to be remarkably accurate. Its error is $O\left(e^{-c\sqrt{N}}\right)$ for some $c > 0$ even when the integrand is infinite at an endpoint of the integration interval.

We shall derive, by methods developed by Frank Stenger and his associates, some related formulas for numerical indefinite integration. The results depend on analyticity of the integrands on the interior of the integration interval, and are valid in the presence of certain endpoint singularities. We standardize on the interval $(-1,1)$ and the lenticular regions Λ_d in the complex plane: Λ_d is bounded by two circular arcs passing through -1 and 1, one arc in the upper half-plane and one in the lower, each arc making an angle of d with the x axis at each point ± 1. Then we have:

Theorem: If for some $d > 0$ $f \le E^1(\Lambda_d)$, (that is, f is analytic in Λ_d and $\int_C |f|$ is bounded along some sequence of contours C that approach $\partial \Lambda_d$) and $f(x) = O\left((1-x^2)^{\alpha-1}\right)$ for some $\alpha > 0$ as $x \to \pm 1$ from inside $(-1,1)$, then

$$\int_{-1}^{x} f(t)\,dt = h \sum_{k=-N}^{N} \psi'(kh) \cdot f(\psi(kh)) \cdot \left(\frac{1}{2} + \sigma\left(\frac{\phi(x)}{h} - k\right)\right) + O\left(e^{-\sqrt{\pi d \alpha N}}\right)$$

uniformly in x. Here ψ and ϕ are the functions $tanh\,(x/2)$ and (its inverse) $log(1+x)/(1-x)$ respectively, and σ is a form of the sine-integral:

P. Keast and G. Fairweather (eds.), Numerical Integration, 105–106.

$$\sigma(x) = \frac{1}{\pi} \int\limits_0^{\pi x} \frac{sint}{t} \, dt \quad .$$

The quantity h depends on N:

$$h(N) = (\frac{\pi d}{\alpha N})^{\frac{1}{2}} + \frac{logN}{2\alpha N} \quad .$$

To use this formula one must have a subroutine for calculating the sine-integral. A second formula that partly avoids this need is:

$$\int\limits_{-1}^x f(t)dt = h \sum_{k=-N}^N \sum_{j=-N}^N \psi'(jh) \cdot f(\psi(jh)) \cdot (\frac{1}{2} + \sigma(k-j)) \cdot sinc\left(\frac{\phi(x)}{h} - k\right)$$

$$+ \left[h \sum_{j=-N}^N \psi'(jh)f(\psi(jh)) \right] \cdot \left[\eta(x) - \sum_{k=-N}^N \eta(\psi(kh))sinc\left(\frac{\phi(x)}{h} - k\right) \right]$$

$$+ 0\left(Ne^{-\sqrt{\pi d \alpha N}} \right) \quad .$$

Here $sinc \, x = (sin \, \pi x)/\pi x$; a simple routine for calculating $\sigma(n)$ for integral n is described.

Graphs of the integration error function, for various sample integrands, are shown. They show that the asymptotic error bound stated above does describe the behavior of the error for moderate values of N ($N < 50$).

ON POSITIVE QUADRATURE RULES

Hans Joachim Schmid
Mathematisches Institut
Univ. Erlangen-Nurnberg
Bismarckstrasse 1 1/2
8520 Erlangen FDR

ABSTRACT

Recently Peherstorfer [1] derived a characterization of positive quadrature rules with knots in (a,b). An equivalent characterization will be presented using the positive definiteness of a matrix of moments. This is the one-dimensional case of a characterization of interpolatory cubature formulae, see, for example, [2] and [3].

Let $q \in \mathbb{P}_k = \mathrm{span}\{1,x,x^2,...,x^k\}$ of degree k be given. We denote by $\Pi_q : \mathbb{P} \to \mathbb{P}_{k-1}: f \to \Pi_q(f)$, $\mathbb{P} = \mathrm{span}\{1,x,x^2,...\}$, the linear projection from $\mathbb{P}$ to $\mathbb{P}_{k-1}$ defined by the unique representation of f as $f = rq + \Pi_q(f)$, $r \in \mathbb{P}$, $\Pi_q(f) \in \mathbb{P}_{k-1}$. For a strictly positive linear functional I on $\mathbb{P}([a,b])$, we define an associated linear functional depending on q by

$$I_q: \mathbb{P}_{2k-1}(\mathbb{R}) \to \mathbb{R}: f \to I_q(f) = I\big(\Pi_q(f)\big) .$$

A positive quadrature rule with k knots of degree m for I is a linear functional

$$Q: \mathbb{P} \to \mathbb{R}: f \to Q(f) = \sum_{i=1}^{k} c_i f(x_i), x_i \in (a,b), c_i > 0 ,$$

such that $Q(f) = I(f)$ for all $f \in \mathbb{P}_m$ and $Q\big(x^{m+1}\big) \neq I\big(x^{m+1}\big)$ hold. Q is generated by the roots of the polynomial $q(x) = (x-x_1)(x-x_2) \cdots (x-x_k)$.

Theorem.

Let s be given, $1 \leq s \leq k+1$, and let $q \in \mathbb{P}_k$ of degree k be orthogonal to $\mathbb{P}_{k-s}$ with respect to I, i.e. $I(fq) = 0$ for all f in $\mathbb{P}_{k-s}$. A positive quadrature rule of degree $2k-s$ with k knots in (a,b) is generated by the roots of q, if and only if I_q is strictly positive on $\mathbb{P}_{2k-1}([a,b])$.

In particular, the quadrature rule of degree $2k-s$ for I is a Gaussian formula for I_q.

P. Keast and G. Fairweather (eds.), Numerical Integration, 107–108.

The characterization can be applied to construct positive quadrature rules of degree $2k-s$ for I by studying the positive definiteness of matrices of moments, associated with I_q. Similarly positive quadrature rules with knots in [a,b] can be obtained.

REFERENCES

1.   F. Peherstorfer: Characterization of quadrature formulae II, SIAM J. Math. Anal. 15 (1984), pp. 1021-1030.

2.   G. Renner: Darstellung von strikt quadratpositiven linearen Funktionalen auf endlichdimensionalen Polynomräumen. Dissertation, Erlangen, 1986.

3.   H.J. Schmid: Interpolatorische Kubaturformeln. Diss. Math. CCXX (1983), pp. 1-122.

QUADRATURE RULES WITH END-POINT CORRECTIONS: COMMENTS ON A PAPER BY GARLOFF, SOLAK AND SZYDELKO

William Squire
Department of Mechanical
and Aerospace Engineering
West Virginia University
Morgantown, WV 26506-6101
U.S.A.

ABSTRACT

Garloff, Solak and Szydelko [1] have proposed a modified trapezoidal rule quadrature with correction terms involving two points outside the range of integration. The rule is exact for cubics. We show:

1) There is an arbitrary factor in their formula but it would require a complex value to make the quadrature exact for quintics.

2) An analogous quadrature can be based on the midpoint rule.

3) By combining the two rules the quadrature is made exact for quintics. Furthermore the procedure is progressive so that the number of divisions can be doubled without wasting function evaluations.

4) The method can be used for integrands which are not real beyond an endpoint such as $x^{1/2}$ and $x \ln x$ which vanish at the endpoints by defining f (endpoint$-x$) as $-f$ (endpoint$+x$).

REFERENCE

1. J. Garloff, W. Solak and Z. Szydelko, New integration formulas which use nodes outside the integration interval, J. Franklin Institute, 321 (1986), pp. 115-126.

P. Keast and G. Fairweather (eds.), Numerical Integration, 109.

COMPARISON OF GAUSS-HERMITE AND MIDPOINT QUADRATURE WITH APPLICATION TO THE VOIGT FUNCTION

William Squire
Department of Mechanical
and Aerospace Engineering
West Virginia University
Morgantown, WV 26506-6101
U.S.A.

ABSTRACT

The abnormal accuracy of the midpoint and trapezoid rules for integrals of the form

$$\int_{-\infty}^{\infty} e^{-t^2} f(t)\, dt$$

and other integrands decaying rapidly for large $|t|$ has been known for some time [1]. Its principal use has been to derive quadrature rules with exponential convergence for finite and semi-infinite ranges.

We will compare the accuracy of the midpoint rule and Gauss-Hermite quadrature for:

$$\int_{-\infty}^{\infty} e^{-t^2} \cos 2\omega t \, dt = \pi^{\frac{1}{2}} e^{-\omega^2} \,, \qquad\qquad\qquad \text{A}$$

$$V(x,y) = \frac{y}{\pi} \int_{-\infty}^{\infty} \frac{e^{-t^2}\, dt}{(x-t)^2 + y^2} = \text{Real part } \frac{i}{\pi} \int_{-\infty}^{\infty} \frac{e^{-t^2}\, dt}{z-t} \qquad\qquad \text{B}$$

where $V(x,y)$ is the Voigt function which is of importance in spectroscopy. The complex integral is equivalent to the complex probability integral and the imaginary part is Dawson's integral.

The integrals are hard to evaluate for x and y small because of the poles at $t = x \pm iy$. Humlecek has proposed a shift to make the integral more tractable.

If

$$\int_{-\infty}^{\infty} \frac{e^{-t^2}\, dt}{z-t} = \sum \frac{W_k}{z-t_k}$$

111

P. Keast and G. Fairweather (eds.), Numerical Integration, 111–112.

where W_k and $\dot{t}_k$ are a set of weights and nodes then

$$e^{-\omega^2}\sum W_k \frac{(-\sin 2\omega t_k + i \cos 2\omega t_k)}{(x+t_k)+i(y+\omega)}$$

will eliminate the difficulty if ω is selected so that the quadrature rule gives the desired accuracy for A.

I have found that for 12 nodes the midpoint rule evaluates A more accurately than Gauss-Hermite quadrature and have just begun to get weights for the complex integrals. The method does give an accurate result for $z = 0$.

REFERENCES

1. E.T. Goodwin, The evaluation of integrals of the form $\int\limits_{-\infty}^{\infty} f(x)e^{-x^2}dx$, Proc. Camb. Phil. Soc. 45 (1949), pp. 241-245.

2. J. Humlecek, An efficient method for evaluation of the complex probability integral: the Voigt function and its derivatives J. Quant. Spectros. Radiat. Transfer., 21 (1979), pp. 309-313.

ON SEQUENCES OF IMBEDDED INTEGRATION RULES

Philip Rabinowitz
Weizmann Institute of Science, Rehovot, Israel.

Jaroslav Kautsky
Flinders University, Bedford Park, S.A., Australia.

Sylvan Elhay
University of Adelaide, Adelaide, S.A., Australia.

John C. Butcher
University of Auckland, Auckland, New Zealand

ABSTRACT

It is shown that given a one-dimensional interpolatory integration rule with positive weights based on a set S of n points, it is possible to find a subset T of S containing n-1 points such that the weights of the interpolatory integration rule based on T are all nonnegative. Consequently, given any one-dimensional positive interpolatory integration rule R, we can generate a sequence of imbedded positive interpolatory rules starting with R. This idea of starting with a rule R and generating a sequence of imbedded rules, which is due to Patterson, can be applied to generate sequences of imbedded multidimensional integration rules. In this case, we do not aim for positive rules but for efficient rules. Some examples are given as to what can be attained.

1.   INTRODUCTION

Several years ago, the first three authors wrote up a paper based on a bit of experimental mathematics. This paper, given in Appendix A, was submitted for publication and returned for revision with the reports of two referees, given in Appendix B. While some of the points raised by referee #2 were valid, his general remark as well as the final remark of referee #1 about the style of the paper were contrary to the intention of the authors who believed that in a paper on experimental mathematics, one should describe in detail the progress of the experiments. As for the main points raised by referee #1, the authors felt that the major interest of the paper was in the conjecture about the existence of sequences of imbedded positive interpolatory integration rules rather than in the calculation and application of such sequences. We leave it to the reader to judge which point of view is more valid. At any rate, the referees' reports reached the first author (PR) when he was visiting at the University of Auckland. When he expressed to his host, Professor Butcher, his dismay at having to

P. Keast and G. Fairweather (eds.), Numerical Integration, 113–139.
© *1987 by D. Reidel Publishing Company.*

revise his paper since he felt the main results were the conjectures, Professor Butcher proceeded to prove the main conjecture. This proof and the corollary are given in Section 2. Thinking about this proof, which is for one-dimensional integration rules, led to some thoughts about multi-dimensional integration rules. This led to the construction of some sequences of imbedded integration rules in 2, 3 and 4 dimensions. These results are given in Section 3.

2. IMBEDDED SEQUENCES OF POSITIVE INTERPOLATORY INTEGRATION RULES

We shall be considering integrals of the form

$$I(kf) = \int_a^b k(x) f(x)dx \ , \ -\infty \le b < a \le \infty \tag{1}$$

where $k(x)$ is such that $I(kf)$ exists for a family of functions F which includes P_n, the set of all polynomials of degree $\le n$, for $n = 1,2,\dots$. If $[a,b]$ is a finite interval and $k \in L_1(a,b)$, then we can take F to be $R[a,b]$, the set of all (bounded) Riemann-integrable functions on $[a,b]$. We shall be interested in approximating $I(kf)$ by interpolatory (product) integration rules (IIR's) of the form

$$Q_n f = \sum_{i=1}^n w_{in} f(x_{in}) \tag{2}$$

where the set of points $X_n = \{x_{in} : i = 1,\dots,n\}$ is specified in advance and the weights $w_{in} = w_{in}(k)$, $i = 1,\dots,n$, are interpolatory. This is equivalent to the condition that $Q_n f = I(kf)$ for all $f \in P_{n-1}$ [7, p.74]. The IIR, $Q_n f$, is said to be based on X_n. An IIR is said to be a positive IIR (PIIR) if $w_{in} > 0$, $i = 1,\dots,n$. $Q_n f$ is said to be of precision p if $Q_n f = I(kf)$ for all $f \in P_p$. If $Q_n f$ is an IIR, then its precision is at least $n-1$. Given two distinct IIR's $Q_n f$ and $Q_m f$ where $Q_n f$ is based on X_n and $Q_m f$, on X_m, we say that $Q_m f$ is (totally) imbedded in $Q_n f$ if $X_m \subset X_n$ and that $Q_m f$ is partially imbedded in $Q_n f$ if $m < n$ and $X_m \cap X_n \ne \phi$. We are now ready to state our main result.

Theorem 1: Given any PIIR $Q_n f$ based on a set X_n, there exists a finite sequence of PIIR's, $\{Q_{n_k} f ; k = 1,\dots,m \le n\}$ such that $Q_{n_1} f = Q_n f$ and such that $Q_{n_{k+1}} f$ is imbedded in $Q_{n_k} f$ and is of precision $n_k - 2$ for $k = 1,\dots,m-1$.

Proof: It suffices to show that for any PIIR based on $X_s = \{x_{is} : i = 1,\dots,s\}$

$$Q_s f = \sum_{i=1}^s w_{is} f(x_{is}) \ , \ w_{is} > 0 \tag{3}$$

there exists an index k such that in the IIR based on the set $\{x_{is} : i = 1,...,s ; i \neq k\}$

$$Q_{s-1}f = \sum_{\substack{i=1 \\ i \neq k}}^{s} w_{i,s-1} f(x_{is}) \tag{4}$$

the weights $w_{i,s-1} \geq 0$. To show the existence of such k, consider the divided difference operator based on X_s

$$D_s f = f[x_{1s},...,x_{ss}] = \sum_{i=1}^{s} \alpha_{is} f(x_{is}) . \tag{5}$$

It is well-known [1, p.123] that $D_s f = 0$ if $f \in P_{s-2}$ and in particular, for $f \equiv 1$, $D_s f = 0$ so that $\sum_{i=1}^{s} \alpha_{is} = 0$. Hence the integration rule

$$Q_s f - K D_s f = \sum_{i=1}^{s} (w_{is} - K\alpha_{is}) f(x_{is}) \tag{6}$$

is exact for all $f \in P_{s-2}$. If we now define k to be any (say the smallest) integer such that

$$w_{ks} / \alpha_{ks} = \min_{\alpha_{is} > 0} w_{is} / \alpha_{is} \equiv K$$

then the integration rule (6) is based on the set $X_s - \{x_{ks}\}$ and is exact for all $f \in P_{s-2}$. Therefore, it is an IIR and hence identical to (4). Clearly, for all i,

$$w_{i,s-1} = w_{is} - K\alpha_{is} \geq 0 .$$

An alternative choice for k is the smallest integer such that

$$w_{ks} / \alpha_{ks} = \max_{\alpha_{is} < 0} w_{is} / \alpha_{is} \equiv -K$$

so that there are in fact at least two sequences satisfying the conditions of the theorem.

A similar proof in a different context appears in Brass [5]. The idea of dropping a point from a Gauss rule and using the IIR based on the remaining points to estimate the error in the Gauss rule was suggested by Berntsen and Espelid [3] and Laurie [12].

Before we state a corollary involving symmetric rules, we require the following definition.

Definition: Let $[a,b]$ be a finite interval or $(-\infty,\infty)$ and let $c = (b + a) / 2$ or 0, respectively. We say that we have a *symmetric situation* if $k(c - x) = k(c + x)$ and the integration rule $Q_n f$ is symmetric, i.e.

$$x_{in} + c = c - x_{n+1-i,n} \text{ and } w_{in} = w_{n+1-i,n} , \quad i = 1,...,n .$$

Note that in a symmetric situation, if n is odd and Q_nf is an IIR, then Q_nf is of precision n.

Corollary: In a symmetric situation, if Q_nf is a (symmetric) PIIR, then there exists a finite sequence of imbedded symmetric PIIR's starting with Q_nf. The precision of each rule is at least $r-3$ where r is the number of points in the previous rule in the sequence.

Proof: (a) $n = 2m + 1$. In this case, D_nf is also symmetric, $\alpha_{in} = \alpha_{n+1-i,n}$. Hence $Q_nf - KD_nf$ is also symmetric and the construction in the proof of Theorem 1 yields a symmetric PIIR with at most $2m - 1$ points for at least one of the choices of k. This rule is of precision $n-2$.

(b) $n = 2m$. In this case, D_nf is antisymmetric, $\alpha_{in} = -\alpha_{n+1-i,n}$, so that the construction given in the proof of Theorem 1 does not yield a symmetric rule. However, we can use the following device. Consider the rule

$$Q_mf = \sum_{i=1}^{m} w_{in} f(x_{in}) \tag{7}$$

approximating the integral $\int_a^c k(x) f(x)dx$, where we assume that the x_{in} are ordered in ascending order. Then Q_mf is exact for all even polynomials in P_{n-2}. We consider now the *null rule* [12]

$$N_mf = \sum_{i=1}^{m} \beta_{im} f(x_{in}), \ \beta_{i1} = 1 \tag{8}$$

with the property that $N_mf = 0$ for all even polynomials in P_{n-4}. We can determine the β_{im} by solving the system of linear equations

$$\sum_{i=2}^{m} \beta_{im} x_{in}^{2j} = -x_{1n}^{2j}, \ j = 0,...,m-2$$

which has a unique solution. Then, as before, $Q_mf - KN_mf$ is exact for all even polynomials in P_{n-4} and we can choose K such that one weight vanishes and all the remaining weights are non-negative. If the index of the vanishing weight is k, then the IIR based on the set $\{x_{in} : i = 1,...,n ; i \neq k, n+1-k\}$ is symmetric, has all non-negative weights and is exact for all $f \in P_{n-3}$, proving our corollary.

Note that we could have proved this second case in other ways. However, we chose this proof since it leads to a method to generate sequences of imbedded multi-dimensional integration rules which are not necessarily interpolatory but do integrate exactly a certain collection of monomials. This is what we study in the next section.

3.   MULTIDIMENSIONAL INTEGRATION RULES

In this discussion of multidimensional integration rules, we shall restrict ourselves to integrals of the form

$$\text{If } = \int_{H_n} f(x)dx \tag{9}$$

where $x = (x_1,...,x_n)$ and H_n is the n-dimensional hypercube

$$H_n : -1 \le x_i \le 1 \, , \, i = 1,...,n .$$

However, many of our results carry over to the more general integral

$$\int_{F_n} w(x) f(x)dx$$

where F_n is a fully-symmetric region and $w(x)$ is a fully-symmetric weight function (For definitions and background reading, see [14] and [7, Section 5.7]). We shall approximate If by fully symmetric integration rules of the form

$$Q_m f = \sum_{i=1}^{m} w_i \sum_{FS} f(y_i) \tag{10}$$

where the symbol $\sum_{FS}$ denotes the sum over $FS(y_i)$, the set of all points of the form $(\pm y_{p(1)},...,\pm y_{p(n)})$, and $\{p(1),...,p(n)\}$ is any permutation of $\{1,...,n\}$. The *generators* y_i will generally have a special form to reduce the number of points in the set $FS(y_j)$. Thus in 3 dimensions, which we shall treat most fully, $Q_m f$ will have the form

$$Q_m f = \sum_{i=1}^{K_0} w_{i0} f(0,0,0) + \sum_{i=1}^{K_1} w_{i1} \sum_{FS} f(y_{i1},0,0)$$

$$+ \sum_{i=1}^{K_2} w_{i2} \sum_{FS} f(y_{i2},y_{i2},0) + \sum_{i=1}^{K_3} w_{i3} \sum_{FS} f(y_{i3},z_{i3},0)$$

$$+ \sum_{i=1}^{K_4} w_{i4} \sum_{FS} f(y_{i4},y_{i4},y_{i4}) + \sum_{i=1}^{K_5} w_{i5} \sum_{FS} f(y_{i5},y_{i5},z_{i5})$$

$$+ \sum_{i=1}^{K_6} w_{i6} \sum_{FS} f(y_{i6},z_{i6},v_{i6}) \tag{11}$$

where $\sum_{j=0}^{6} K_j = m$ and $v_{ik} > z_{ik} > y_{ik} > 0$.

Given a rule (10) with m generators specified in advance, we can usually choose the weights w_i to integrate exactly m monomials of the form

$$x^j = x_1^{j_1} \dots x_n^{j_n}, \ j_1 \geq j_2 \geq \dots \geq j_n \geq 0 \ , \ j_i \ \text{even}.$$

We shall usually assume an ordering in the vectors $j = (j_1,\dots,j_n)$ as follows: $j < k$ if

$|j| < |k|$ or if $|j| = |k|$ and $j_1 = k_1, \dots, j_p = k_p$, $j_{p+1} < k_{p+1}$ where $|j| = \sum_{i=1}^{n} j_i$.

Then, given the m generators y_i , $i = 1,\dots,m$, we set up the m linear equations corresponding to the first m monomials specified by the first m vectors j_e,

$$\sum_{i=1}^{m} w_i \sum_{FS} y_i^{j_e} = I(x^{j_e}) \quad e = 1,\dots,m \tag{12}$$

and solve for the w_i , assuming that the matrix is nonsingular. This corresponds to the one-dimensional IIR except that, in contrast to the one-dimensional case, the matrix may be singular. Returning to the three-dimensional example, there are certain constraints on the K_j which insure that the matrix is nonsingular [11,14].

As in the one-dimensional case, a particular choice of generators may integrate exactly many more monomials and these are the ones most sought after and most difficult to obtain. In the literature, integration rules of type (10) with arbitrary m are not usually studied. Rather, integration rules of precision p are studied, which means rules which integrate exactly all monomials x^j such that $|j| \leq p$. It is not at all clear that the use of such rules of precision p is the best strategy [6]. However, almost all known rules (10) were computed to be of precision p [8,9,10,14,18]. The computation of economical rules of high precision, $p > 11$, is quite a complicated problem and the presently known rules [9,10] are far from optimal. At any rate, we should consider the more general case of arbitrary m.

Given a set of generators, $\{y_i , i = 1,\dots,m\}$ such that the system (12) has a solution, we can always obtain a null rule

$$N_m f = \sum_{i=1}^{m} \beta_i \sum_{FS} f(y_i)$$

such that $N_m f = 0$ if $f(x) = x^{j_e}$, $e = 1,\dots,m-1$. Then we have a theorem corresponding to Theorem 1.

Theorem 2: Given a rule $Q_m f$ with m generators and positive weights which is exact for the first m monomials, then there exists a rule with $m - 1$ generators and non-negative weights which is exact for the first $m - 1$ monomials. Consequently we can find a sequence of imbedded positive integration rules starting with $Q_m f$.

However, this theorem is not as useful as Theorem 1. First, there are not that many positive integration rules, especially for high precision and large dimension and those that exist, such as the product Gauss rules, require an enormous amount of function evaluations. Second, the rules that do exist are not that accurate so that the loss of accuracy caused by negative weights is usually not critical provided that the magnitudes of these weights are of reasonable size. Third, not all generators require the same number of function evaluations so that it is more efficient to drop a more expensive generator rather than a less expensive one, by not requiring that the weights of the resulting rule remain non-negative.

We shall therefore attempt to generate sequences of imbedded integration rules by attempting to remove as expensive a generator as possible from the previous rule. In contrast to the case in Theorem 2, we will not always be able to remove any generator we wish since the coefficient of that generator in the null rule may vanish. Hence our strategy will be to remove the most expensive generator with a non-vanishing coefficient in the null rule. It may be that we can remove any one of several equally expensive generators. In that case, our criterion will be to remove that generator for which the weights in the resulting rule are closest to positivity.

A more haphazard way of going about this, but one which allows one to remove several generators simultaneously, thus allowing one to produce rules of a specified precision, is to set up a system of equations (12) without one or several generators and try to solve it. If it has a solution, fine; if not, we try to remove a different group of generators. Usually, if we succeed with a certain group of generators, we will succeed with another group of generators similar in structure to the successful one in which case we again have a choice which we resolve as above. We illustrate this procedure in the following example.

Example 1.

Consider the Stenger 3-dimensional rule of precision 11 given in [17,18]. It has 13 generators listed in Table 1 and requires 151 function evaluations. Given these generators, one should be able to calculate the weights by solving the linear system given by the first 13 equations in (12). However, the matrix of this system is singular. On the other hand, in the 16×13 matrix arising from the system given by all 16 equations in (12) which must be satisfied for the rule to be of precision 11, the rank is 13 and we can determine the weights by QR decomposition. We must also have recourse to some sophisticated methods of numerical linear algebra when we try to compute an imbedded rule of precision 9 which should involve 11 generators. For when we set up the 11×13 matrix corresponding to the first 11 equations in (12), it turns out to be deficient with rank 10. Fortunately, this matrix augmented by the right hand side of (12) also has rank 10 so that the system is consistent and consequently we can get a solution with 10 rather than 11 generators. It turns out that we can remove only two of the three expensive generators. The 'best' rule is that obtained using the generators (1-6, 10-13) and requires 91 function evaluations. For precision 7, we need only 7 generators and the set (1-5, 10, 11) is best. For precision 5, we need only the four generators (1-3, 11) and for precision 3, the two generators (1,2). Thus we have a complete sequence of imbedded rules, each of higher precision than the previous one. We do not give the weights in these rules since they can be computed quite easily.

We remark that Berntsen [3] suggested the use of two embedded rules of precision p - 2 together with a sequence of rules of precision p-4,p-6,... to estimate the error of a rule of precision p . For the Stenger rule, we have given all the rules required except a second one of precision p - 2 . But this is also available using the set of generators (1-7, 9-12) or, if we want rules with less generators in common, (1-5, 7, 9-12) .

Table 1. The Generators in the Stenger 3-dimensional Rule of Precision 11.

i	y_i	i	y_i
1	$(0,0,0)$	8	$(y_{13},z_{13},0)$
2	$(y_{11},0,0)$	9	$(y_{23},z_{23},0)$
3	$(y_{21},0,0)$	10	(y_{14},y_{14},y_{14})
4	$(y_{31},0,0)$	11	(y_{24},y_{24},y_{24})
5	$(y_{12},y_{12},0)$	12	(y_{34},y_{34},y_{34})
6	$(y_{22},y_{22},0)$	13	(y_{15},y_{15},z_{15})
7	$(y_{32},y_{32},0)$		

$$y_{11} = y_{12} = y_{13} = y_{23} = y_{14} = y_{15} = .9324695142$$
$$y_{21} = y_{22} = z_{13} = y_{24} = z_{15} = .6612093865$$
$$y_{31} = y_{32} = z_{23} = y_{34} = .2386191860$$

Since the weights in Stenger's rule are not all positive, Theorem 2 is not applicable. Furthermore Stenger's rule is not very efficient requiring 151 function evaluations in contrast to Espelid's rule using only 89 points [8,p.59]. However, it has the merit of being the first non-product rule of degree 11 for the hypercube H_n for $n = 2,3,...,20$. In fact it is the only one mentioned in [18] and [14]. Improved rules for the general hypercube are given in Keast [10] and in Genz and Malik [9].

If we start with an optimal or near-optimal rule, we cannot expect to generate a sequence of rules with precision going down in steps of 2. However, if we reduce our requirement to partially imbedded sequences, we can still get someting useful as in the following two-dimensional example.

Example 2.
    Consider the 48-point two-dimensional integration rule of precision 15 computed by Rabinowitz and Richter [16]. It has 9 generators as follows:

$$(y_{i1},0), (y_{i2},y_{i2}), (y_{i3},z_{i3}) , i = 1,2,3 .$$

If we remove one generator, we cannot get a rule of precision 9 since there are 9 monomials of degree ≤ 8. However, if we add the generator $(0,0)$, then we can get a rule of precision 9 with 41 points. In fact we can get 3 distinct rules depending upon which generator (y_{i3},z_{i3}) we remove. Subsequently we can get imbedded rules of degree 7, 5 and 3 choosing the less expensive generators, the choice depending on which weights have the most desirable properties.

A less satisfactory situation occurs when we start with the three-dimensional rule of degree 9 computed by Mantel and Rabinowitz [14] as in the following example.

Example 3.
    The optimal three-dimensional rule of degree 9 with all points inside the cube has 57 points. One such rule is given in [14]; others of a different structure appear in [8]. None of these is positive. A positive rule with 58 points is given in [8]. The

generators in the rule under consideration are as follows:

$$(0,0,0), \; (y_{11},0,0), \; (y_{21},0,0), \; (y_{12},y_{12},0), \; (y_{13},z_{13},0), \; (y_{14},y_{14},y_{14}) \; .$$

If we remove one generator, we have 5 parameters at our disposal while there are 7 monomials of degree ≤ 6. Hence we cannot compute an imbedded rule of precision 7. It is easy to compute one of precision 5 by using the first four generators. However this is not too satisfactory since it yields zero for any monomial which has $x^2y^2z^2$ as a factor. On the other hand, if we use generators 1-3 and 6, we get a rule of precision 5 which is not only less expensive but also integrates exactly the monomial x^4y^2. By using 5 generators, we can determine the weights so that the rule will still be of precision 5 but will in addition be exact for the monomial $x^2y^2z^2$. Finally, we can do what we did in Example 2 and add an inexpensive generator, but this time one involving a free parameter. Then the number of unknowns grows to 7 and we can try to determine this parameter and the weights so as to integrate exactly all monomials of degree ≤ 6 thus yielding a partially imbedded rule. However, we must be careful in the choice of the form of the generator. Thus the addition of a generator of the form $(\alpha,0,0)$ does not work since the resulting system of equations has no solution. However the addition of a generator of the form (α,α,α) leads to a successful computation. We determine the value of α by binary search. For any choice of α, we solve the first 6 equations in (12) for the weights w_i, $i = 1,...,6$ and substitute in the last equation in (12) to compute the residual. This yields a pair of rules of degrees 9 and 7 using a total of 65 points. The value of $\alpha = y_{24}$ as computed is

$$y_{24} = .5476041077 \; .$$

Other 9/7 pairs with 65 points are given in [4], as well as one with 64 points. These authors have used much more sophisticated techniques to get their results in contrast to the relatively simple-minded approach used here.

As a final example, we give a partially imbedded pair of four-dimensional rules.

Example 4.
    The following 57 point four-dimensional rule of precision 7 has been computed by Mantel and Rabinowitz.

$$Q_4 f \; = \; \sum_{i=1}^{4} w_i \sum_{FS} f(y_i)$$

where $y_1 = (0,0,0,0)$

$y_2 = (\alpha,0,0,0)$ $\alpha \; = \; .9866146547$

$y_3 = (\beta,\beta,\beta,0)$ $\beta \; = \; .8495213324$

$y_4 = (\gamma,\gamma,\gamma,\gamma)$ $\gamma \; = \; .5054081059$

and the weights can be computed from (12) and are positive. To get a rule of precision 5, we need 4 parameters. If we drop one generator, we can only integrate exactly 3 monomials. By dropping y_3 and adding an inexpensive generator of the form $(a,0,0,0)$ for any value of a in $(0,1]$, we can determine a rule of precision 5. Hence we can choose a in such a way that the resulting integration rule will be exact in addition for one monomial of degree 6, which must be x_1^6. The resulting rule with 4 generators is partially imbedded in Q_4f and yields a 7/5 pair requiring 65 function evaluations. The value of a is .3857458432 and one of the weights is negative.

The number of points in this pair compares favourably with that in the pair by Phillips [15] which uses 73 points. On the other hand, Genz and Malik [9] have a pair requiring only 57 points. However, the rule of precision 7 is not positive and the rule of degree 5 is less accurate than that in our pair in the sense that it integrates exactly fewer monomials.

4.   CONCLUDING REMARKS

For fully symmetric rules, a rule with m generators exact for the first m monomials corresponds to a one-dimensional IIR based on a set with m points or more exactly, in a symmetric situation, to a (symmetric) IIR determined by $[(m+1)/2]$ points. In both cases, if we start with a positive rule, we can find a sequence of imbedded positive rules starting with the given rule. However, in contrast to the one-dimensional case, it may not always be possible to find a rule based on an arbitrary subset of the m generators, since the matrix in (12) may be singular. The ordering of the monomials determines the coefficients in the rules and a different ordering may give quite different rules. This problem has been avoided in the literature by dealing only with rules of a certain precision. However, it is our feeling that a rule which integrates the first p monomials is also useful even though it lies between two rules of precision $q - 2$ and q. We also believe that it is more important to integrate a monomial of the form, say, $x^2 y^2 z^2$, rather than one of the form x^6, and the remarks in [6] seem to support this belief. In our examples, we have retained to a great degree the notion of rules of precision q since this has been the standard practice and since Berntsen [2] has based his cautious adaptive approach on rules of various precisions. However, in Example 3 and 4, we have also indicated this new approach which should be investigated more fully.

REFERENCES

1.   K. Atkinson, 'An Introduction to Numerical Analysis', Wiley, New York, 1978.

2.   J. Berntsen, 'Cautious adaptive numerical integration over the 3-cube', Report No. 17, Dept. of Informatics, Univ. of Bergen, 1985.

3.   J. Berntsen and T.O. Espelid, 'On the use of Gauss quadrature in adaptive automatic integration schemes', *BIT* **24** (1984) 239-242.

4.   J. Bernsten and T.O. Espelid, 'On the construction of higher degree three dimensional imbedded integration rules', Report No.16, Dept. of Informatics, Univ. of Bergen, 1985.

5.   H. Brass, 'Eine Fehlerabschätzung für positive Quadraturformeln', *Numer. Math.* **47** (1985) 395-399.

6. R. Cranley and T.N.L. Patterson, 'The evaluation of multidimensional integrals', *Computer J.* **11** (1968) 102-110.

7. P.J. Davis and P. Rabinowitz, 'Methods of Numerical Integration', Second Ed., Academic Press, New York, 1984.

8. T.O. Espelid, 'On the construction of good fully symmetric integration rules', Report No.13, Dept. of Informatics, Univ. of Bergen, 1984.

9. A.C. Genz and A.A. Malik, 'An imbedded family of fully symmetric numerical integration rules', *SIAM J. Numer. Anal.* **20** (1983) 580-588.

10. P. Keast, 'Some fully symmetric quadrature formulae for product spaces', *J. Inst. Math. Appl.* **23** (1979) 251-264.

11. P. Keast and J.N. Lyness, 'On the structure of fully symmetric multidimensional quadrature rules', *SIAM J. Numer. Anal.* **16** (1979) 11-29.

12. D.P. Laurie, 'Practical error estimation in numerical integration', *J. Comp. Appl. Math.* **12** & **13** (1985) 425-431.

13. J.N. Lyness, 'Symmetric integration rules for hypercubes III. Construction of integration rules using null rules', *Math. Comp.* **19** (1965) 625-637.

14. F. Mantel and P. Rabinowitz, 'The application of integer programming to the computation of fully symmetric integration formulas in two and three dimensions', *SIAM J. Numer. Anal.* **14** (1977) 391-425.

15. G.M. Phillips, 'Numerical integration over an N-dimensional rectangular region', *Computer J.* **10** (1967/68) 297-299.

16. P. Rabinowitz and N. Richter, 'Perfectly symmetric two-dimensional integration formulas with minimal numbers of points', *Math. Comp.* **23** (1965) 765-779.

17. F. Stenger, 'Tabulation of certain fully symmetric numerical integration formulas of degree 7, 9 and 11', *Math. Comp.* **25** (1971) 935 and Microfiche Supplement.

18. A.H. Stroud, 'Approximate Calculation of Multiple Integrals', Prentice-Hall, Englewood Cliffs, N.J., 1971.

APPENDIX A

SOME CONJECTURES CONCERNING IMBEDDED SEQUENCES OF POSITIVE INTERPOLATORY INTEGRATION RULES

Philip Rabinowitz
Weizmann Institute of Science, Rehovot, Israel.

Jaroslav Kautsky
Flinders University, Bedford Park, S.A., Australia.

Sylvan Elhay
University of Adelaide, Adelaide, S.A., Australia.

ABSTRACT

On the basis of numerical experiments, we conjecture that for any interpolatory integration rule with positive weights, there exists an interpolatory rule based on the set of all its points less one with non-negative weights. Consequently, each such rule generates a finite imbedded sequence of positive interpolatory rules. With very few exceptions, the number of points in the rules in such sequences increases arithmetically.

In a note in 1968, Patterson [10] proposed the use of sequences of interpolating integration rules based on subsets of certain Gauss and Lobatto integration points. Such sequences will henceforth be called sequences of Gauss-based integration rules (GBIR's) even though they may be based on different sets of integration points. In these sequences, each rule is an extension of the previous rule in that it uses all the points of the previous rule, or equivalently, each rule is imbedded in its successor. Such a sequence is called an imbedded sequence. We shall be interested mainly in consecutive sequences of GBIR's, that is, sequences which contain with each n-point rule $Q_n f$, the rule $Q_{n-1} f$ obtained by dropping one point. We shall call $Q_{n-1} f$ the (immediate) predecessor of $Q_n f$.

In the specific cases considered by Patterson, he started with an n-point rule, $n = 2^r + 1$, with the points denoted by x_j, $j = 1,...,n$, $-1 \leq x_n < x_{n-1} < ... < x_2 < x_1 \leq 1$. He then formed the r subsets

$$S_i = \{ x_{2^i(j-1)+1} : j = 1,2,...,2^{r-i}+1 \}, \quad i = 1,2,...,r$$

by successively deleting every second point from the previous subset. For each of the subsets S_i, he computed the interpolatory weights w_j in the numerical integration rule

$$\int_{-1}^{1} f(x)dx = \sum_{x_j \in S_i} w_j f(x_j) + Ef$$

by numerical integration of the Lagrangian interpolating coefficients

$$e_j(x) = \frac{\Pi_i(x)}{\Pi_i'(x_j)(x-x_j)} \quad , \quad x_j \in S_i \ , \quad \Pi_i(x) = \prod_{x_k \in S_i} (x-x_k) \ ,$$

using a Gauss rule of sufficiently high order. It turned out that the resulting rules were all positive integration rules (PIR's), i.e. integration rules in which all the weights are positive. The precision of the rule based on the points in S_i was $2^{r-i}+1$ because of symmetry and not 2^{r-i} as stated by Patterson, where a rule is said to be of precision p if it integrates correctly all polynomials of degree $\leq p$ but not all polynomials of degree $p+1$. The points in $S_i - S_{i+1}$ interlace those of S_{i+1}, i.e. between any two points of S_{i+1}, there is a point of $S_i - S_{i+1}$.

Patterson computed such sequences of GBIR's starting with G_{33} and G_{65}, the Gauss rules with 33 and 65 points, respectively, and with the 65-point Lobatto rule, L_{65}. He implicitly states that their use is indicated when we are integrating functions about which we are reasonably certain that an n-point rule will give the desired accuracy. In this case, we may save on function evaluations by approximating the integral with rules based on the sets S_i without wasting any previously computed function values. If the results of two (or three, if we are conservative) successive approximations agree to within the desired accuracy, we stop and accept the last approximation; otherwise, we go on until we reach the final Gauss or Lobatto rule upon which we can rely. Of course, there is always the danger of false convergence or early termination, especially with low tolerances, a well-documented phenomenon ([1, pp.317-319]). However, as Lyness [7] has pointed out, this is almost inevitable and is one of the occupational hazards of automatic numerical integrators. Various methods have been proposed to alleviate this hazard, but it is inherent in any process which uses information at a finite set of points to approximate a number which is the limit of an infinite process.

Returning now to Patterson, it is clear why he chose a 65-point Lobatto rule as his base rule since he wanted all the rules in the sequence to have the interlacing property and the Lobatto property, namely, that they include the endpoints, $x = \pm 1$. However, the choice of a 33- or 65-point Guass rule as a base rule is not so reasonable. A more reasonable choice is $n = 2^r - 1$, in which case

$$S_i = \{ \ x_{2^i j} \ , \ j = 1,2,...,2^{r-i}-1 \ \} \ , \quad i = 1,2,...,r-1 \ .$$

The reason this choice is preferable is that the resulting sets S_i are closer to the sets corresponding to the Gauss rules with the same number of points. To illustrate this, we list in Table 1 the points $x_{33,4j+1}$, x_{9j}, $x_{31,4j+1}$ and x_{7j}, $j = 1,...,4$ where x_{mk} denotes the point x_k in the rule G_m. Furthermore, the magnitude of the differences between the points $x_{33,4j+1}$ and x_{9j} are much larger than the differences between $\bar{x}_{33,4j+1}$ and $\bar{x}_{9j}$, the corresponding Lobatto points, also listed in Table 1. One possible reason for Patterson's choice may have been that for the same value of n, the programs in the Gauss and Lobatto cases are identical, the only difference being in the input data. We remark that the choice $n = 2^r - 1$ also leads to a sequence of interlacing PIR's. The weights for these rules and for all other rules computed in the course of this work were computed using the program COWIQ [2] which is based on [6]. Other possibilities would have been the method used by Patterson mentioned above and the one proposed by Gustafson [4].

Now this idea of Patterson never caught on since in the same issue of

Mathematics of Computation, Patterson published a second paper [9] in which he introduced the Gauss-Kronrod-Patterson (G-K-P) sequence of integration rules. In this sequence, he starts with G_3, the three-point Guass rule of precision 5, extends it to K_7, the seven-point Kronrod rule of precision 11, and continues to extend K_7 with P_{15}, the 15-point Patterson rule of precision 23. Subsequently, he extends the rule P_{2^n-1} of precision $3 \cdot 2^{n-1}-1$ to the rule $P_{2^{n+1}-1}$ of precision $3 \cdot 2^n-1$, $n = 4,5,6$. The G-K-P sequence is also a sequence of interlacing imbedded PIR's. If we compare it up to P_{63} with the GBIR sequence based on G_{63}, we see that the precisions of G-K-P are 5, 11, 23, 47 and 95 while the corresponding precisions in the GBIR sequence for rules with the same number of points are 3, 7, 15, 31 and 125. Thus, all rules up to the ultimate rule are more accurate in the G-K-P sequence. In addition, the G-K-P sequence is open-ended whereas in the GBIR sequence, we have to specify the base rule in advance. This accounts for the neglect of GBIR's.

However, GBIR sequences have a role to play in cases where G-K-P sequences are unavailable provided that the weights in such rules are positive, or at least not of large magnitudes. In our subsequent discussion, we shall be concerned mainly with PIR's in view of their theoretical and practical importance. One case in which such GBIR's may be useful is in Gauss-Laguerre integration. If we consider the integration rule

$$\int_0^\infty e^{-x}f(x)dx = \sum_{i=1}^n w_i f(x_i) + C_n f^{(2n)}(\xi) , \quad 0 < \xi < \infty , \qquad (1)$$

then there do not exist any PIR's with positive points which are Kronrod extensions [5]. Hence, a sequence corresponding to the G-K-P sequence does not exist in this case. Therefore, if we could have at our disposal a sequence of positive GBIR's based on a Laguerre rule of sufficiently high precision, we would have a useful and economical tool to evaluate the integrals in (1).

Accordingly, we attempted to find such sequences based on Gauss-Laguerre rules. Because of accuracy considerations, we were forced to work with relatively low precision rules, namely GL_{20} and GL_{21}. The way we decided on these values was as folows. We evaluated the points and weights of GL_n for some large value of n using the numerically stable Golub-Welsch algorithm [3]. We then recomputed the weights using the interpolatory weight program COWIQ. If these agreed with the Golub-Welsch weights, then we were satisfied that COWIQ was sufficiently accurate for the value of n; otherwise, we reduced n and repeated the computation.

Having chosen an initial rule, we adopted the following procedure after several false starts. We dropped a point from our rule and computed the interpolatory weights for the resulting set of points. If they were all positive, we had a new imbedded PIR and proceeded to the next stage; otherwise, we dropped a different point. Had we failed to find a PIR after stepping through all the points in the initial rule, we would have exited with an indication of failure; otherwise, we were in stage 2. In general, when we entered stage k, we had already obtained a sequence of imbedded PIR's with $r = n-1,...,n-k+1$ points. We then proceeded as before with the possibility of either reaching stage k+1 or ending in failure. Had we ended in failure, we could have tried to backtrack. However, this never occurred in any of our experiments. In each case, we were able to proceed from any stage to the next one until we reached the final stage - a one-point PIR. We list the indices of the points in the order in which they were removed for the rules GL_{20} and GL_{21}. Our first conjecture is that such consecutive sequences exist for all Gauss-Laguerre rules.

$$GL_{20}: \quad 20,19,18,17,16,15,14,13,11,9,12,7,5,2,10,6,3,8,4,1 \; .$$
$$GL_{21}: \quad 21,20,19,18,17,16,15,14,12,10,8,13,6,4,11,2,9,7,3,5,1 \; .$$

Note that when read from right to left, as in Hebrew, the sequences give the indices of the points in the order of building up a (k+1)-point PIR from a k-point PIR. Note also that such sequences are not unique. A trivial example of this is that the last two indices in any sequence may always be interchanged since a one-point rule is always a PIR so that if we have a two-point PIR, we can drop either of the points.

We tested the rules we computed on a variety of functions. We give results for four functions in Table 2. The first function $e^x / (1 + x^2)$ is an example of the misuse

of Laguerre integration. We have an integral $\quad Ig = \int_0^\infty g(x)dx \quad$ where $g(x)$ decreases

slowly. We write the integrand as $e^{-x}e^x g(x) = e^{-x} f(x)$ and apply Laguerre integration to $f(x)$. (In early tabulations of Laguerre integration rules, for example [12] and [13], $w_k e^{x_k}$ is tabulated along with w_k and x_k to facilitate this computation when done by hand.) Since $f(x)$ does not usually exhibit polynomial-like behaviour on $[0,\infty)$, this approach may fail to give good accuracy. For such functions, the GBIR sequence based on a Laguerre rule does not converge. The reason is that the points added on at the end of the sequence are those points in the Laguerre rule which are most distant from the origin and for such functions, these points made a non-negligible contribution to the quadrature sum.

The other three integrals are of the form $\quad If = \int_0^\infty e^{-x} f(x)dx \quad$ where $f(x)$ is

bounded. In these cases, points at a distance from the origin make a negligible contribution to the quadrature sum since the corresponding weights are so small. Thus, in all three cases we have convergence. However, the convergence is *not* to If but to $GL_{20}f = If$ to reasonable accuracy. Hence, if we are interested in 10 significant figures, we can stop at n = 18 (or 19) with two (or three) successive approximations agreeing to 10 figures. For $f(x) = \sin 5x$, convergence after 17 steps is to the incorrect value .29615678 which is also the value of $GL_{20}f$. For $f(x) = (1 + .02x)^{-1}$, we have very rapid convergence to the correct value. It will generally be the case that when we have rapid convergence, it will be to the true value of the integral. In both this case and the case $f(x) = \cos x$, the use of a sequence of GBIR's results in a considerable saving in function evaluations. Thus, for $f(x) = \cos x$, Stroud and Secrest [14, p.54] list approximations to If by GL_4f, GL_8f, $GL_{12}f$ and $GL_{16}f$. They achieved 10 figure accuracy with $GL_{16}f$ at a cost of 40 function evaluations as against 17 needed here. Furthermore, to verify that they had indeed achieved this accuracy, they would have had to evaluate $GL_{20}f$ making a total of 60 function evaluations as against 18 (or 19) using our method. Similarly, in the case $f(x) = (1 + .02x)^{-1}$, we need only 7 function values to get the aswer correct to 10 figures and one (or two) more to verify it. Had we used successive Laguerre integrations with say 2, 3, 4 points, we would have reached the correct value of GL_4f using a total of 9 function evaluations but would have needed to compute GL_5f to verify it for a total of 14 function evaluations. Of course, had we gone up by two's starting with GL_2f, we would have required only 11 function values all told; on the other hand, had be started with GL_1f and gone up by two's, we would have needed a total of 16 values. In general, the question of determining an optimal (or near-optimal) sequence of non-imbedded integration rules is an intractable problem since it is so dependent on the nature of the integrand. The use of a sequence of imbedded rules sidesteps this problem, with the proviso in the cases of GBIR's that the base rule is sufficiently accurate.

Encouraged by these results, we proceed to investigate the Gauss-Hermite case

$$\int_{-\infty}^{\infty} e^{-x^2} f(x) dx = \sum_{i=1}^{n} w_i f(x_i) + C_n f^{(2n)}(\xi) \quad , \quad -\infty < \xi < \infty. \quad (2)$$

This case is a little better than the Laguerre case in that Kronrod extensions exist for some values of n, namely n = 1,2,4 . (The Kronrod extension of GH_1 is GH_3.) Aside from these cases, it has been shown [5] that no other positive Kronrod extension exists. Since no Hermite-Kronrod-Patterson sequence has been computed, the use of sequences of GBIR's is called for in the Hermite case too.

The Hermite rules, as well as the Gauss and Lobatto rules, are cases of what we shall call a symmetric situation, i.e. we have an integration rule of the form

$$\int_{-a}^{a} w(x) f(x) dx = \sum_{i=1}^{n} w_i f(x_i) + Ef$$

where a may be infinite, $w(x)$ is an even function and the weights w_i and points x_i satisfy

$$w_i = w_{n-i+1} \ , \ x_i = -x_{n-i+1} \ , \ i = 1,2,..., \left[\frac{n}{2}\right] \ .$$

In a symmetric situation, it makes sense to generate symmetric GBIR's. Thus, starting with an n-point rule, our strategy is to remove a symmetric pair of points and check if the remaining points generate a PIR. If so, we enter the next stage; otherwise, we try another symmetric pair. Otherwise, the procdure is similar to that employed in the Laguerre case.

In the Hermite case, accuracy considerations led us to start with the rules GH_{36} and GH_{37}. For these rules we list the sequences of indices, in the order in which they were removed, which resulted in PIR's. (We only give the index, m_k, of one of the pair; the index of the other is $n+1-m_k$).

$$GH_{36} : 1,2,3,4,5,6,7,9,11,8,13,10,15,12,16,17,14,18 .$$

$$GH_{37} : 1,2,3,4,5,6,7,9,11,8,13,15,17,10,18,12,14,16,19 .$$

As in the Laguerre case, our second conjecture is that for every Hermite rule there exists such a consecutive sequence of symmetric imbedded PIR's. In Table 3, we give several examples of the use of these sequences in the computation of integrals of the form in (2). Here we have convergence in all four cases but only in two cases is it to the true value, since in the other cases, $GH_{37}f$ is not accurate to any degree. Note that in the case $f(x) = \cos 4x^2$, we would have had early termination had we requested a low tolerance, say .005 .

After our success with the Gauss-Laguerre and Gauss-Hermite cases, we returned to the Gauss (= Gauss-Legendre) case to see if we could compute a sequence of positive GBIR's in which the number of points increases arithmetically rather than geometrically as in the case of the Patterson sequence, our point of departure. In fact, in all cases of imbedded sequences known to the authors such as the G-K-P sequence, the sequence of trapezoidal rules with $2^k + 1$ points, $k = 0,1,2,...$ and the Clenshaw-Curtis sequence with $2^k n + 1$ points, $k = 1,2,...$, the number of points increases geometrically. Our efforts were crowned with success, and in some cases, gave us a very orderly pattern of removal of points from the base rule. Thus, starting with a Gauss rule with $2^r - 1$ points, the order of removal of the pairs of points is given by the following sequence of indices, where again we give only one of the pair:

$$2^{r-1}\text{-}1(\text{-}2)1, 2^{r-1}\text{-}2(\text{-}4)2, 2^{r-1}\text{-}4(\text{-}8)4,... \ .$$

We observed this pattern for $r = 4,5,6$ and conjecture that it holds for all values of r. Similarly, we have the following sequence based on a Gauss rule with $2^r + 1$ points, the Patterson case:

$$2^{r-1}(\text{-}2)2,3, 2^{r-1}\text{-}1(\text{-}4)7,5, 2^{r-1}\text{-}3(\text{-}8)13,... \ .$$

We give below some other sequences although we did not attempt to find any more patterns:

$$G_{13} : \ 6, 4, 2, 1, 5, 3, 7 \ .$$
$$G_{14} : \ 3, 5, 1, 7, 2, 4, 6 \ .$$
$$G_{16} : \ 4, 6, 2, 8, 1, 5, 3, 7 \ .$$
$$G_{18} : \ 4, 2, 6, 8, 1, 5, 7, 3, 9 \ .$$
$$G_{19} : \ 9, 7, 5, 3, 1, 8, 4, 6, 2, 10 \ .$$

As mentioned above, these sequences are not unique.

We next restricted our attention to the case $n = 15$ for which the pattern is given by

$$G_{15} : \ 7, 5, 3, 1, 6, 2, 4, 8 \ .$$

and tried to see if this pattern carried over to the Gegenbauer or Ultraspherical case, $w(x) = (1 - x^2)^\alpha$, $\alpha > -1$. It turned out that for certain values of α, namely $\alpha = -.51, -.5, -.49, .49$, the G_{15} sequence was successful, yielding a sequence of symmetric positive GBIR's. The rule for $\alpha = -.5$ is the Gauss-Chebyshev rule of the first kind

$$\int_{-1}^{1} (1 - x^2)^{-1/2} f(x)dx \ = \ \frac{\pi}{n} \sum_{i=1}^{n} f\left(\cos \frac{(2i - 1)\pi}{2n}\right) + C_n f^{(2n)}(\xi), \ -1 < \xi < 1 . \ (3)$$

For $\alpha = -.99$ and for $\alpha = .51$, the G_{15} sequence did not work; however, in both cases, other successful sequences were found. For $\alpha = .5$, we had a special situation. We were unable to find any sequence of symmetric positive GBIR's in which the number of points increases arithmetically. This was so since when we removed a pair of points, the weights corresponding to some of the remaining points vanished. We shall treat this situation in some detail later.

We next tested the G_{15} sequence for some Jacobi weights,

$$w(x) = (1 - x)^\alpha (1 + x)^\beta , \ \alpha, \beta > -1$$

even though these do not generate symmetric rules for $\alpha \neq \beta$. For $\alpha = .5$, $\beta = -.5$, the sequence yielded PIR's whereas for $\alpha = 0$, $\beta = 1$, it failed. We then tried to find a good sequence of indices for the Jacobi (0,1) case. Since this is not a symmetric rule, we decided to proceed by removing one point at a time as in the Laguerre case. And indeed we found the following sequence of positive GBIR's:

$$GJ_{15}{}^{(0,1)} : \; 2, 8, 6, 10, 4, 12, 14, 1, 15, 7, 3, 11, 9, 5, 13 \; .$$

We then returned to the Gauss case to see if we could find a similar sequence and again our efforts were crowned with success with the sequence

$$G'_{15} : \; 8, 6, 10, 4, 12, 2, 14, 1, 15, 5, 11, 7, 9, 3, 13 \; .$$

For a symmetric base rule with an odd number of points, such sequences are not of much practical value since the precision of a symmetric interpolatory rule with an odd number of points $2n - 1$ is the same as that of a rule with $2n$ points. However, the existence of such sequences is of theoretical interest and we conjecture that they exist for all Gauss rules.

A final experiment in this group gave us somewhat of a surprise. We tried to generate a sequence of PIR's based on the 15 point Gauss-Chebyshev rule (3). Removing points as in the sequence G'_{15}, things proceedly smoothly for a while, and every alternate rule was a symmetric rule. However, we reached a stage where, after removing a point and then the symmetric point, we found ourselves with negative weights, whereas removing a different point resulted in positive weights. Thus, in this case, symmetry was not retained even though we have a symmetric weight function. Nevertheless, positive sequences of GBIR's existed in this case too, one of which is listed below:

$$GC_{15} : \; 8, 6, 10, 4, 12, 2, 14, 9, 5, 15, 3, 13, 1, 11, 7 \; .$$

As a result of all these experiments, we formulated the following conjecture:
For any Gaussian integration rule with respect to any weight function $w(x) \geq 0$ such that $\int_a^b w(x)dx > 0$ and $\int_a^b w(x)x^k \, dx < \infty$, $k = 0,1,2,...$, of the form

$$\int_a^b w(x) \, f(x) \, dx \; = \; \sum_{i=1}^{n} w_i \, f(x_i) \; + \; C_n \, f^{(2n)}(\xi) \, , \quad -\infty \leq a < \xi < b \leq \infty \qquad (4)$$

we can find a sequence of positive integers $m_1, m_2, ..., m_n$ such that the interpolatory integration rules based on $x_{m_1}, ..., x_{m_k}$, $1 \leq k \leq n$ have non-negative weights for all k. Furthermore, if we have a symmetric situation, then we can find a sequence of positive integers $m_1, m_2, ..., m_{[n/2]}$ such that the interpolatory integration rules based on $\pm x_{m_1}, ..., \pm x_{m_k}$ if n is even and on $x_{m_1} = 0, \pm x_{m_2}, ..., \pm x_{m_k}$ if n is odd, $1 \leq k \leq [n/2]$, have non-negative weights for all k.

This conjecture is weaker than the individual conjectures we made for the Laguerre, Hermite and Gauss cases in that it only postulates non-negative weights rather than positive weights. However, this is the best we can do in the general case since we have found a counterexample to the stronger conjecture. On the other hand, this counterexample was one of the few cases where we could *prove* our conjecture analytically. Again we defer a discussion of this case, namely the Gauss-Chebyshev rule of the second kind,

$$\int_{-1}^{1} (1 - x^2)^{1/2} f(x)dx = \sum_{i=1}^{n} w_i f\left(\cos \frac{i\pi}{n+1}\right) + C_n f^{(2n)} (\xi), \quad -1 < \xi < 1 \qquad (5)$$

since it is connected with a stronger conjecture which implies the one above concerning Gaussian rules in general. We arrived at this stronger conjecture by noticing that in the course of numerical experiments in generating sequences of GBIR's by removing points, we never had to backtrack, i.e. if starting with an n-point rule, we removed a point x_p so that the resulting n-1 point rule had all positive weights, then this n-1 point rule always had an n-2 point positive predecessor. (If the resulting n-1 point rule had s vanishing weights, we continued with the positive rule with n-1-s points.) Thus, once we found a positive predecessor, this predecessor also had a positive predecessor and we did not have to look for a second positive predecessor at a previous stage. This led us to formulate the following conjecture:

Given an n-point interpolatory PIR, there always exists an n-1 point interpolatory non-negative predecessor, and in the symmetric case, there always exists a n-2 point symmetric non-negative predecessor.

A corollary of this conjecture states that given any positive interpolatory integration rule, there exists a sequence of positive GBIR's with that rule as the base rule. However, the number of points in these rules need not increase arithmetically, as our discussion of (5) will show.

We tested this conjecture on a variety of PIR's including closed and open Newton-Cotes rules, the 15 point Patterson rule, Chebyshev equal-weighted rules, rules based on Kronrod points, Fejér rules, Clenshaw-Curtis rules, Lobatto rules and Radau rules. In every case, we were able to generate a non-negative predecessor as well as a positive not necessarily consecutive sequence of GBIR's. Even in a symmetric situation we succedded in generating non-negative immediate predecessors. In fact, if n is odd and there exists a symmetric positive predecessor, then by removing one of the pair, we arrive at a non-negative predecessor since the weight of the symmetric point vanishes and the remaining weights are the same as when we remove a symmetric pair.

In the general Lobatto case, the rules in the sequence of positive GBIR's did not always have the Lobatto property although in the Patterson case, $n = 2^r + 1$, our experiments indicate that they do. On the other hand, in all experiments on Radau rules, all the rules in the sequence did have the Radau property.

In addition to the case of rule (5), we found other examples of non-existence of positive immediate predecessors. Two examples are GH_{11} and the closed 4-point Newton-Cotes rule. In the Hermite rule we got negative weights when we removed the point $x_6 = 0$ and removal of any other point resulted in a vanishing weight for the symmetric point. In the Newton-Cotes rule with points $\{1, {}^1/_3, -{}^1/_3, -1\}$, the removal of any point left a pair of points which were the points of a 2-point Radau rule, either left hand or right hand, so that the weight of the third point vanished.

We now discuss at length the case of the rule (5). In this rule, the points are the zeros of the Chebyshev polynomial of the second kind $U_n(x) = \sin(n + 1)\theta / \sin\theta$,

$x = \cos\theta$. To study the effect of removing a point or a pair of symmetric points, we must consider separately the cases n odd $= 2m + 1$ and n even $= 2m$. Consider first the case $n = 2m + 1$. If we drop the point x_{2k+1} (and x_{n-2k}), then among the remaining points, we have all the points $\cos(2p\pi \,/\, 2m+2) = \cos(p\pi \,/\, m+1)$, $p = 1,...,m$ which are the zeros of $U_m(x)$. Consequently, the interpolatory rule based on the set of $2m$ (or $2m-1$) points is again a rule (5) with $n = m$ in which m weights are positive and the remaining weights are zero. This is a well known result in the theory of Gaussian integration, namely, that if we augment the set of points of an n-point Gaussian rule (4) with $m \leq n$ additional points in the interior of the integration interval, then the interpolatory weights of the n original points are the (positive) Gaussian weights while the weights of the m additional points vanish. A similar situation holds with rules of Lobatto and Radau type with $m \leq n$ -2 and $m \leq n$ - 1 respectively. If we drop the point x_{2k} (and x_{n-2k+1}), then among the remaining points, we have the points $\cos((2p+1)\pi \,/\, 2m+2)$, $p = 0,...,m$ which are all the zeros of $T_{m+1}(x)$, the Chebyshev polynomial of the first kind. The weight w_j corresponding to one of the remaining points x_j of the form $\cos(2p\pi \,/\, 2m+2)$ is given by

$$C \int_{-1}^{1} (1 - x^2)^{1/2} T_{m+1}(k)\, P(x)dx \;\; = \;\; C \int_{-1}^{1} T_{m+1}(x)\, (1 - x^2)^{-1/2}\, (1 - x^2)\, P(x)dx \;\; = 0$$

where $P(x)$ is the product $\prod(x - x_k)$ taken over the set of remaining points excluding the point x_j and hence is a polynomial of degree $m-2$ (or $m-3$). Thus, in any case we end up with vanishing weights.

Consider now the case $n = 2m$. If we drop the point x_{2k} (and x_{n-2k+1}) then among the remaining points are all the points $\cos(2p-1)\pi \,/\, 2m+1$, $p = 1,...,m$ which are the zeros of the Jacobi polynomial $P_m^{(-1/2,1/2)}(x)$. The weight w_j corresponding to one of the remaining points x_j is given by

$$C \int_{-1}^{1} (1 - x^2)^{1/2} P_m^{(-1/2,1/2)}(x)\, P(x)dx \;\; = \;\; C \int_{-1}^{1} \sqrt{\frac{1 - x}{1 + x}}\; P_m^{(-1/2,1/2)}(x)(1+ x)P(x)dx = 0$$

since as before $P(x)$ is of degree $m-2$ (or $m-3$). If we drop the point x_{2k+1} (and x_{n-2k}) we have a similar situation with $P_m^{(1/2,-1/2)}(x)$. Thus, in both cases, we end up with vanishing weights. To summarize this discussion, we see that both when n is odd and when n is even, we cannot have a positive immediate (symmetric) predecessor; furthermore, when n is odd, we have *proved* the existence of a non-negative immediate (symmetric) predecessor and when $n = 2^r$ - 1 , a sequence of positive GBIR's. This is nothing knew since Monegato [8] has already demonstrated the existence of this sequence in his study of Kronrod extensions.

Another case involving Kronrod extensions is the Lobatto-Chebyshev rule of the first kind

$$\int_{-1}^{1}(1- x^2)^{-1/2} f(x)dx \;\; = \;\; \frac{\pi}{n} \sum_{i=0}^{n}{}'' f\left(\cos \frac{i\pi}{n}\right) + C_n f^{(2n)}(\xi) , \;\; 1 < \xi < 1 . \qquad (6)$$

where the double prime indicates that the first and last terms in the sum are to be halved. When n is even, $n = 2m$, then it is known [8] that (6) is the Kronrod extension both of (6) and of (3) with $n = m$. From this it follows that if we drop one point or a pair of symmetric points from (6) with $n = 2m$, then we get among the remaining points, the points of either (6) or (3) with $n = m$. For these points, the weights will be positive while for the rest, the weights will vanish. Here then is another case for which we have proven our conjecture and again, for $n = 2^r$, we have a sequence of positive GBIR's all of the form (6).

Similarly, for the Radau-Jacobi rules

$$\int_{-1}^{1} \sqrt{\frac{1+x}{1-x}}\, f(x)\, dx \;=\; w_0\, f(1) \;+\; \sum_{i=1}^{n} w_i\, f\!\left(\cos \frac{i\pi}{n+1}\right) + C_n\, f^{(2n+1)}(\xi), \;\; -1 < \xi < 1$$

$$\int_{-1}^{1} \sqrt{\frac{1-x}{1+x}}\, f(x)\, dx \;=\; w_0\, f(-1) \;+\; \sum_{i=1}^{n} \hat{w}_i\, f\!\left(\cos \frac{i\pi}{n+1}\right) + \hat{C}_n\, f^{(2n+1)}(\xi), \;\; -1 < \xi < 1$$

with positive weights w_i and $\hat{w}_i$, $i = 0,...,n$. If n is odd, $n = 2m + 1$, and we remove a point $\cos \frac{(2k+1)\pi}{n+1}$, $k = 0,...,m$, then among the remaining points, we will have the points of the same rule with $n = m$ so that the weights of these points will be positive while the weights of the other points will vanish. Hence, if we start with $2^r - 1$ points, we will get a sequence of positive GBIR's.

We know of one case in which we can prove the existence of a positive immediate predecessor. This is the case of the rule (3) with $n = 2m + 1$. If we remove the point $x_{m+1} = 0$, the resulting interpolatory rule has precision $2m - 1$ since it will not integrate exactly the functions $T_{2m+1}(x) / x$ and $\prod(x - x_j) = T_{2m+1}(x) / 2^{2m}x$, where the product is taken over all the remaining points. By corollary 2 in [11], the rule based on these points is positive if and only if there exists a polynomial

$$P_{2m}(z) \;=\; \sum_{k=0}^{2m} a_k\, z^k, \quad a_{2m} = 1$$

with all its zeros in the open unit circle $U : |z| < 1$ such that

$$\prod(x - x_j) \;=\; 2^{-2m+1} \sum_{k=0}^{2m} a_k\, T_k(x) . \qquad (7)$$

Since $\prod(x - x_j) = T_{2m+1}(x) / 2^{2m}x$ and $x = T_1(x)$, (7) is equivalent to $2 \sum_{k=0}^{2m} a_k T_k(x) = T_{2m+1}(x) / T_1(x)$. Multiplying through by $T_1(x)$, using the fact that

$$2\, T_1 T_k = T_{k+1} + T_{k-1} , \;\; k > 0 \;\; \text{and} \;\; 2\, T_1 T_0 = 2\, T_1 .$$

and equating coefficients of the T_k, we find that

$$P_{2m}(z) \;=\; z^{2m} - z^{2m-2} + ... + (-1)^{m-1} z^2 + (-1)^m \frac{1}{2}$$

$$=\; \frac{z^{2m+2} + (-1)^m}{z^2 + 1} \;-\; (-1)^m \frac{1}{2} .$$

Since all the zeros of $P_{2m}(z)$ are in U, we have proved the existence of a PIR in this case.

In conclusion, we shall list again our various conjectures.

1.   For every Laguerre and Radau rule, there exists a consecutive sequence of positive GBIR's. This may also hold for generalized Laguerre rules $(w(x) = x^{\alpha} e^{-x}$, $\alpha > -1)$ and for Radau-Laguerre rules.

2.   For every Hermite rule, there exists a consecutive sequence of symmetric positive GBIR's.

3.   For every Gauss and Lobatto rule, there exists consequtive sequences of positive GBIR's and of symmetric positive GBIR's. This may also hold for Gauss- and Lobatto-Gegenbauer rules $(w(x) = (1 - x^2)\alpha$, $\alpha > -1)$ with the exception of $\alpha = \frac{1}{2}$ in the Gauss case and $\alpha = -\frac{1}{2}$ in the Lobatto case.

4.   Every positive (symmetric) interpolatory rule has an immediate non-negative (symmetric) predecessor. Consequently, for every such rule, there exists a sequence of positive (symmetric) GBIR's.

TABLE 1 : Comparison of values of Gauss and Lobatto points x_{mj} and $\bar{x}_{mj}$ belonging to G_m and L_m with the corresponding points belonging to a GBIR with m points.

j	$x_{33,4j+1}$	x_{9j}	$x_{31,4j}$	x_{7j}	$\bar{x}_{33,4j+1}$	$\bar{x}_{9j}$
1	.997	.968	.931	.949	-	-
2	.902	.836	.716	.742	.917	.900
3	.682	.613	.388	.406	.699	.677
4	.366	.324	-	-	.377	.363

TABLE 2 : Numerical evaluation of $\int_0^\infty e^{-x} f(x)dx$ by a consecutive sequence of positive interpolatory rules with k points, k = 2(1)20 , based on the Laguerre rule GL_{20} .

k\f	$e^x(1 + x^2)^{-1}$	cos x	sin 5x	$(1 + .02x)^{-1}$
2	1.261	.35	.59	.98064
3	1.487	.39	.57	.9807562
4	1.488	.39	.56	.9807562
5	1.449	.508	-.75	.980755506
6	1.533	.516	-.12	.9807554977
7	1.509	.50095	.73	.9807554965
8	1.508	.50104	.68	.9807554965
9	1.509	.49913	.17	.9807554965
10	1.523	.50020	.38	.9807554965
11	1.527	.499976	.307	.9807554965
12	1.526	.499977	.303	.9807554965
13	1.532	.5000040	.2955	.9807554965
14	1.537	.49999965	.2962	.9807554965
15	1.541	.500000017	.296155	.9807554965
16	1.545	.4999999996	.29615682	.9807554965
17	1.548	.5000000000	.29615678	.9807554965
18	1.551	.5000000000	.29615678	.9807554965
19	1.554	.5000000000	.29615678	.9807554965
20	1.557	.5000000000	.29615678	.9807554965
Exact	$\pi/2$	.5	.19230769	.9807554965

TABLE 3 : Numerical evaluation of $\int_{-\infty}^{\infty} e^{-x^2} f(x)dx$ by a consecutive symmetric sequence of positive interpolatory rules with k points, $k = 3(2)37$, based on the Hermite rule GH_{37} .

k\f	$\sin^2 x$	$(x^2 - 1)|x||\log|x||$	$x^2 e^x$	$\cos 4x^2$
3	.586	.013	1.47	.1065
5	.537	.110	1.66	.1156
7	.553	.116	1.713	.1104
9	.56014	.568	1.706876	.1158
11	.560204	.578	1.70690664	.1160
13	.56020235	.566	1.706906845	.1135
15	.56020234	.565	1.706906845	.1134
17	.5602022526	.405	1.706906846	.5859
19	.5602022597	.466	1.706906846	.8059
21	.5602022594	.438	1.706906846	.6949
23	.5602022594	.435	1.706906846	.6818
25	.5602022594	.4331	1.706906846	.6735
27	.5602022594	.43325	1.706906846	.67414
29	.5602022594	.4332412	1.706906846	.674110
31	.5602022594	.433241447	1.706906846	.67411118
33	.5602022594	.4332414444	1.706906846	.6741111694
35	.5602022594	.4332414444	1.706906846	.6741111695
37	.5602022594	.4332414444	1.706906846	.6741111695
Exact	.5602022594	.5	1.706906846	.6880216950

REFERENCES

1. P.J. Davis and P. Rabinowitz, 'Methods of Numerical Integration', Academic Press, New York, 1975.

2. S. Elhay and J. Kautsky, 'COWIQ and SIWIQ: Fortran subroutines for the weights of interpolatory quadratures', Tech. Rep. TR82-08, Dept. of Computing Science, Univ. of Adelaide, South Australia, 1982.

3. G.H. Golub and J.H. Welsch, 'Calculation of Gauss quadrature rules', *Math. Comp.* **23** (1969) 221-230.

4. S.-A. Gustafson, 'Algorithm 417. Rapid computation of weights of interpolatory quadrature rules', *CACM* **14** (1971) 807.

5. D.K. Kahaner and G. Monegato, 'Nonexistence of extended Gauss-Laguerre and Gauss-Hermite quadruatre rules with positive weights', *ZAMP* **29** (1978) 983-986.

6. J. Kautsky and S. Elhay, 'Calculation of the weights of interpolatory quadratures', *Numer. Math.* **40** (1982) 407-422.

7. J. Lyness, 'When not to use an automatic quadrature routine', *SIAM Rev.* **25** (1983) 63-87.

8. G. Monegato, 'A note on extended Gaussian quadrature rules', *Math. Comp.* **30** (1976) 812-817.

9. T.N.L. Patterson, 'The optimum addition of points to quadrature formulae', *Math. Comp.* **22** (1968) 847-856.

10. T.N.L. Patterson, 'On some Gauss and Lobatto based integration formulae', *Math. Comp.* **22** (1968) 877-881.

11. F. Peherstorfer, 'Characterization of positive quadrature formulas', *SIAM J. Math. Anal.* **12** (1981) 935-942.

12. P. Rabinowitz and G. Weiss, 'Tables of abscissas and weights for numerical evaluation of integrals of the form $\int_0^\infty e^{-x} x^n f(x)dx$ ', *MTAC* **13** (1959) 285-294.

13. H.E. Salzer and R. Zucker, 'Table of the zeros and weight factors of the first fifteen Laguerre polynomials', *Bull. Amer. Math. Soc.* **55** (1949) 1004-1012.

14. A.H. Stroud and D.H. Secrest, 'Gaussian Quadrature Formulas', Prentice-Hall, Englewood Cliffs, NJ, 1966.

15. I.P. Mysovskih, A special case of quadrature formulae containing preassigned nodes, (Russian), Vesci. Akad. Nauuk BSSR Ser. Fiz.-Tehn. Navuk, No. 4, 1964, 125-127.

APPENDIX B

REFEREE #1

Referee's report on the paper "Some conjectures concerning imbedded sequences of positive interpolatory integration rules" by P. Rabinowitz, J. Kautsky & S. Elhay.

It seems to me that the paper should have two major objectives:

1.   The construction of special sequences of interpolatory rules with positive weights, as those proposed by the authors.

In general, it is not a simple matter to construct sequences of rules with positive weights; there is no underlying theory to accomplish this. The authors conjecture that it is always possible to construct an imbedded sequence of rules with this property if one starts with the nodes of a Gaussian rule and proceeds by successively deleting points. I feel that this point could (and should) be made in fewer pages. Also the references to Patterson's work should be shortened, since the authors' argumentation is on weak grounds. Personally, I believe that the strategy proposed by Patterson of doubling the points at each step is a fairly sound one. Otherwise, how does one know how many points to add in order to obtain a more trustworthy estimate? Doesn't adding only one or two, or three, points increase the danger of false convergence? Doubling seems to be a safer choice, although a bit expensive. Concerning their conjecture, the authors should *briefly* report the testing they have done, i.e., the cases they have examined.

2.   The second part of the paper should be devoted to the problem of constructing, and listing, sequences of imbedded rules in the Laguerre and Hermite cases, since in these cases Kronrod extensions do not exist (except for the first few values of n). However, it is not clear how these rules would compare, from a practical point of view, with sequences of Gaussian rules, or with the extensions proposed by Kahaner et al. in *SIAM J. Stat. Comput. 5* (1984), 42-55.

Furthermore, how does one know which Gaussian rule will assure the desired accuracy? If one chooses the number of points too high, isn't there a chance that the predecessor rules, especially the first half, or two-thirds, for example, will give a much lower accuracy? This point might be discussed a little more fully. For example, is it true that the predecessor rules, in spite of their degree of precision being only half that of a Gaussian rule, give an accuracy comparable to that of the corresponding Gaussian rule? Or something of the sort.

The authors might wish to consider the remarks above, but at any rate ought to compress considerably the description (not the results) of their experiments.

REFEREE #2

Referee's remarks on P-4686,
"SOME CONJECTURES CONCERNING IMBEDDED SEQUENCES ETC."
    by P. Rabinowitz, et al.

As a general remark, I would like to see the paper better structured into sections with appropriate titles. At the same time, the exposition should be tightened up; it is much too rambling.

(There followed a few specific requests for changes the majority of which have been made.)

ASYMPTOTIC EXPANSIONS AND THEIR APPLICATIONS IN NUMERICAL
INTEGRATION

Elise de Doncker-Kapenga
Computer Science Department
Western Michigan University
Kalamazoo, MI 49008
U.S.A.

ABSTRACT. Numerical integration may require expensive calculations, especially for high
dimensionality or when the integrand depicts some kind of singular behaviour. In this paper
we shall discuss how the knowledge and use of the underlying quadrature error functional
expansions contributes to a satisfactory treatment of some of these problems. Particular
applications include the construction of quadrature rules of a specific degree of precision,
as well as the treatment of some types of integrand singularities on the boundaries of the
integration domain by means of extrapolation or transformation methods. A generalization
of the expansions to non-integer mesh ratios will allow extrapolation on a sequence of
product offset trapezoidal rule sums which is invariant with respect to the group of the
affine transformations of the N-simplex onto itself.

1. INTRODUCTION AND SUMMARY

Given an Euler-Maclaurin asymptotic series for the error functional, successive terms of the
expansion can be eliminated by the use of extrapolation methods.

Linear extrapolation applied to a sequence of offset trapezoidal rule sums results in a
Romberg type method. For the N-simplex, an extension to include rules with half-integer
mesh ratios leads to the construction of a set of minimum point formulas [5]. The definitions
and the main result are reviewed in section 2. They will also be needed in section 3, for
deriving expansions valid when integrand singularities are present.

To the expansions satisfied by certain classes of singularities over the N-cube or the
N-simplex, linear extrapolation can be applied using modified Romberg integration or the
method devised by Lyness [20] (which involves solving a linear system). For each term to be
eliminated, knowledge of its dependence on the mesh ratio is mandatory (but its coefficient is
not required). Hollosi and Keast [16] implement a nonlinear scheme (Brezinski's E-algorithm
[1]) to calculate integrals over a triangle, where the integrand has a vertex singularity of
type r^ρ (and r is the radial spherical coordinate in a coordinate system which has that
vertex at the origin). The exponent ρ has to be supplied to the program.

P. Keast and G. Fairweather (eds.), Numerical Integration, 141–151.
© *1987 by D. Reidel Publishing Company.*

Lyness [19] and Lyness and Monegato [26] provide the error expansions for integration over the N-cube and the N-simplex respectively, of functions with vertex singularities of type r^ρ. For the N-simplex, the quadrature error involved is that of a set of N! quadrature rules (each corresponding to one of the simplices in a simplex subdivision of the N-cube [21]).

In [23] [24] an expansion is constructed for the error incurred over the unit square when the integrand has boundary line singularities of the form $x^\lambda y^\nu$ apart from the r^ρ vertex singularity. Similar results can be derived for the triangle and for half-integer mesh ratios. These will be given here, as well as some numerical results. Furthermore, an extension to higher dimensions covering cases with (only) the line singularities will be outlined. This constitutes section 3.

Applications in automatic integration, where transformation methods or nonlinear extrapolation by means of Wynn's ε-algorithm [34] [35] are used, will be discussed in section 4. These methods work for a large class of problems without reference to the specific integrand behaviour.

2. PRODUCT OFFSET TRAPEZOIDAL RULES AND EXPANSIONS VALID WHEN THE INTEGRAND IS SMOOTH

Offset trapezoidal rules play an important role in the construction of asymptotic expansions, since μ^N-copies of more general quadrature rules (of polynomial degree ≥ 0) can be written as a weighted sum of μ^N-panel offset trapezoidal rules, and the error expansions can be derived correspondingly. We use the following definition of the one-dimensional offset trapezoidal rule [5].

Definition 1. Let $\mu > 0$, $0 \leq a \leq b \leq 1$ and $-1 \leq \alpha \leq 1$. The μ-panel offset trapezoidal rule operator $R^{[\mu;\alpha]}[a,b]$ is given by

$$R^{[\mu;\alpha]}[a,b]f = \frac{1}{\mu} \sum_{q=-\infty}^{+\infty} \theta_q f(x_q) \tag{1}$$

with

$$x_q = \frac{q - 1 + t_\alpha}{\mu},$$
$$t_\alpha = \frac{1 + \alpha}{2}$$

$$\tag{2}$$

and

$$\theta_q = H(b - x_q) - H(a - x_q),$$

where H represents Heaviside's function, with $H(0)$ defined to be $\frac{1}{2}$.

For $a \leq b$ this means that $\theta_q = 1$ for $a < x_q < b$, $\frac{1}{2}$ when $x_q = a < b$ or $a < b = x_q$, 0 if $a = b$ and 0 in all other cases. When μ is an integer, $\mu = m$, Definition 1 coincides with that given by Lyness and Puri [27].

The product offset trapezoidal rule is formed by using the 1D rule operator in an iterated way. For the (unit) simplex (S_N) and the hypercube (H_N) this yields

Definition 2. *Let $\mu > 0$ and $-1 \leq \alpha_i \leq 1$, $i = 1, ..., N$. The product offset trapezoidal rules for H_N and for S_N are given by*

$$R^{[\mu;\vec{\alpha}]}(H_N) = \prod_{i=1}^{N} R_{x_i}^{[\mu;\alpha_i]}[0,1] \tag{3}$$

and

$$R^{[\mu;\vec{\alpha}]}(S_N) = R_{x_1}^{[\mu;\alpha_1]}[0,1] R_{x_2}^{[\mu;\alpha_2]}[0, 1 - x_1] \; ... \; R_{x_N}^{[\mu;\alpha_N]}[0, 1 - \sum_{i=1}^{N-1} x_i] \tag{4}$$

respectively, where each $R_{x_i}^{[\mu;\alpha_i]}$ represents a 1D offset trapezoidal rule of the form (1), operating in the x_i direction.

Examples of the quadrature rule sums for a triangle include

$$R^{[\frac{1}{2};\vec{0}]}(S_2)f \; = \; 0$$
$$R^{[\frac{3}{2};\vec{0}]}(S_2)f \; = \; \frac{4}{9}f(\frac{1}{3}, \frac{1}{3})$$
$$R^{[\frac{5}{2};\vec{0}]}(S_2)f \; = \; \frac{4}{25}(f(\frac{1}{5}, \frac{1}{5}) + f(\frac{3}{5}, \frac{1}{5}) + f(\frac{1}{5}, \frac{3}{5})) \tag{5}$$

The expansion for the quadrature error

$$E^{[\mu;\vec{\alpha}]}(\mathcal{D})f = R^{[\mu,\vec{\alpha}]}f - I(\mathcal{D})f \tag{6}$$

where $I(\mathcal{D})f$ represents the integral over $\mathcal{D}$ and $\mathcal{D} = S_N$ or H_N, is given by

Theorem 1. *Let $f \in C^p(\mathcal{D})$, $\mathcal{D} = S_N$ or H_N. Then*

$$E^{[\mu;\vec{\alpha}]}(\mathcal{D})f = \sum_{s=1}^{p-1} \frac{B_s(\mu; \mathcal{D}, \vec{\alpha}, f)}{\mu^s} + \mathcal{O}(\frac{1}{\mu^p}), \tag{7}$$

where the dependence of B_s on μ is only through its fractional part. The expansion terminates when f is a polynomial and it is even when $|\alpha_i| = 0$ or 1 for $i = 1, 2, ..., N$.

If the product rules are formed using only mid-point ($\alpha_i = 0$) or end-point ($|\alpha_i| = 1$) offset trapezoidal rules, they are found [5] [8] to satisfy an even Euler-Maclaurin expansion for integer or half-integer μ, and can be used as basic rules for Romberg integration (if the integrand has a sufficient number of continuous derivatives). For S_N this allows the construction of rules with some nice features. With the mesh ratios μ_k, $k = 0, 1, ...$, where $\mu_k = k + \frac{1}{2}$ if N is even and $\mu_k = k+1$ if N is odd, the basic simplex product mid-point rules are affine invariant. In the family constituted by their diagonal Romberg table elements of (odd) degrees d, the elements of degrees $d \leq N + 1$ coincide with the minimum point formulas currently known [15].

For H_N the generalization to the μ^N-copy $Q^{(\mu)}(H_N)$ of a rule $Q(H_N)$ of polynomial degree ≥ 0 (and having no points outside H_N) is straightforward. For S_N a similar generalization requires a quadrature rule $Q(T(\vec{\gamma}))$ for each of the simplices $T(\vec{\gamma})$, $\vec{\gamma} \in$ the permutation group $\wp_N$ of N elements, occurring in a subdivision of H_N into $N!$ simplices of equal volume [21]. Assume that each $Q(T(\vec{\gamma}))$ is exact for constants and does not have nodes outside $T(\vec{\gamma})$. Then its μ^N-copy can be written as a linear combination of μ^N-panel simplex product trapezoidal rules apart from, possibly, $\frac{1}{\mu^N}$ times a weighted sum of function values at some vertices $\vec{v}^{(k)}$ of $T(\vec{\gamma})$. If $Q(T(\vec{\gamma}))$ has weights $w_q(\vec{\gamma})$ and nodes $\vec{t}_{\tilde{\alpha}_q(\vec{\gamma})}$, this is of the form

$$Q^{(\mu)}(T(\vec{\gamma}))f = \sum_{q=1}^{n(\vec{\gamma})} w_q(\vec{\gamma}) R^{[\mu;\tilde{\alpha}_q(\vec{\gamma})]}(S_N)f - \frac{1}{\mu^N} \sum_{k=1}^{N+1} a_k(\vec{\gamma}) f(\vec{v}^{(k)}). \tag{8}$$

The second summation on the right is only present if $Q(T(\vec{\gamma}))$ has nodes at some but not all the vertices of $T(\vec{\gamma})$. Let Q represent the quadrature rule set $\{Q(T(\vec{\gamma})), \vec{\gamma} \in \wp_N\}$. The error expansion for $Q^{(\mu)}(S_N)f$ can then be obtained via the expansions of the constituting trapezoidal rule sums. It emerges that results similar to those of Lyness [21] can also be developed when μ is not restricted to being an integer.

3. SINGULAR INTEGRANDS

3.1 QUADRATURE ERROR EXPANSIONS FOR FUNCTIONS WITH VERTEX AND EDGE SINGULARITIES

Lyness [19] [20] handles the integration over H_N of functions singular at a vertex, which have no other singularities within or on the boundaries of H_N. The singular factor (r_ρ) is called *homogeneous of degree ρ (about the origin)*. The following definition is from Lyness and Monegato [26].

Definition 3. *The function $r_\rho(\vec{x})$ is homogeneous of degree ρ about the point $\vec{p}$ if*

$$r_\rho(\beta(\vec{x} - \vec{p})) = \beta^\rho r_\rho(\vec{x} - \vec{p}), \tag{9}$$

for $\vec{x} \neq \vec{p}$ and $\beta \in \Re$.

This behaviour includes singularities of type r^ρ. Note that, for example, the function $(\sum_{i=1}^{N} c_i x_i)$, $c_i > 0$, $i = 1, 2, ..., N$, also belongs to this class [19]. The result also allows the derivation of quadrature error expansions for functions with singularities of type $r^\rho \log^k r$, $k \geq 0$ integer.

Lyness and Monegato [26] present the analogue of the theory for S_N; they consider integrands homogeneous about one or more vertices.

Similar results can be shown to hold for μ^N-copies where μ is not necessarily integer [8]. The coefficients in the resulting expansions are functions of the fractional part of μ.

For the square (H_2), singular factors of the form $x^\lambda y^\nu r_\rho$, with singularities along two sides and a homogeneous behaviour at the vertex where they intersect, are dealt with in [23] [24]. One of the principal results in that work is the following.

Theorem 2. *Let* $f(x,y) = x^\lambda y^\nu r_\rho$, $Q_{\alpha\beta}$ *a one-point rule for* H_2, $m > 0$ *integer and* $p \neq \lambda + \nu + \rho + 2$. *Then*

$$Q_{\alpha\beta}^{(m)}(H_2)f = I(H_2)f + \sum_{s=0}^{p-1} \frac{B_s}{m^s} + \frac{A_{\lambda+\nu+\rho+2} + C_{\lambda+\nu+\rho+2}\log_2 m}{m^{\lambda+\nu+\rho+2}} +$$

$$+ \sum_{t=0}^{p-[\lambda]-2} \frac{E_{\lambda+1+t}^{(1)}}{m^{\lambda+1+t}} + \sum_{t=0}^{p-[\nu]-2} \frac{E_{\nu+1+t}^{(2)}}{m^{\nu+1+t}} + \mathcal{O}(m^{-p}), \tag{10}$$

where the coefficients are independent of m.

Using this, an expansion is also derived for functions obtained by appending to $f(x,y)$ a smooth factor $g(x,y)$ (having a specified number of integrable partial derivatives with respect to x and y), by using Taylor series expansions of $g(x,y)$ around $x = 0$ and $y = 0$. The result is of a form not unlike (10), with the terms in $1/m^{\lambda+\nu+\rho+2}$ and $(\log_2 m)/m^{\lambda+\nu+\rho+2}$ replaced by summations over l of terms in $1/m^{\lambda+\nu+\rho+2+l}$ and $(\log_2 m)/m^{\lambda+\nu+\rho+2+l}$ respectively [24].

It can be verified, using techniques similar to those of Lyness and Monegato [26], that $f(x,y)g(x,y)$ enjoys an expansion of the same form (10) when integrated over S_2, using for Q a simplex product offset trapezoidal rule or a quadrature rule pair. Furthermore, the result in Theorem 2 holds for non-integer μ, with coefficients that are functions of the fractional part of μ. The extrapolation examples of the next section are based on the expansions for $E^{[\mu,\bar\alpha]}(\mathcal{D})fg$, with $\mathcal{D} = S_2$ or H_2, μ integer or half-integer and $|\alpha_i| = 0$ or 1, $i = 1, 2$. The expansions are of the form outlined above (for integer μ), but the coefficients are functions of the fractional part of μ. For the present choice of the $|\alpha_i|$, the B_s coefficients drop out for odd s.

To conclude this section, we shall state the result of a generalization of the work in [23] involving the integration over H_2 of functions with line singularities $x^\lambda y^\nu$, to the N-dimensional cube.

Theorem 3. *Let* $f(\vec{x}) = \prod_{i=1}^k x_i^{\lambda_i} g(\vec{x})$, *where* $g(\vec{x})$ *is assumed analytic in all its variables,* $k \leq N$, $\lambda_i > -1$, $Q_{\bar\alpha}(H_N)$ *is a one-point rule for* H_N *and* $m > 0$ *integer. Then*

$$Q_{\bar\alpha}^{(m)}(H_N)f(\vec{x}) \sim I(H_N)f + \sum_{s\geq 1} \frac{B_s}{m^s}$$

$$+ \sum_{s=1}^k \sum_{(i_1,i_2,...,i_s)\in C_s} \sum_{t\geq 0} \frac{E_{\lambda_1+\lambda_2+...+\lambda_s+t+s}^{(s)}}{m^{\lambda_1+\lambda_2+...+\lambda_s+t+s}} \tag{11}$$

where the coefficients are independent of m *and* C_s *is the set of the combinations of the first* k *integers taken* s *at a time.*

The proof is long and we shall not give it here. Roughly, by the use of subtraction functions, the problem for k is reduced to the problem for $k - 1$ ... and so on until $k = 1$. The latter problem is solved in a way similar to the 2-dimensional case in [23] (again using subtraction functions and also the formalism of Lyness and McHugh [25]). Note that, if the function $g(\vec{x})$ is known to have a specific number of continuous partial derivatives with respect to the x_i, $i = 1, 2, ..., N$, an expansion with a remainder term of the proper order and bounds on the summations in (11) can be given.

3.2 EXTRAPOLATION EXAMPLES

In this section we shall give numerical results for the following examples [8].

$$I_1 = C_1 \int_{S_2} x^{-\frac{1}{5}} y^{-\frac{1}{5}} (x+y)^{-\frac{1}{5}} \, dx \, dy$$

$$I_2 = C_2 \int_{H_2} x^{-\frac{1}{5}} y^{-\frac{1}{5}} (x+y)^{-\frac{1}{5}} \, dx \, dy$$

$$I_3 = C_3 \int_{S_2} \frac{(x+y)^{-\frac{1}{2}}}{(1+x+y)^2} \, dx \, dy \tag{12}$$

where C_1, C_2 and C_3 are normalization constants ($I_1 = I_2 = I_3 = 1$).

Linear extrapolation will be performed on quadrature sums resulting from the product offset trapezoidal rule with $|\alpha_i| = 1$, used with either the geometric sequence of mesh ratios $\mathcal{G} = \{m_k = 2^k\}_{k=0,1,...}$ or Bulirsch' sequence [2] $\mathcal{F} = \{1, 2, 3, 4, 6, 8, 12, ...\}$ (introduced for reasons involving the stability of the computations). The properties of these sequences in relation to the stability of the extrapolation processes is studied also by Lyness [20] [22] and by Hollosi and Keast [16]. Furthermore we shall apply the product offset mid-point rule ($|\alpha_i| = 0$) with the sequence $\mathcal{H}_{\frac{1}{2}} = \{\mu_k = k + \frac{1}{2}\}_{k=0,1,...}$.

As the first extrapolation method we perform "modified Romberg integration" [3] [13] to the integral I_1, using $\mathcal{G}$ and the product end-point trapezoidal rule:

$$T_0^k = R^{[2^k; \bar{1}]}(S_2) f, \quad k = 0, 1, ...$$

$$T_l^k = \frac{2^{\gamma_l} T_{l-1}^{k+1} - T_{l-1}^k}{2^{\gamma_l} - 1}, \quad k = 0, 1, ...; l = 0, 1, ... \tag{13}$$

where the γ_l are the successive exponents of $\frac{1}{m}$ in the error expansion,

$$\{\gamma_l\}_{l=0,1,...} = \{0.8, 1.4, 1.8, 2, 2.8, 3.8, 4, 4.8, 5.8, ...\} \tag{14}$$

The errors $E_l^k = |T_l^k - 1|$ are displayed in Table 1.

Next we will extrapolate by solving a linear system [20], which allows more general sequences of mesh ratios (and more general expansions).

TABLE 1. $E_l^k = |T_l^k - 1|$

l m_k	0	1	2	3	4	5	6
1	1.00						
2	.848	.642					
4	.581	.220	.370 10^{-1}				
8	.367	.787 10^{-1}	.772 10^{-2}	.407 10^{-2}			
16	.223	.287 10^{-1}	.176 10^{-2}	.638 10^{-3}	.505 10^{-3}		
32	.133	.106 10^{-1}	.435 10^{-3}	.101 10^{-3}	.783 10^{-4}	.672 10^{-5}	
64	.779 10^{-1}	.395 10^{-2}	.113 10^{-3}	.169 10^{-4}	.111 10^{-4}	.156 10^{-6}	.668 10^{-6}
128	.453 10^{-1}	.148 10^{-2}	.302 10^{-4}	.156 10^{-5}	.423 10^{-7}	.334 10^{-7}	.102 10^{-7}

TABLE 2. Errors $E_{0,p;0} = |T_{0,p;0} - 1|$

		errors			number of points used			condition numbers		
		$E_{0,3;0}$	$E_{0,5;0}$	$E_{0,7;0}$	$T_{0,3;0}$	$T_{0,5;0}$	$T_{0,7;0}$	$\kappa_{0,3;0}$	$\kappa_{0,5;0}$	$\kappa_{0,7;0}$
	G	$.407\ 10^{-2}$	$.672\ 10^{-5}$	$.885\ 10^{-8}$	36	528	8256	15	33	43
I_1	F	$.782\ 10^{-2}$	$.865\ 10^{-4}$	$.455\ 10^{-6}$	22	67	229	70	566	1808
	$H_{1/2}$	$.384\ 10^{-3}$	$.526\ 10^{-4}$	$.222\ 10^{-5}$	10	35	84	37	1021	25190
	G	$.165\ 10^{-2}$	$.818\ 10^{-5}$	$.259\ 10^{-9}$	81	1089	16641	15	33	43
I_2	F	$.323\ 10^{-2}$	$.685\ 10^{-4}$	$.816\ 10^{-6}$	37	121	433	70	566	1808
	$H_{1/2}$	$.287\ 10^{-2}$	$.106\ 10^{-4}$	$.135\ 10^{-5}$	27	86	197	37	1021	25190
	G	$.374\ 10^{-3}$	$.698\ 10^{-9}$	$.232\ 10^{-8}$	36	528	8256	5.0	6.7	7.7
I_3	F	$.323\ 10^{-2}$	$.685\ 10^{-4}$	$.816\ 10^{-6}$	22	67	229	22	81	180
	$H_{1/2}$	$.358\ 10^{-2}$	$.135\ 10^{-3}$	$.687\ 10^{-5}$	10	35	84	13	165	2597

The error expansion can be represented by

$$R^{[\mu;\vec{a}]}(\mathcal{D})f = a_0\phi_0(\mu) + a_1\phi_1(\mu) + \ldots + a_q\phi_q(\mu) + \mathcal{O}(\phi_{q+1}(\mu)); \qquad (15)$$

here the $\phi_l(\mu)$, $l = 0, 1, \ldots, q+1$ are functions of μ such as $\mu^{-\gamma_l}$ and they satisfy

$$\lim_{\mu\to\infty} \frac{\phi_{l+1}(\mu)}{\phi_l(\mu)} = 0. \qquad (16)$$

The quadrature rule sum corresponding to μ_s is denoted by $T_{s,0;0}$. The extrapolation based on the quadrature rule sums for $\{\mu_k, \mu_{k+1}, \ldots, \mu_{k+q}\}$ is carried out by solving the $(q+1) \times (q+1)$ system

$$T_{s,0;0} = \sum_{l=0}^{q} T_{k,q;l}\phi_l(\mu_s), \quad s = k, k+1, \ldots, k+q \qquad (17)$$

for the $T_{k,q;l} \approx a_l$, $l = 0, 1, \ldots, q$, in particular

$$T_{k,q;0} \approx a_0 = I(\mathcal{D})f. \qquad (18)$$

For the examples under study, the error expansions involved have only terms in $\frac{1}{\mu^{\gamma_l}}$. The exponents are

$$\{\gamma_l\}_{l=0,1,\ldots} = \{0.8, 1.4, 1.8, 2, 2.8, 3.8, 4, 4.8, 5.8, \ldots\}, \quad for\ I_1\ and\ I_2,$$
$$\{1.5, 2, 2.5, 3.5, 4, 4.5, 5.5, 6, \ldots\}, \quad for\ I_3. \qquad (19)$$

Table 2 displays the errors $E_{0,q;0} = |T_{0,q;0} - 1|$ for $q = 3, 5, 7$, the number of evaluation points used by $T_{0,q;0}$ and the corresponding condition numbers $\kappa_{0,q;0}$ [20] for I_1, I_2, I_3; $\mathcal{G}$, $\mathcal{F}$ and $\mathcal{H}_{\frac{1}{2}}$.

Overall, the geometric sequence gives the most accurate results but is far more expensive than the others. The sequence $\mathcal{H}_{\frac{1}{2}}$ is the least expensive. Examination of the condition numbers reveals that the expected roundoff error in the present calculations is up to $1\frac{1}{2}$ figures more for $\mathcal{H}_{\frac{1}{2}}$ than for $\mathcal{F}$, which is not highly significant. Apart from its low cost, an important advantage of the $\mathcal{H}_{\frac{1}{2}}$ is that the quadrature rule sums produced with the product mid-point rule as the basic rule are invariant with respect to the group of all affine transformations of S_2 onto itself. As a consequence, the evaluation points can be generated easily and economically.

4. APPLICATIONS IN AUTOMATIC INTEGRATION

For applications to automatic integration we advocate the use of nonlinear extrapolation by means of the ε-algorithm of Wynn [34] [35]. It was incorporated in a non-adaptive integration procedure by Kahaner [18] and in a local adaptive procedure by Genz [14]. Genz also justified its use for sequences with a widely applicable type of error expansion. A global adaptive algorithm was developed and is applied by several of the (1D) integrators in the subroutine package QUADPACK [28]. The algorithm was modified for integration over a triangle [9] [10] and over a set of triangles [11]. A parallelization has also been studied [12].

Another application of asymptotic error expansions involves transformation methods, yielding an alternative approach to dealing with boundary singularities. In 1D, a nonlinear transformation makes the transformed integrand and all its derivatives vanish at the end-points (0 and 1, say) of the transformed range. Let the transformed integrand be $g(t)$, satisfying

$$g^{(s)}(0) = g^{(s)}(1) = 0, \quad s = 0, 1, \ldots \tag{20}$$

and

$$g \in C^p[0, 1] \tag{21}$$

for a sufficiently large p. An approximation to $I[0,1]g$ can then be obtained by using the end-point trapezoidal rule $R^{[m;1]}[0,1]$, for which the classical Euler-Maclaurin formula is known to reduce (because of (20) and (21)) to a remainder term of order $\frac{1}{m^p}$. This is generally true for arbitrary mesh ratios $\mu > 0$ [8]. A transformation of this nature is the IMT method [17]. It was implemented in 1D non-adaptive integration routines using a Chebyshev approximation of the transformation function [6] [7]. A 2D extension, for computing improper integrals over a bounded or unbounded planar region, was given in [29].

The double-exponential formulas [33] map the original integration range onto $(-\infty, +\infty)$ and evaluate the resulting integral with use of a trapezoidal rule over the infinite range. In order to keep the truncation error small (dominated by the error of the quadrature formula), the transformation $x = \varphi(u)$ should be such that the transformed integrand $g(u)$ decays very rapidly as $|u| \to \infty$. Takahasi and Mori [33] propose the use of a transformation such that $g(u)$ behaves asymptotically as $exp(-\frac{\pi}{2}exp|u|)$ (*double-exponential*) as $u \to \pm\infty$, and suggest the map

$$\varphi(u) = tanh(\frac{\pi}{2}sinh\, u). \tag{22}$$

A comparison of double-exponential transformations and the IMT formula was done in [30]. In [31] a method for integrating over the N-dimensional sphere was obtained by deriving a set of suitable transformations to be applied in the radial coordinate of the spherical coordinate system. The transformations implemented allow integrating over a sphere when the integrand has singularities on the surface and/or in the center. A shortcoming which these methods have in common is that underflow and overflow easily occur because the transformed points are very close to the boundaries. This can usually be taken care of by careful programming of the integrand function.

REFERENCES

1. BREZINSKI, C.: 'A general extrapolation algorithm', *Numer. Math.*, **35** (1980), 175-187.

2. BULIRSCH, R.: 'Bemerkungen zur Romberg–Integration', *Numer. Math.*, **6** (1964), 6-16.

3. DAVIS, P.J. and RABINOWITZ, P.: *Methods of Numerical Integration*, Academic Press Inc., London (1975).

4. DE DONCKER, E.: 'An adaptive extrapolation algorithm for automatic integration', *SIGNUM Newsl.*, **13** (1978), 12-18.

5. DE DONCKER, E.: 'New Euler-Maclaurin Expansions and their application to quadrature over the n-dimensional simplex', *Math. Comp.*, **33** (1979), 1003-1018.

6. DE DONCKER, E. and PIESSENS, R.: 'Automatic computation of integrals with singular integrand over a finite or an infinite range', *Computing*, **17** (1979), 265-279.

7. DE DONCKER, E. and PIESSENS, R.: 'A simple quadrature code for integrands with end point singularities', *Report TW 29*, Computer Science Dept., K.U. Leuven (1979).

8. DE DONCKER, E.,: 'Numerical integration and asymptotic expansions', *Ph.D. Thesis*, K.U. Leuven (1980).

9. DE DONCKER, E. and ROBINSON, I.: 'An algorithm for automatic integration over a triangle using non-linear extrapolation', *ACM Trans. Math. Softw.* **10** (1984), 1-16.

10. DE DONCKER, E. and ROBINSON, I.: 'TRIEX : Integration over a triangle using non-linear extrapolation', *ACM Trans. Math. Softw.*, **10** (1984), 17-22.

11. DE DONCKER-KAPENGA, E. and KAHANER, D.K.: 'Adaptive integration over a triangulated region', in preparation.

12. DE DONCKER, E. and KAPENGA, J.A.: 'A parallelization of adaptive integration methods', *Proceedings of the NATO Advanced Research Workshop on Numerical Integration*, D. Reidel Publ. Comp. (1987).

13. FOX, L.: 'Romberg integration for a class of singular integrands', *Comp. J.*, **10**, 1 (1967), 87-93.

14. GENZ, A.: 'The approximate calculation of multidimensional integrals using extrapolation methods', *Ph.D. Thesis*, Univ. of Kent, Canterbury (1975).

15. GRUNDMANN, A. and MÖLLER, H.M.: 'Invariant integration formulas for the n-simplex by combinatorial methods', *SIAM J. Numer. Anal.*, **15** (1978), 282-290.

16. HOLLOSI, M. and KEAST, P.: 'Numerical integration of singular functions over a triangle', *Report 1985CS-3* , Mathematics, Statistics and Computing Science, Dalhousie University (1985).

17. IRI, M., MORIGUTI, S. and TAKASAWA, Y.: 'On a certain quadrature formula'(in Japanese), 'Kokyuroku' of Research Institute for Math. Sciences, **19** (1970), 82-118.

18. KAHANER, D.: 'Numerical quadrature by the ε-algorithm', *Math, Comp.*, **26** (1972), 689-693.

19. LYNESS, J.N.: 'An error functional expansion for N-dimensional quadrature with an integrand function singular at a point', *Math. Comp.*, **20** (1976), 346-364.

20. LYNESS, J.N.: 'Applications of extrapolation techniques to multi-dimensional quadrature of some integrand functions with a singularity', *J. Comp. Phys.*, **20** (1976), 1-21.

21. LYNESS, J.N.: 'Quadrature over a simplex: part 2. A representation for the error functional', *SIAM J. Numer. Anal.*, **15** (1978), 870-887.

22. LYNESS, J.N.: 'Extrapolation methods for multi-dimensional quadrature of singular and regular integrand functions', *Report ANL/MCS-TM-45*, Argonne National Laboratories (1985).

23. LYNESS, J.N. and DE DONCKER-KAPENGA, E.: 'On quadrature error expansions - part 1', *Journal of Computational and Applied Mathematics*, to appear.

24. LYNESS, J.N. and DE DONCKER-KAPENGA, E.: 'On quadrature error expansions - part 2', in preparation.

25. LYNESS, J.N. McHUGH, B.J.J.: 'On the remainder term in the N-dimensional Euler-Maclaurin expansion', *Numer. Math.*, **15** (1970) 333-344.

26. LYNESS, J.N. and MONEGATO, G.: 'Quadrature error expansions for the simplex when the integrand has singularities at vertices', *Math. Comp.*, **34** (1980), 213-225.

27. LYNESS, J.N. and PURI, K.K.: 'The Euler-Maclaurin expansion for the simplex', *Math. Comp.*, **27** (1973), 273-293.

28. PIESSENS, R., DE DONCKER-KAPENGA, E., ÜBERHUBER, C.W. and KAHANER, D.K.: '*QUADPACK - A subroutine package for automatic integration*', Springer-Verlag Series in Computational Mathematics 1, Berlin, Heidelberg, New York, London (1983).

29. ROBINSON, I., and DE DONCKER, E.: 'Automatic computation of improper integrals over a bounded or unbounded planar region', *Computing*, **27** (1981), 253-284.

30. ROOSE, D.: 'Eén- en meerdimensionale integratie met behulp van transformatietechnieken', *Thesis Eng. Comp. Sc.*, K.U. Leuven (1978).

31. ROOSE, D. and DE DONCKER, E.: 'Automatic integration over a sphere', *Journal of Computational and Applied Mathematics*, **7** (1981), 203-224.

32. SAG, T.W. and SZEKERES, G: 'Numerical evaluation of high-dimensional integrals', *Math. Comp.*, **18** (1964), 245-253.

33. TAKAHASI, H. and MORI, M.: 'Double exponential formulas for numerical integration', *Publ. RIMS, Kyoto Univ.*, **9** (1974), 721-741.

34. WYNN, P.: 'On a device for computing the $e_m(S_n)$ transformation', *Mathematical Tables and Aids to Computing*, **10** (1956), 91-96.

35. WYNN, P.: 'Acceleration techniques in numerical analysis, with particular reference to problems in one independent variable', *Proc. IFIP Congress 1962*, North-Holland, Amsterdam (1963), 149-156.

CONSTRUCTION OF THREE-DIMENSIONAL CUBATURE FORMULAE WITH POINTS ON REGULAR AND SEMI-REGULAR POLYTOPES

Ann Haegemans and Ronald Cools
Department of Computer Science
University of Leuven
Celestijnenlaan 200A
B-3030 Heverlee
Belgium

ABSTRACT. A method of constructing three-dimensional cubature formulae of arbitrary degree for some three-dimensional regions is presented. We search for formulae with points as vertices of a series of regular polytopes. The radii of the circumscribing spheres and the weights (the same for all points on the same polytope) can be computed from a nonlinear system of equations. Using polynomials that are invariant under the symmetries of the polyhedral group the system reduces to a set of systems analogous to the system of a one-dimensional quadrature problem. Formulae of degree 3, 5, 7, 9, 11, 13, ... are computed for the sphere and the entire three-dimensional region with certain weight functions.

1. INTRODUCTION

A. Haegemans [6] and R. Cools and A. Haegemans [3] constructed cubature formulae for two dimensional regions, using points on regular polygons. The aim of this paper is to extend these ideas for constructing cubature formulae for some three-dimensional regions. However, while there are infinitely many regular polygons, there are only a limited number of regular polytopes in three dimensions. Using the theory of reflection groups and invariant polynomials we are able to construct cubature formulae of arbitrary degree.

2. TRANSFORMATION GROUPS AND INVARIANT POLYNOMIALS

2.1 Introduction

Let G be a group of linear transformations acting on a finite dimensional vector space V over a field K. A set $\Omega \subset V$ is said to be *invariant with respect to a group* G, if $g(\Omega) = \Omega$ for any $g \in G$. A function $\theta(x)$, defined on V is said to be

P. Keast and G. Fairweather (eds.), Numerical Integration, 153–163.

invariant with respect to the group G if $\theta(g(x)) = \theta(x)$ for any $g \in G$ and $x \in V$.
If $\theta(x)$ is a polynomial function, $\theta(x)$ is called an *invariant polynomial with respect to* G.

Polynomials $\psi_1(x), \ldots, \psi_r(x)$ are called *algebraically dependent* if there is a polynomial $\psi(x)$ in r variables with complex coefficients, not all zero, such that $\psi(\psi_1(x), \ldots, \psi_r(x)) \equiv 0$. Otherwise $\psi_1(x), \ldots, \psi_r(x)$ are called *algebraically independent*.

A linear transformation acting on the n-dimensional vector space V is said to be a *reflection* if it fixes an $n-1$ dimensional hyperplane. G is a *reflection group* if it is generated by reflections. The set of points $g(a)$, where $a \in V$ is fixed, and g runs through all elements of a group G, is called a G-*orbit* containing the point a.

Let $\chi_1(x), \ldots, \chi_l(x)$ be invariant polynomials of G. $\chi_1(x), \ldots, \chi_l(x)$ forms an *integrity basis* for the invariant polynomials of G $\Leftrightarrow$ any polynomial invariant under G is a polynomial in $\chi_1(x), \ldots, \chi_l(x)$. Each polynomial χ_i is called a *basic invariant polynomial of* G.

THEOREM 1. *Let G be a finite reflection group acting on the n-dimensional vector space V over a field K of characteristic 0. The invariant polynomials of G have an integrity basis consisting of n homogeneous elements which are algebraically independent over K.*

$\square$

This theorem was proved by Chevalley [2].

THEOREM 2. *Let G be a finite group acting on the n-dimensional space V. Let $\chi_1, \ldots, \chi_n$ be homogeneous polynomials forming an integrity basis for the invariant polynomials of G. Let $d_1, \ldots, d_n$ be the respective degrees of $\chi_1, \ldots, \chi_n$. Then*

$$\prod_{i=1}^{n} d_i = |G| \quad , \quad \sum_{i=1}^{n} (d_i - 1) = r$$

where $r = $ number of reflections in G.

$\square$

A proof of this theorem can be found in Flatto [5].

2.2 Invariant polynomials for polyhedral groups

Let G_C be the symmetry group of the cube with vertices

$$(\pm \frac{\sqrt{3}}{3} , \pm \frac{\sqrt{3}}{3} , \pm \frac{\sqrt{3}}{3}) \tag{1}$$

This symmetry goup consists of 48 transformations including 9 reflections. G_C is a reflection group.
A G_C-orbit consists of 48 points :

$$(\pm kr , \pm lr , \pm \sqrt{1-k^2-l^2}\, r) + \text{permutations} \tag{2}$$

with

$$0 \leqslant |k| \leqslant 1 \qquad 0 \leqslant |l| \leqslant 1 \qquad 0 \leqslant k^2+l^2 \leqslant 1$$

at distance r form the origin. We use from now on an abbreviated notation for equation (2)

$$(kr , lr , \sqrt{1-k^2-l^2}\, r)_{FS} \tag{3}$$

Giving special values to k and/or l one can obtain different orbits :

Type 6 : 48 points
k and l arbitrary

$$(kr , lr , \sqrt{1-k^2-l^2}\, r)_{FS} \tag{4}$$

Type 5 : 24 points
k arbitrary , $l = 0 , 0 \leqslant |k| \leqslant 1$

$$(kr , 0 , \sqrt{1-k^2}\, r)_{FS} \tag{5}$$

Type 4 : 24 points
$k = l$,arbitrary , $\quad 0 \leqslant |k| \leqslant \dfrac{\sqrt{2}}{2}$

$$(kr , kr , \sqrt{1-2k^2}\, r)_{FS} \tag{6}$$

Type 3 : 12 points
$k = l = \dfrac{\sqrt{2}}{2}$

$$(\frac{\sqrt{2}}{2} r , \frac{\sqrt{2}}{2} r , 0)_{FS} \tag{7}$$

Type 2 : 8 points : the cube
$k = l = \dfrac{\sqrt{3}}{3}$

$$(\frac{\sqrt{3}}{3} r , \frac{\sqrt{3}}{3} r , \frac{\sqrt{3}}{3} r)_{FS} \tag{8}$$

Type 1 : 6 points : the octahedron
$k = l = 0$

$$(0 , 0 , r)_{FS} \tag{9}$$

Let G_D be the symmetry group of the dodecahedron with vertices

$$(\pm \frac{\sqrt{3}}{3} , \pm \frac{\sqrt{3}}{3} , \pm \frac{\sqrt{3}}{3})$$

$$(0 , \frac{\sqrt{3}}{3\tau} , \frac{\tau\sqrt{3}}{3}) + \text{cyclic permutations} \tag{10}$$

with $\tau = \dfrac{1+\sqrt{5}}{2}$.

This symmetry group consists of 120 transformations including 15 reflections. G_D is a reflection group.

In the following, we use instead of

$$(\pm kr , \pm lr , \pm \sqrt{1-k^2-l^2}\, r) + \text{cyclic permutations} \tag{11}$$

the following notation

$$(kr , lr , mr)_{SC} \quad \text{with} \quad m = \sqrt{1-k^2-l^2} \tag{12}$$

A G_D-orbit consists of 120 points. As above we can give special values to k and/or l and then we obtain orbits with a different number of points.

Type 5 : 120 points
k and l arbitrary

$$(kr , lr , mr)_{SC}$$

$$((\frac{k\tau}{2} + \frac{l}{2\tau} + \frac{m}{2})r , (\frac{k}{2\tau} + \frac{l}{2} - \frac{m\tau}{2})r , (\frac{k}{2} - \frac{l\tau}{2} - \frac{m}{2\tau})r)_{SC}$$

$$((\frac{k\tau}{2} + \frac{l}{2\tau} - \frac{m}{2})r , (\frac{k}{2\tau} + \frac{l}{2} + \frac{m\tau}{2})r , (\frac{k}{2} - \frac{l\tau}{2} + \frac{m}{2\tau})r)_{SC} \tag{13}$$

$$((\frac{k\tau}{2} - \frac{l}{2\tau} + \frac{m}{2})r , (\frac{k}{2\tau} - \frac{l}{2} - \frac{m\tau}{2})r , (\frac{k}{2} + \frac{l\tau}{2} - \frac{m}{2\tau})r)_{SC}$$

$$((\frac{k\tau}{2} - \frac{l}{2\tau} - \frac{m}{2})r , (\frac{k}{2\tau} - \frac{l}{2} + \frac{m\tau}{2})r , (\frac{k}{2} + \frac{l\tau}{2} + \frac{m}{2\tau})r)_{SC}$$

with

$$0 \leqslant |k| \leqslant 1 \qquad 0 \leqslant |l| \leqslant 1 \qquad 0 \leqslant k^2+l^2 \leqslant 1$$

at distance r form the origin.

Type 4 : 60 points
k arbitrary , $l = 0$

$$(kr , 0 , mr)_{SC}$$

$$((\frac{k\tau}{2} + \frac{m}{2})r , (\frac{k}{2\tau} - \frac{m\tau}{2})r , (\frac{k}{2} - \frac{m}{2\tau})r)_{SC} \tag{14}$$

$$((\frac{k\tau}{2} - \frac{m}{2})r , (\frac{k}{2\tau} + \frac{m\tau}{2})r , (\frac{k}{2} + \frac{m}{2\tau})r)_{SC}$$

Type 3 : 30 points

$k = l = 0$

$$(0 , 0 , r)_{SC}$$

$$(\frac{r}{2} , \frac{r\tau}{2} , \frac{r}{2\tau})_{SC} \tag{15}$$

Type 2 : 20 points : the dodecahedron
$$k = 0 \ , l = \frac{\sqrt{3}}{3\tau}$$

$$(\frac{r\sqrt{3}}{3} , \frac{r\sqrt{3}}{3} , \frac{r\sqrt{3}}{3})_{SC}$$

$$(0 , \frac{r\sqrt{3}}{3\tau} , \frac{r\tau\sqrt{3}}{3})_{SC} \tag{16}$$

Type 1 : 12 points : the icosahedron
$$k = 0 \ , l = \frac{\sqrt{\tau}}{\sqrt{\sqrt{5}}}$$

$$(0 , r\frac{\sqrt{\tau}}{\sqrt{\sqrt{5}}} , \frac{r}{\sqrt{\tau}\sqrt{5}})_{SC} \tag{17}$$

Using Theorem 1 , Theorem 2 and the results of Coxeter [4] we obtain the following degrees for the basic invariant polynomials :
For the symmetry group G_C of the cube :

$$d_1 = 2 \quad d_2 = 4 \quad d_3 = 6 \tag{18}$$

and for the symmetry group G_D of the dodecahedron :

$$d_1 = 2 \quad d_2 = 6 \quad d_3 = 10. \tag{19}$$

Using the results of Coxeter [4] and Neutsch [8] we can compute the basic invariant polynomials :
For the symmetry group G_C of the cube :

$$\chi_1 = x^2 + y^2 + z^2 = \phi_{C2}$$
$$\chi_2 = x^2y^2 + y^2z^2 + z^2x^2 = \phi_{C4} \tag{20}$$
$$\chi_3 = x^2y^2z^2 = \phi_{C6}$$

and for the symmetry group G_D of the dodecahedron :

$$\chi_1 = x^2 + y^2 + z^2 = \phi_{D2}$$
$$\chi_2 = 4x^2y^2z^2 + \frac{1}{\tau}(x^2y^4 + y^2z^4 + z^2x^4)$$
$$\qquad\qquad - \tau \ (x^4y^2 + y^4z^2 + z^4x^2) = \phi_{D6} \tag{21}$$

$$\chi_3 = \sqrt{5} \, (x^4 + y^4 + z^4 - 2\,x^2 y^2 - 2\,y^2 z^2 - 2\,z^2 x^2)\times$$

$$\times [\, (\tau^6 - \frac{1}{\tau^6})\,x^2 y^2 z^2 + \frac{1}{\tau^2}\,(\,x^4 y^2 + y^4 z^2 + z^4 x^2)$$

$$- \tau^2 (x^2 y^4 + y^2 z^4 + z^2 x^4)] = \phi_{D10}$$

3. INVARIANT CUBATURE FORMULAE

We are concerned with determining the points (x_i, y_i, z_i) and weights w_i in a cubature formula

$$\sum_{i=1}^{M} w_i \, f(x_i, y_i, z_i) \tag{22}$$

which is an approximation of

$$\iiint_R w(x,y,z) \, f(x,y,z) \, dx \, dy \, dz = I[f] \tag{23}$$

A cubature formula is said to be *invariant with respect to a transformation group* G, if the domain of integration R and the weight function $w(x,y,z)$ are invariant with respect to G, and if the set of points is a union of G-orbits, where the points of one and the same orbit have the same coefficient w_i.

The vector space of all polynomials in x, y and z of degree $\leqslant n$ is denoted by P_n. A cubature formula which is exact for all $P \in P_m$ but not for all $P \in P_{m+1}$, is said to have *degree m*.

THEOREM 3: Let the formula (22) be invariant with respect to G. In order for the formula (22) to be exact for all polynomials in P_m it is necessary and sufficient that (22) is exact for all those polynomials which are invariant with respect to G.

□

This theorem is proved by Sobolev [9].

A basic cubature rule operator $Q_i(k,l,r)$ is defined as

$$Q_i(k,l,r)\,f = \sum_{j=1}^{N} f(x_j, y_j, z_j) \qquad r = \sqrt{x_j^2 + y_j^2 + z_j^2} \tag{24}$$

Where the sum runs over the N different points of an orbit and $i = 1, \cdots, 6$ for G_C and $i = 1, \cdots, 5$ for G_D depending on the type of the orbit.

We search for cubature formulae of the form

$$I[f] \simeq \sum_{i=1}^{m} \sum_{j=1}^{M_i} w_{ij} Q_i(k_{ij}, l_{ij}, r_{ij})\,f + w_0\,f(0,0,0) \tag{25}$$

where $m = 5$ or 6 for G_C and G_D respectively. M_i is the number of orbits of type$_i$ and each orbit is fully determined by k_{ij}, l_{ij}, r_{ij} and w_{ij} is the corresponding weight. Since the location of the points is symmetric with respect to the $xy-$, the $yz-$ and the $zx-$ plane we only have to consider cubature formulae of odd degree. Requiring that the cubature formula is exact for all invariant polynomials up to a certain degree leads to a nonlinear system that can be reduced to a set of systems analogous with the system of a one-dimensional quadrature problem. We will explain this by an example : degree 9 for the symmetry G_C

We set

$$\sum_{j=1}^{M_i} w_{ij} \, r_{ij}^{2k} = W_i \, R_i^k \qquad i = 1,2,3,4,5,6.$$

Requiring that the formula (25) is exact for the following invariant polynomials up to degree 9

$$\theta_{01} = \phi_{2C}^0$$

$$\theta_{21} = \phi_{2C} / 6$$

$$\theta_{41} = \phi_{2C}^2 / 6$$

$$\theta_{42} = \phi_{4C} / 8$$

$$\theta_{61} = \phi_{2C}^3 / 6$$

$$\theta_{62} = \phi_{2C} \, \phi_{4C} / 8$$

$$\theta_{63} = (\phi_{2C} \, \phi_{4C} - 9 \, \phi_{6C}) / 12$$

$$\theta_{81} = \phi_{2C}^4 / 6$$

$$\theta_{82} = \phi_{2C}^2 \, \phi_{4C} / 8$$

$$\theta_{83} = (\phi_{2C}^2 \, \phi_{4C} - 9 \, \phi_{2C} \, \phi_{6C}) / 12$$

$$\theta_{84} = (\phi_{2C}^2 \, \phi_{4C} + 3 \, \phi_{2C} \, \phi_{6C} - 4 \, \phi_{4C}^2) / 24$$

gives us the following system :

$$w_0 + 6W_1 + 8W_2 + 12W_3 + 24W_4 + 24W_5 + 48W_6 = I[\theta_{01}]$$

$$W_1R_1 + \frac{4}{3}W_2R_2 + 2W_3R_3 + 4W_4R_4 + 4W_5R_5 + 8W_6R_6 = I[\theta_{21}]$$

$$W_1R_1^2 + \frac{4}{3}W_2R_2^2 + 2W_3R_3^2 + 4W_4R_4^2 + 4W_5R_5^2 + 8W_6R_6^2 = I[\theta_{41}]$$

$$\frac{1}{3}W_2R_2^2 + \frac{3}{8}W_3R_3^2 + 3a(k)W_4R_4^2 + 3d(k)W_5R_5^2 + 6p(k,l)W_6R_6^2 = I[\theta_{42}]$$

$$W_1R_1^3 + \frac{4}{3}W_2R_2^3 + 2W_3R_3^3 + 4W_4R_4^3 + 4W_5R_5^3 + 8W_6R_6^3 = I[\theta_{61}]$$

$$\frac{1}{3}W_2R_2^3 + \frac{3}{8}W_3R_3^3 + 3a(k)W_4R_4^3 + 3d(k)W_5R_5^3 + 6p(k,l)W_6R_6^3 = I[\theta_{62}]$$

$$\frac{1}{4}W_3R_3^3 + 2b(k)W_4R_4^3 + 2e(k)W_5R_5^3 + 4q(k,l)W_6R_6^3 = I[\theta_{63}]$$

$$W_1R_1^4 + \frac{4}{3}W_2R_2^4 + 2W_3R_3^4 + 4W_4R_4^4 + 4W_5R_5^4 + 8W_6R_6^4 = I[\theta_{81}]$$

$$\frac{1}{3}W_2R_2^4 + \frac{3}{8}W_3R_3^4 + 3a(k)W_4R_4^4 + 3d(k)W_5R_5^4 + 6p(k,l)W_6R_6^4 = I[\theta_{82}]$$

$$\frac{1}{4}W_3R_3^4 + 2b(k)W_4R_4^4 + 2e(k)W_5R_5^4 + 4q(k,l)W_6R_6^4 = I[\theta_{83}]$$

$$c(k)W_4R_4^4 + h(k)W_5R_5^4 + 2s(k,l)W_6R_6^4 = I[\theta_{84}]$$

with

$$a(k) = k_4^2(2 - 3k_4^2)$$
$$b(k) = 2k_4^2(1 - 3k_4^2)^2$$
$$c(k) = 2k_4^2(1 - 3k_4^2)^2(1 - 2k_4^2)$$
$$d(k) = k_5^2(1 - k_5^2)$$
$$p(k,l) = k_6^2 + l_6^2 - k_6^4 - l_6^2k_6^2 - l_6^2$$
$$\vdots$$

Choosing $W_5 = W_6 = 0$, $M_4 = 1$, $M_3 = 1$, $M_2 = 1$, $M_1 = 2$ we can solve the system if we choose k_4 properly so that $w_0 = 0$. The formula so obtained has 56 points. More details of setting up the system and solving it will be given in a report [1].

4. RESULTS

We construct cubature formulae for the following regions :

C_3:   the cube $\{(x,y,z): -1 \leqslant x \leqslant 1, -1 \leqslant y \leqslant 1, -1 \leqslant z \leqslant 1\}$ with weight function $w(x,y,z) = 1/8$ (only for the symmetry group G_C.)

S_3:   the sphere $\{(x,y,z): x^2 + y^2 + z^2 \leqslant 1\}$ with weight function
$$w(x,y,z) = \frac{3}{4\pi}$$

$E_3^{r^2}$:   the entire three-dimensional region with weight function
$$w(x,y,z) = \frac{1}{\pi\sqrt{\pi}}e^{-(x^2+y^2+z^2)}$$

E_3^r : · the entire three-dimensional region with weight function
$$w(x,y,z) = \frac{1}{8\pi\sqrt{\pi}}e^{-(x^2+y^2+z^2)^{1/2}} \, .$$

Up to now we have the following results :

degree	#	G_C				#	G_D		
		C_3	S_3	$E_3^{r^2}$	E_3^r		S_3	$E_3^{r^2}$	E_3^r
1	1	PI	PI	P	P	1	PI	P	P
3	6	PI	PI	P	P	12	PI	P	P
5	14	PI	PO	P	P	13	PI -	P	P
7	27	PO	PO	P	P	33	PI	P	P
9	56	NO	PI	/	/	45	PI	P	P
11	77	/	PO	P	/	86	PI	P	P
13	?					137	PO	P	P

means the number of points in the formula. P denotes that all weights are positive, N denotes that at least one weight is negative. I means that all points are inside the region, while O means that at least one orbit is outside the region (for C_3 and S_3).

Most of the formulae of degree 1 to 7 are known in the literature(see e.g. Stroud [10]).

The formulae of degree 9 with the symmetry G_D were also constructed by Konjaev [7].

The formulae of degree 9 form with symmetry G_C and the formulae of degree 11 and 13 are new.

In Table I we give a degree 11 formula for S_3 and in Table II a degree 13 formula for $E_3^{r^2}$.

Region : S_3
Symmetry group : G_D
Degree : 11
Number of points : 86

type	radius	weight	#
1	0.262968935070	0.522108688506(-2)	12
1	0.567588940403	0.230662886046(-1)	12
1	0.914681642532	0.124977409505(-1)	12
2	0.806174562986	0.174461676499(-1)	20
3	0.976829142867	0.538850832401(-2)	30

Table I : 86-point formula of degree 11 for S_3.

Region : $E_3^{r^2}$
Symmetry group : G_D
Degree : 13
Number of points : 137
$k_4 = 0.192750641161$

type	radius	weight	#
	0.0	0.102744273286	1
1	0.971696497325	0.506525883155(-1)	12
1	0.183853327608(1)	0.469594215398(-2)	12
1	0.420015937983(1)	0.350694085436(-6)	12
2	0.169021665105(1)	0.107122025012(-1)	20
2	0.325461244319(1)	0.136081776301(-4)	20
4	0.254950975680(1)	0.309215652903(-3)	60

Table II : 137-point formula of degree 13 for $E_3^{r^2}$.

5. CONCLUSION

Using the method described above we can (theoretically) find cubature formulae
of arbitrary degree. We found some new formulae with "very few" points.
However, there is not a straightforward method: for each degree one has to start
again and look at the equations. The solutions are not always real and even in the
case of real solutions, the points are not always inside the region (C_3 , S_3) and
the weights are not always positive.

Acknowledgements : The authors wish to thank the National Fund for Scientific research (Belgium) and the NATO for the financial support for the presentation of this paper at the NATO Advanced Research Workshop in Halifax, Canada, August 11-15, 1986, on Numerical Integration : Recent Developments, Software and Applications, co-directed by Professors P. Keast and G. Fairweather.

6. REFERENCES

[1] M. Beckers and A. Haegemans, 'Construction of three-dimensional invariant cubature formulae' *Report TW 85* (K. U. Leuven, 1986).

[2] C. Chevalley, 'Invariants of finite groups generated by reflections', *Amer. J. Math.* 77 (1955) pp 778-782.

[3] R. Cools and A. Haegemans, 'Automatic computation of points and weights of cubature formulae for circular symmetric planar regions', *Report TW 77* (K. U. Leuven, 1986).

[4] H. S. M. Coxeter, 'The product of the generators of a finite group generated by reflections', *Duke Math. J.* 18 (1951) pp 765-782.

[5] L. Flatto, 'Invariants of finite reflection groups', *L'enseignement Math.* 24 (1978) pp 237-292.

[6] A. Haegemans, 'Circularly symmetrical integration formulae of degree $2d-1$ for two-dimensional circularly symmetrical regions', *BIT* 16 (1976) pp 52-59.

[7] S. I. Konjaev, 'Ninth-order quadrature formulae invariant with respect to the icosahedral group', *Soviet Math. Dokl.* 18 (1977) pp 497-501.

[8] W. Neutsch, 'Optimal spherical design and numerical integration on the sphere', *Journal of Comp. Phys.* 51 (1983) pp 313-325.

[9] S. L. Sobolev, 'Formulas of mechanical cubature on the surface of a sphere' (in Russ.), *Sibirsk Mat. Z.* 3 (1962) pp 769-796.

[10] A. H. Stroud, 'Approximate calculation of multpile integrals', (Prentice-Hall, Englewood Cliffs, N.J., 1971).

CONSTRUCTION OF SEQUENCES OF EMBEDDED CUBATURE FORMULAE FOR CIRCULAR SYMMETRIC PLANAR REGIONS

Ronald Cools and Ann Haegemans
Department of Computer Science
University of Leuven
Celestijnenlaan 200A
B3030 Heverlee

ABSTRACT. The construction of a cubature formula requires the solution of a large system of nonlinear equations for determining the knots and weights. The number of equations and unknowns can be reduced by imposing some structure on the formula. We construct cubature formulae for circular symmetric regions, with knots on regular polygons. Due to our special structure, we obtain a reduction of the number of equations and unknowns and the systems of nonlinear equations can be solved automatically and quickly. In this paper an algorithm is described and used to construct cubature formulae and sequences of embedded cubature formulae.

1. INTRODUCTION

One of the objections to the use of Gauss quadrature formulae for approximating one-dimensional integrals, is that when one proceeds from the m-point formula G_m to the n-point formula G_n, $n > m$, all function values needed for G_m are discarded by G_n. This objection can be overcome if one uses so called extended-Gauss-formulae of which the Kronrod-rules are well known.

The same objection can be made against most cubature formulae in 2 dimensions. In order to reduce the amount of work, i. e. the number of function evaluations, we are concerned in this paper with the construction of cubature formulae of the form

$$I[f] = \iint_R w(x,y)f(x,y)dxdy \simeq \sum_{i=1}^{N} u_i f(x_i,y_i) + \sum_{i=N+1}^{N+M} u_i f(x_i,y_i) \quad (1)$$

which have a higher precision than the given formula

$$I[f] \simeq \sum_{i=1}^{N} w_i f(x_i,y_i). \quad (2)$$

The construction of such pairs of cubature formulae was studied before in [1].

P. Keast and G. Fairweather (eds.), Numerical Integration, 165–172.

A cubature formula which is exact for all polynomials of degree lower than or equal to m, but not for all polynomials of degree $m + 1$, is said to have degree m.

A sequence of cubature formulae is called economic if no other sequence of the same degrees is known which uses fewer knots.

A region is called circular symmetric if the region R and the weight function $w(x,y)$ remain unchanged under the linear transformation $x \leftarrow x\cos\theta + y\sin\theta$, $y \leftarrow y\cos\theta - x\sin\theta$ where θ is arbitrary. We only consider circular symmetric regions and we are specially interested in the following regions:

S_2:   the circle $\{(x,y) : x^2 + y^2 \leqslant 1\}$ with weight function $w(x,y) = 1$

$E_2^{r^2}$:   the entire two-dimensional plane with weight function
$w(x,y) = e^{-(x^2+y^2)}$.

E_2^r: the entire two-dimensional plane with weight function
$w(x,y) = e^{-(x^2+y^2)^{1/2}}$.

To construct cubature formulae for these regions, we only need moments of the form $\mu_{2l} = I[r^{2l}]$, with $r^2 = x^2 + y^2$. For circular symmetric regions, these can be written as one-dimensional integrals

$$\mu_{2l} = I[r^{2l}] = \int_0^R \overline{w}(r) r^{2l+1} dr. \tag{3}$$

2. CUBATURE FORMULAE WITH A SPECIAL STRUCTURE

We search for cubature formulae of the form

$$I[f] \simeq \sum_{i=1}^{nmax} \sum_{t=1}^{a_i} w_{it} Q_{N_i}(r_{it}, \alpha_i) f + w_0 f(0,.) \tag{4}$$

with

$$Q_N(r, \alpha)f = \frac{1}{N} \sum_{j=0}^{N-1} f(r, \alpha + \frac{2\pi j}{N})$$

and f a function of the polar coordinates r and θ. a_i is the number of N_i-gons, each N_i-gon is fully determined by one vertex (r_{it}, α_i), and w_{it} is a weight. We search for cubature formulae with knots on regular N_n-gons where

$$N_1 = A.2 \qquad \alpha_1 = 0$$
$$N_n = A.2^{n-1} \qquad \alpha_n = \frac{\pi}{N_n}, n = 2, \ldots, nmax \quad \text{with } A \in \mathbb{N}_0\backslash\{1\} \tag{5}$$

$\mathbb{N}_0\backslash\{1\}$ is the set of positive integers greater than 1.

Using invariant polynomials with respect to reflection groups (See [4]), in [2] we proved

THEOREM 1. *A cubature formula (4) with the knots as specified in (5) has degree* $m = 2k - 1$ *if and only if the formula is exact for*

$$r^{2l} \qquad\qquad l = 0, \ldots, k-1, \tag{6}$$

and

$$r^{N_n + 2l}(1 - \cos N_n \theta) \quad , \quad n = 2, \ldots, nmax \tag{7}$$
$$\text{and } l = 0, \ldots, k-1-\frac{N_n}{2}$$

with

$$nmax = 1 \qquad\qquad \text{if } k - 1 < A$$
$$nmax = 2 + \left\lfloor \log_2\left(\frac{k-1}{A}\right) \right\rfloor \quad \text{if } k - 1 \geqslant A$$

$\square$

In [2] we then showed that demanding that a cubature formula of the form (4) is exact for the polynomials (6) and (7) is equivalent to solving quadrature problems according to algorithm 1.

---

1. $j = nmax - 1$

2. While $j \geqslant 1$, solve the following system of nonlinear equations (all the unknowns appear only in the left-hand side):
$$2 \sum_{t=1}^{a_{j+1}} w_{j+1,t} r_{j+1,t}^{N_{j+1}+2l} = \mu_{N_{j+1}+2l} - \sum_{i=j+2}^{nmax} \sum_{t=1}^{a_i} w_{it} r_{it}^{N_{j+1}+2l} \tag{8}$$
$$l = 0, \ldots, k - 1 - \frac{N_{j+1}}{2}$$
and modify $j : j \leftarrow j - 1$.

3. Solve the following system of nonlinear equations (all the unknowns appear only in the left-hand side):
$$\sum_{t=1}^{a_1} w_{1t} r_{1t}^{2l} = \mu_{2l} - \sum_{i=2}^{nmax} \sum_{t=1}^{a_i} w_{it} r_{it}^{2l} \quad , l = 0, \ldots, k - 1 \tag{9}$$

Algorithm 1

In [2] we studied the case where all the w_{it} and r_{it} in the left-hand side of equations (8) and (9) are unknown. Such formulae we call Gauss-cubature formulae. It is then possible to express $nmax$ and the number of knots as a function of A and m. It is also posible to compute the value of the structure parameter A for which the number of knots becomes minimal. A FORTRAN program for automatic computing such formulae, CISYR, was also developed.

In this paper we study the case where the cubature formula has a certain number of preassigned knots.

3. CONSTRUCTION OF EMBEDDED CUBATURE FORMULAE

3.1 From low degree to high degree

Following the same idea as in Kronrod-Patterson schemes, we want to use all knots of a known cubature formula (2) of degree $2k - 1$ in a cubature formula (1) of degree $2(k + l) - 1$. We assume that both cubature formulae are of the form (4) with knots as in (5). It doesn't matter how the low degree formula is constructed; it may also be constructed by adding knots to a lower degree formula.

Let

$$\sum_{i=1}^{nl} \sum_{t=1}^{a_i} w_{it} Q_{N_i}(r_{it}, \alpha_i) f + w_0 f(0, .)$$

be a cubature formula of degree $2k - 1$ with

$$nl = 1 \qquad\qquad\qquad \text{if } k - 1 < A$$

$$nl = 2 + \left\lceil \log_2 \left\lceil \frac{k-1}{A} \right\rceil \right\rceil \quad \text{if } k - 1 \geqslant A$$

Let

$$\sum_{i=1}^{nl} \sum_{t=1}^{a_i} u_{it} Q_{N_i}(r_{it}, \alpha_i) f + \sum_{i=1}^{nh} \sum_{t=1}^{b_i} v_{it} Q_{N_i}(s_{it}, \alpha_i) f + u_0 f(0, .) \qquad (10)$$

be a cubature formula of degree $2(k + l) - 1$ with

$$nh = 1 \qquad\qquad\qquad\qquad \text{if } k + l - 1 < A$$

$$nh = 2 + \left\lceil \log_2 \left\lceil \frac{k+l-1}{A} \right\rceil \right\rceil \quad \text{if } k + l - 1 \geqslant A$$

Demanding that the cubature formula (10) is exact for the polynomials (6) and (7) of degree $\leqslant 2(k + l) - 1$, is equivalent to solving (extended-)quadrature problems according to algorithm 2, which is a modification of algorithm 1.

We will always assume that:

C1: the number of equations in the quadrature problems (11) is even and

C2: (the number of equations in the extended–quadrature problem (12)) minus (the number of preassigned radii) is even.

More formally: we assume that

C1: $k + l - N_j/2$ is even for $nl < j \leqslant nh$ and

C2: $k + l - N_j/2 - a_j$ is even for $1 < j \leqslant nl$.

In these cases the systems of nonlinear equations can easily be solved. Note that if $nl = nh$, the nonlinear equations (12) for $j = nh = nl$ correspond to an extended–quadrature problem with respect to the one-dimensional integral (3).

1. $j = nh$

2. While $j > nl$, solve the following quadrature problem (all the unknowns appear only in the left-hand side):

$$2 \sum_{t=1}^{b_j} v_{jt} s_{jt}^{N_j+2l} = \mu_{N_j+2l} - \sum_{i=j+1}^{nh} \sum_{t=1}^{b_i} v_{it} s_{it}^{N_j+2l}, \quad (11)$$

$$l = 0, \ldots, k + l - 1 - \frac{N_j}{2}$$

and modify $j : j \leftarrow j - 1$.

3. While $j > 1$, solve the following extended-quadrature problem (all the unknowns appear only in the left-hand side):

$$2 \sum_{t=1}^{b_j} v_{jt} s_{jt}^{N_j+2l} + 2 \sum_{t=1}^{a_j} u_{jt} r_{jt}^{N_j+2l} \quad (12)$$

$$= \mu_{N_j+2l} - \sum_{i=j+1}^{nl} \sum_{t=1}^{a_i} u_{it} r_{it}^{N_j+2l} - \sum_{i=j+1}^{nh} \sum_{t=1}^{b_i} v_{it} s_{it}^{N_j+2l},$$

$$l = 0, \ldots, k + l - 1 - \frac{N_j}{2}$$

and modify $j : j \leftarrow j - 1$.

4. Solve the following extended-quadrature problem:

$$\sum_{t=1}^{a_1} u_{1t} r_{1t}^{2l} + \sum_{t=1}^{b_1} v_{1t} s_{1t}^{2l} = \mu_{2l} - \sum_{i=2}^{nl} \sum_{t=1}^{a_i} u_{it} r_{it}^{2l} - \sum_{i=2}^{nh} \sum_{t=1}^{b_i} v_{it} s_{it}^{2l}, \quad (13)$$

$$l = 0, \ldots, k - 1.$$

Algorithm 2

From one-dimensional integration we know the following

THEOREM 2. *Given a n-point Gauss-quadrature formula of degree $2n - 1$, then $n + 1$ is the minimum number of new points we need to add to obtain a quadrature formula of degree greater than $2n - 1$.*

$\square$

This theorem was proved by Monegato [5]. When $j = nl = nh$, theorem 2 gives us some constraints on the choice of k, l and A.

We consider only in detail the case where the low degree cubature formula is a Gauss-cubature formula.

When is $nl = nh$?

$$nl = nh \Leftrightarrow 2(k + l) - 1 \leqslant 2A2^{nl-1}$$

$$\Leftrightarrow l \leqslant A2^{nl-1} - k + \frac{1}{2}$$

$$\Leftrightarrow l \leqslant A2^{nl-1} - k$$

a. $nl = nh = 1$.

The cubature formula of degree $2k - 1$ consists of $[k/2]$ radii. We obtain the degree $2(k + l) - 1$ cubature formula if we can find a solution of $k + l$ nonlinear equations. Of these $k + l$ equations, $[k/2]$ are needed to determine the weights of the preassigned radii. We then still have $k + l - [k/2]$ equations for the new radii and weights. Thus, there are $\left| \dfrac{k + l - [k/2]}{2} \right|$ new radii. Theorem 2 shows us that

$$\left| \frac{k + l - [k/2]}{2} \right| > \left| \frac{k}{2} \right| \tag{14}$$

b. $nl = nh \geqslant 2$.

The number of nonlinear equations to obtain the new radii and weights, is $k + l - A2^{nl-2}$. Of this, $\dfrac{k}{2} - A2^{nl-3}$ are needed to determine the weights of the preassigned radii. Thus, there are $\dfrac{k}{2} - A2^{nl-3} + l$ equations to determine the new radii and weights. Theorem 2 shows us that

$$\frac{1}{2} \left| \frac{k}{2} - A2^{nl-3} + l \right| > \frac{k}{2} - A2^{nl-3} \iff l > \frac{k}{2} - A2^{nl-3} \tag{15}$$

3.2 From high degree to low degree

If a cubature formula of degree $2k - 1$ is known, it is always possible to construct a lower degree interpolary cubature formula using a subset of the knots of the given formulae. This requires only the solution of a system of linear equations. If the high degree formulae is a Gauss-cubature formulae of the form (4) with knots as specified in (5), the low degree interpolatory cubature can have at most degree $k - 1$ if k is even and degree k if k is odd.

4. RESULTS

4.1 The low degree formula is a Gauss-cubature formula (2).

In spite of the constraints C1,C2,(14) and (15), there are still a lot of possible combinations of k , l and A. In many cases, the additional knots became complex. For S_2, most of the real solutions have knots outside the region. In [3] we give tables with possible combinations of k , l and A and tables with the real pairs with all knots inside the region. In Table I, we give an economic pair for S_2. Note that sometimes it is more economic to use 2 formulae (2) than to use an embedded formula (1). Note also that in many cases it is more economic to construct a formula of degree $2k - 1$ by discarding knots of a formula of degree $2(k + l) - 1$.

4.2 The low degree formula is an embedded–cubature formula (1).

Naturally, we assume that the low degree formula has real knots that are inside the region. Real solutions now become very rare. At the moment, the only formulae we know are:

For E_2^r :

>   degree 5-9-19 with 7+18+61 knots
>   degree 5-9-27 with 7+18+133 knots

For $E_2^{r^2}$:

>   degree 17-23-39 with 71+84+238 knots
>   degree 17-23-43 with 71+84+294 knots

None of these triplets is an economic sequence.

5. CONCLUSION

The special structure that we used is very well suited to constructing Gauss-cubature formulae for circular symmetric regions: high degree formulae can be computed automatically and quickly. The special structure is not well suited to construct sequences of cubature formulae: a lot of formulae have complex knots or the difference between the low and high degree becomes large.

6. REFERENCES

[1] R. Cools and A. Haegemans, 'Optimal Addition of Knots to Cubature Formulae for Planar Regions', *Numer. Math.* **49**, pp 269-274 (1986) .

[2] R. Cools and A. Haegemans, 'Automatic computation of knots and weights of cubature formulae for circular symmetric planar regions', *Report TW 77* (K. U. Leuven, 1986).

[3] R. Cools and A. Haegemans, 'Tables of sequences of cubature formulae for circular symmetric planar regions', *Report TW 83* (K. U. Leuven, 1986).

[4] L. Flatto, 'Invariants of finite reflection groups', *L'enseignement Math.* 24 (1978) pp 237-292.

[5] G. Monegato, 'An overview of results and questions related to Kronrod schemes', *Numerische Integration*, ed G. Hammerlin, (Basel,1979) pp 231-240.

Table I : Radii and weights of an embedded pair of cubature formulae for S_2.
(low degree = 17 , high degree = 27)

```
                    DEGREE 17 WITH  71 KNOTS
------------------------------------------------------------------------
                    R( I )                           W( I )
------------------------------------------------------------------------
N = 14, ALPHA = PI/N:
             0.9428090415820633658 7D+00    0.3198670545750828227320-01
------------------------------------------------------------------------
N = 14, ALPHA = 0:
             0.36287539183788280420D+00     0.47165090102593798184D-01
             0.62628572445574393350D+00     0.65878332316776996083D-01
             0.82511131069746725821D+00     0.59342041742643147520D-01
             0.98898634548766119135D+00     0.11569638455880253082D-01
------------------------------------------------------------------------
N =  1, ALPHA = 0:
             0.00000000000000000000D+00     0.11840734028811499207D+00
------------------------------------------------------------------------


                    DEGREE 27 WITH 183 KNOTS
------------------------------------------------------------------------
                    R( I )                           W( I )
------------------------------------------------------------------------
N = 14, ALPHA = PI/N:
             0.9428090415820633658 7D+00    0.15206272800073253972D-01
             0.68450601500701400185D+00     0.30359783588322466825D-01
             0.84402412935255954835D+00     0.24026694045280758112D-01
             0.99048704201934852797D+00     0.56817401865177141636D-02
------------------------------------------------------------------------
N = 14, ALPHA = 0:
             0.36287539183788280420D+00     0.32201285769753704457D-01
             0.62628572445574393350D+00    -0.26024231540795000648D-01
             0.82511131069746725821D+00     0.20235553301535040341D-01
             0.98898634548766119135D+00     0.94695655858931660527D-02
             0.16070100264144024694D+00     0.14754471738696702298D-01
             0.55890063474200112629D+00     0.54517003750985255996D-01
             0.70231201601793654000D+00     0.27923774056626528388D-01
             0.92555858903287597858D+00     0.17729425675526136217D-01
             0.99803053279792988415D+00    -0.16818637020019234268D-02
------------------------------------------------------------------------
```

Acknowledgements : The authors wish to thank the NATO for the financial support for the presentation of this paper at the NATO Advanced Research Workshop in Halifax, Canada, August 11-15, 1986, on Numerical Integration : Recent Developments, Software and Applications, co-directed by Professors P. Keast and G. Fairweather.

ON THE CONSTRUCTION OF HIGHER DEGREE THREE DIMENSIONAL EMBEDDED INTEGRATION RULES

Jarle Berntsen and Terje O. Espelid
Department of Informatics
Universitetet I Bergen
Allegt. 55
N-5000 Bergen, Norway

ABSTRACT

We want to approximate $I[f] \equiv \int_{R_3} f(x)\,dx$ for a given function $f(x)$ defined on a fully symmetric (FS)-region, R_3, in E_3. Let $x = \{x_1, x_2, ..., x_N\}$ be a fully symmetric set of points in R_3. Let $Q_A[f]$ and $Q_B[f]$ be two FS-rules of degree $2m+1$ and $2m-1$ respectively approximating $I[f]$ and defined as follows

$$Q_A[f] \equiv \sum_{i=1}^{N} a_i f(x_i) \quad (\text{degree } 2m+1)$$

$$Q_B[f] \equiv \sum_{i=1}^{N} b_i f(x_i) \quad (\text{degree } 2m-1) .$$

Suppose furthermore that $a_i \neq b_i$, $i = 1(1)N$. We want to construct **a pair of rules** in order to be able to estimate the error in $Q_A[f]$ as an approximation to $I[f]$. This is useful, for example, in automatic quadrature. The following four strategies for constructing such pairs of FS-rules in three dimensions will be discussed (m fixed)

Strategy 1.

Construct both rules simultaneously. This implies solving a large set of nonlinear equations.

Strategy 2.

Construct rule Q_A first and then rule Q_B by adding points if necessary.

Strategy 3.

Construct rule Q_A first and then rule Q_B without adding points (Q_B embedded in Q_A).

Strategy 4.

Construct rule Q_B first and then rule Q_A by adding points. This strategy is similar to Kronrod's idea in one dimension.

P. Keast and G. Fairweather (eds.), Numerical Integration, 173–174.

In any of these four strategies we want to minimize the total number of points, N, used. In addition we will prefer to have a pair of rules where Q_A is a good rule, that is $a_i > 0$, $i = 1(1)N$. (Note $x_i \in R_3$, $i = 1(1)N$).

Any FS-set X can be associated with a set of structure parameters: in three dimensions this can be written $(K_0, K_1, K_2, K_3, K_4, K_5, K_6)$. Here K_i is an integer indicating how many generators of type i there are in X. For definitions, see Mantel and Rabinowitz. This implies that

$$N = K_0 + 6K_1 + 12K_2 + 24K_3 + 8K_4 + 24K_5 + 48K_6$$

Consistency conditions for the pair of rules can then be developed following the idea of Mantel and Rabinowitz giving an integer programming problem to be solved (one for each strategy). In the next step, for a given set of structure parameters, one has to solve the non-linear equations following the strategy in question. Next best solutions may also be of interest. We have succeeded in constructing several new pairs of rules, with rule Q_A good, for the 3-cube, i.e. $9/7(1,2,1,0,2,1,0)65$ and $11/9(1,3,2,0,3,2,0)115$. A few tests of these pairs of rules in adaptive schemes will also be presented.

REFERENCES

1. J. Berntsen and T.O. Espelid, On the construction of higher degree three dimensional embedded integration rules. Reports in Informatics, No. 16, 1985, Department of Informatics, University of Bergen.

2. T.O. Espelid, On the construction of good fully symmetric integration rules. Reports in Informatics, No. 13, 1984, Department of Informatics, University of Bergen.

3. F. Mantel and P. Rabinowitz, The application of integer programming to the computation of fully symmetric integration formulas in two and three dimensions. SIAM J. Numer. Anal., 14 (1977), pp. 391-425.

FULLY SYMMETRIC INTEGRATION RULES FOR THE UNIT FOUR-CUBE

T. Sorevik and T.O. Espelid
Department of Informatics
Universitetet I Bergen
Allegt. 55
N-5000 Bergen, Norway

ABSTRACT

For multidimensional integration, there exist some well-known families of fully-symmetric rules of different degrees and dimensions, for example, Phillips (1967), Stenger (1971), Keast (1979), Genz and Malik (1980, 1983). In two and three dimensions, there also exist several rules of different degrees and over different regions. The only rules for the 4-cube we have been able to find are one degree 7 rule by Stroud (1967) and one degree 5 rule by Stroud and Goit (1968). In this talk we will discuss efficient rules of degrees 7 and 9.

The main tools we are using are the consistency conditions for rules to be of a particular degree developed by Keast and Lyness (1979). We are mainly interested in good rules (rules with all the evaluation points inside the 4-cube, and all the weights positive). In a recent paper, Espelid (1984) has shown how to use the theory of moments for polynomials in constructing good fully-symmetric rules (FSG-rules) and has used the technique in 3 dimensions. We will use the same technique in 4 dimensions. Using this technique we were able to construct the minimum point, FSG-rules for all degrees up to and including degree 9. We were also able to construct pairs of rules of degrees 9 and 7 and of degrees 7 and 5. We will also present some tests on some of the new rules of degrees 9 and 7 and compare them with ADAPT and the degree 9 version of Genz and Malik's "embedded family".

REFERENCES

1. T.O. Espelid, On the construction of good fully symmetric integration rules, Report in informatics, No. 13, (1984), Dept. of Informatics, University of Bergen, Norway.

2. A.C. Genz and A.A. Malik, Remarks on algorithm 006: An adaptive algorithm for numerical integration over an N-dimensional rectangular region, J. Comput. Appl. Math., 6 (1980), pp. 295-302.

P. Keast and G. Fairweather (eds.), Numerical Integration, 175–176.

3. A.C. Genz and A.A. Malik, An imbedded family of fully symmetric numerical integration rules, SIAM J. Numer. Anal., 20 (1983), pp. 580-588.

4. P. Keast, Some fully symmetric quadrature formulae for product spaces, J. Inst. Maths. Applics. 23 (1979), pp. 251-264.

5. P. Keast and J.N. Lyness, On the structure of fully-symmetric multidimensional quadrature rules, SIAM J. Numer. Anal., 16 (1979), pp. 11-29.

6. G.M. Phillips, Numerical integration over an N-dimensional rectangular region, Comput. J., 10 (1967), pp. 297-299.

7. F. Stenger, Tabulation of certain fully symmetric formulae of degree 7, 9, and 11, Math. Comp., 25 (1971), Microfiche Supplement.

8. A.H. Stroud, Some seventh degree integration formulas for symmetric region, SIAM J. Numer. Anal., 4 (1967), pp. 37-44.

9. A.H. Stroud and E.H. Goit, Jr., Some extensions of integration formulas, SIAM J. Numer. Anal., 5 (1968), pp. 243-251.

ON THE CONSTRUCTION OF CUBATURE FORMULAE WITH FEW NODES USING GROEBNER BASES

H.Michael Möller
FB Mathematik und Informatik
FernUniversität Hagen, Postfach 940
5800 Hagen 1, F.R. Germany

ABSTRACT. One method for constructing cubature formulae of a given degree of precision consists of using the common zeros of a finite set F of polynomials as nodes. The formula exists if and only if F is an H-basis and some well-defined orthogonality conditions hold. Groebner bases and especially Buchberger's algorithm for their computation allow an effective calculation of H-bases and easy proofs and generalizations of known methods based on H-bases. Groebner bases, a powerful tool in Computer Algebra for analyzing ideals and solving systems of algebraic equations, allow in addition the calculation of the common zeros of the polynomials in F also in cases, where the number of unknowns is different from the number of equations.

1. INTRODUCTION

Given a fixed n-dimensional integral I ,

$$I(f) = \int_B f(x)w(x)dx \quad , \quad B \subseteq \mathbb{R}^n \ , \ w(x) \geq 0 \ \forall x \in B$$

such that $I(p) < \infty$ for all $p \in P := \mathbb{R}[x_1,\ldots,x_n]$, and given a positive integer d , we are interested in constructing cubature formulae (c.f.) of degree d ,

$$I(f) = \sum_{k=1}^{N} A_k f(y_k) + R(f) \quad \text{with} \quad R\big|_{P_d} = 0 \quad ,$$

where P_d denotes the linear space of all $p \in P$ with $\deg p \leq d$. We concentrate our interest in obtaining c.f. with few nodes, i.e. with N near to the known lower bound

$$N \geq \dim P_s \quad \text{with} \quad s = [\tfrac{d}{2}] \ ,$$

or to the lower bound, see [9] ,

$$N \geq \dim P_{s-1} + [\tfrac{s}{2}] \quad \text{with} \quad d = 2s-1 \quad \text{and} \quad n=2 \quad .$$

177

P. Keast and G. Fairweather (eds.), Numerical Integration, 177–192.
© 1987 by D. Reidel Publishing Company.

A given set of points $\{y_1,\ldots,y_N\} \subset \mathbb{R}^n$ is only under strong constraints a set of nodes of a cubature formula of degree d. These constraints can be investigated by studying

$$\mathcal{a} = \{ f \in P \mid f(y_k) = 0, \ k=1,\ldots,N \},$$

or at least by studying $\mathcal{a} \cap P_d$. The latter is a finite dimensional vectorial space, and using linear algebra techniques the constraints are analyzed by Stroud and other authors, see [28]. But $\mathcal{a}$ is an ideal (and called in that context a nodal ideal). This allows the use of ideal theoretic results such as Max Noether's theorem, Macaulay's h-bases (for nodal ideals), Hilbert function, real ideals etc., see [9,14,22,28].

The ideal theoretic approach for constructing c.f. consists in testing a finite set of polynomials for being an H-basis, and the calculation of their common zeros. In case the zeros are real and simple and the polynomials satisfy some well-defined orthogonality conditions a cubature formula of degree d with these common zeros as nodes exists. This approach was used for instance in [15].The main difficulties in that paper were raised by the H-basis tests, such that some steps of the proofs were only indicated. Similar problems occured in [20,26], where H-bases were at least implicitly computed.

Starting with a parameter depending set of polynomials, the H-basis test leads to a number of nonlinear equations for the parameters. This and the fact, that the H-basis contains usually more than n elements causing the failure of numerical methods for the computation of the zeros suggests the employment of formula manipulating systems.

Such systems, nowadays also called Computer Algebra Systems (CAS), and the corresponding theory had a rapid development during the last twenty years; for a survey see [2], for recent contributes [3,4]. Many CAS contain packages for solving systems of algebraic equations, like REDUCE, SCRATCHPAD II, and MAPLE. They are based on Buchberger's algorithm for computing Groebner bases. For a given system with only a finite number of solutions, these packages produce one equation in one variable, say in x_1, at least one eq. in the variables x_1,x_2, at least one in x_1,x_2,x_3 etc., such that by numerical methods and substitutions all solutions can be determined.

The surprizing fact is, that Groebner bases are also a very practical tool for the theoretical investigation of polynomial sets. Under mild restrictions Groebner bases are also H-bases. In this case, Buchberger's algorithm produces H-bases and hence the Groebner basis test imply the H-basis test. In this paper, we provide using Groebner bases simple proofs of known theorems [15,20,26] for the construction of bivariate c.f. and also proofs for the n-dimensional generalizations of these results.

2. CUBATURE FORMULAE AND H-BASES

A cubature formula of degree d with weights A_k and nodes y_k exists if and only if for a linear basis $\{p_1,\ldots,p_D\}$ of P_d

$$(2.1) \quad I(p_i) = \sum_{k=1}^{N} A_k p_i(y_k), \quad i=1,\ldots,D = \binom{d+n}{n},$$

holds. These moment fitting equations can be considered as a nonlinear system of equations for the unknowns A_k and y_k. Because of the large number of equations even for moderate d and n, this method of constructing c.f. is rarely employed except when the integral is invariant under a group of affine linear transformations. In this case, choosing sets of points invariant under the transformations, so called orbit sets, and all points in an orbit set with the same weight, then a theorem of Sobolev [7] states, that a considerably simpler system of nonlinear equations has to be solved in place of (2.1). This is the construction principle for fully symmetric c.f. (invariant under all permutations of $x_1,\ldots,x_n$ and all sign changes $x_i \longrightarrow -x_i$) and for c.f., which are invariant under all mappings of the simplex onto itself. For a collection of invariant c.f. and references see [8]; for a good introduction to invariant c.f. see also [23].

However the gain in having simpler moment fitting equations sometimes causes the c.f. have too many nodes. Even if one has an invariant cubature formula with a minimal number of nodes w.r.t. all c.f. for the same integral, same degree and same symmetries, there may exist a cubature formula of the same degree for the same integral with less nodes. For instance, a tetrahedral invariant cubature formula of degree 6 needs at least 12 nodes [12], whereas Rasputin [25] constructed a cubature formula of degree 6 for the same integral with the minimal number of nodes 10. Similarly for the integral over the square exactly two c.f. of degree 9 with the minimal number of nodes 17 exist but fully symmetric c.f. require in this case at least 20 nodes. It is to be expected that for large degrees d the discrepancy between the minimal number of nodes considering all c.f. of degree d for the given integral and the minimal number of nodes considering only all c.f. having the same (large number of) symmetries increases strongly. Unfortunately only a small number of c.f. is known up to now with the minimal number of nodes w.r.t. all c.f. of the same degree.

In our effort to solve (2.1) directly, we observe that these fitting equations are linear in A_k. This allows us to split the problem. By a standard linear algebra argument, we know that weights $A_k \in \mathbb{R}$ exist satisfying (2.1) if and only if

$$(2.2) \quad p \in P_d, \quad p(y_k) = 0, \quad k=1,\ldots,N \implies I(p) = 0$$

holds; and in that case, given the nodes y_k, (2.1) is a linear system of equations for determining the weights.

Denoting by a_d the linear space of all polynomials of degree $\leq d$ vanishing at $y_1,\ldots,y_N$, (2.2) holds if and only if $I(p_i) = 0$ for all p_i in a given linear basis of a_d and then the weights in (2.1) can be chosen, such that at most

$$(2.3) \quad \binom{d+n}{n} - \dim a_d$$

of them are non-zero [28].

In the univariate case a_d has the linear basis $\{\omega, x\omega, \ldots, x^{d-N}\omega\}$, where ω is a polynomial of degree N vanishing at $y_1, \ldots, y_N$. Then (2.2) holds if and only if $I(p\omega) = 0$ for all $p \in P_{d-N}$. This orthogonality condition and the requirement, that the zeros of ω are real and simple, allows the construction of a numerical integration formula of degree d, and any such formula can be obtained by this method. For generalizing it to the multivariate case, we need some definitions.

DEFINITION 1. $f \in P$ is d-orthogonal, if $gf \in P_d \Rightarrow I(gf)=0$.

DEFINITION 2. Let a be an ideal. $\{f_1, \ldots, f_s\} \subset a$ is an H-basis of a if for all $f \in a$ polynomials $g_1, \ldots, g_s$ exist, such that

$$(2.4) \quad f = \sum_{i=1}^{s} g_i f_i \quad , \quad \deg(g_i f_i) \leq \deg f.$$

Macaulay [13] introduced the notion of H-bases and showed, that any ideal possesses an H-basis. These bases are sometimes also called Macaulay bases or, as in [14], canonical bases.

Using these definitions, one easily gets the following [14].

THEOREM 1. Let $y_1, \ldots, y_N \in \mathbb{R}^n$, $a = \{f \in P \ / \ f(y_k)=0, \ k=1,\ldots,N\}$. Then the following conditions are equivalent.

i) A cubature formula of degree d with nodes $y_1, \ldots, y_N$ exists.

ii) Let $\{f_1, \ldots, f_s\}$ be an H-basis of a. Then $f_1, \ldots, f_s$ are d-orthogonal.

PROOF
"$\Rightarrow$" Let $gf_i \in P_d$. Then $I(gf_i) = \sum_{k=1}^{N} A_k g(y_k) f_i(y_k)=0$ since $f_i \in a$.

"$\Leftarrow$" Let $f \in a_d$. Then with g_i as in (2.4) $I(f)= \Sigma I(g_i f_i)=0$.

Using ii) $\rightarrow$ i) of the theorem, c.f. of degree d can be constructed by choosing an appropriate set of d-orthogonal polynomials $F = \{f_1, \ldots, f_s\}$. If $f_1, \ldots, f_s$ have only a finite number of common zeros, say $y_1, \ldots, y_N$, all of them real and simple, then by standard arguments of algebraic geometry the ideal a generated by F is indeed a nodal ideal

$$a = \{f \in P \ / \ f(y_k) = 0, \ k=1,\ldots,N \}.$$

If F is in addition an H-basis, then the theorem guarantees the existence of a cubature formula of degree d with $y_1, \ldots, y_N$ as nodes.

The critical point is the test of F for being an H-basis of an ideal a, because then it is easy by algebraic methods to test whether a is a nodal ideal:

As shown in [16], $f_1, \ldots, f_s$ have only a finite number of common zeros if and only if they have only affine common zeros (for this equivalence we need, that F is an H-basis!), i.e. if and only if the homogeneous polynomials $\varphi_1, \ldots, \varphi_s$

$$\varphi_i(x_1, \ldots, x_n) := \lim_{t \to 0} t^{\deg f_i} f_i(\frac{x_1}{t}, \ldots, \frac{x_n}{t}), \quad i=1,\ldots,s ,$$

have only the point $(0,0,\ldots,0)$ as common zero. The finiteness of the number of common zeros can also be tested by means of the Hilbert function

$$H(k,a) := \dim P_k - \dim(a \cap P_k) , \quad k=0,1,2,\ldots .$$

If and only if $H(k,a) = H(K,a)$ for all $k \geq K$ holds for a (sufficiently large) K, then the polynomials in a have exactly $H(K,a)$ common zeros, counted with corresponding multiplicities. In case a is a nodal ideal and the Hilbert function is known, $H(d,a)$ gives by (2.3) an upper bound for the number of non-zero weights of the cubature formula. The Hilbert function can be determined by calculating the numbers $\dim(a \cap P_k)$ or in case a Groebner basis of a is known [18].

The simplicity of the common zeros of $f_1,\ldots,f_s$ is easily tested. A common zero y^* is simple if and only if

$$\text{rank} \left. \left(\frac{\partial f_i}{\partial x_k}(y^*) \right) \right|_{\substack{i=1,\ldots,s \\ k=1,\ldots,n}} = n .$$

An alternative method for testing a to be a nodal ideal is given by Schmid[27]. In case the polynomials in a have only a finite number of common zeros, the notion of nodal ideal and real ideal coincide. The test for real ideals requires the computation of the representers of all equivalence classes in P/a and the test of a matrix for positive definiteness.

In a special instance the H-basis test is simple.

EXAMPLE (Max Noether's theorem)
If $f_1,\ldots,f_n$ have only affine common zeros, then $F = \{f_1,\ldots,f_n\}$ is an H-basis. If their common zeros are simple, say $y_1,\ldots,y_N$, then

$$a = \{ f \in P / \quad f(y_k) = 0 , \quad k=1,\ldots,N \}$$

is generated by F and $N = \Pi \deg f_i$ holds.

In the bivariate case, Max Noether's theorem was used by Mysovskih and by Stroud[28] for constructing cubature formulae. Very recently it was also used by Cools and Haegemans in [6]. The formulation presented here was given in [16].

In general, if a polynomial set $F = \{f_1,\ldots,f_r\}$ has to be tested for being an H-basis, it is sufficient to consider polynomials $f = \Sigma\, g_i f_i$ with $\deg(f) < \max \deg(g_i f_i)$. If every such f has a representation $f = \Sigma\, g'_i f_i$ with $\max \deg(g'_i f_i) < \max \deg(g_i f_i)$, then F is by induction an H-basis.

This idea was used in [15] for the bivariate case. Starting with r lin. independent $(2s-1)$-orthogonal polynomials of degree s , $f_1,\ldots,f_r$, condition (ii) in [15,p.189] guaranteed, that among the polynomials of degree $s+k$

$$x^{k-j} y^j f_i , \quad j=0,\ldots,k, \ i=1,\ldots,r ,$$

$s+k+1$ linear independent polynomials exist constituting together with a basis of $a \cap P_{s+k-1}$ a basis of $a \cap P_{s+k}$, $k \in \mathbb{N}$. On the other hand, condition (i) and (iii) in [15,p.189] required the existence of some dependence relations (syzygies) among polynomials

$$(2.5) \quad \sum_{j=0}^{k} \sum_{i=1}^{r} \lambda_{ij} x^{k-j} y^{j} f_i = \sum_{j=1}^{k} \sum_{i=1}^{r} \mu_{ij} x^{k-j} y^{j-1} f_i$$

for $k=1$ and $k=2$. Multiplications with power products $x^l y^m$ gave syzygies of higher degrees. A careful counting of linearly independent polynomials and syzygies showed, that for any $f = \Sigma \ g_i f_i$ with $\deg(f) < \max \deg(g_i f_i)$ a syzygy $\Sigma \ g_i f_i = \Sigma \ g_i' f_i$ with $\max \deg(g_i' f_i)$ $< \max \deg(g_i f_i)$ exists which is a combination of appropriate multiples of the syzygies (2.5). As indicated above, this shows that F is an H-basis.

This procedure together with the condition, that the common zeros are simple and real, provides c.f. of degree $2s-1$ with the minimal number of nodes $\dim P_{s-1} + [s/2]$. For moderate s , say $s \leq 4$, it can be applied successfully . For larger s , it may happen, that for the given integral the conditions of [15,p.189] hold only for polynomials with non-real coefficients or even hold for no set of polynomials [24].

Constructive conditions for the existence of bivariate c.f. with a minimal number of nodes are given in the following two theorems. Theorem 2 was first stated and proved in [20], theorem 3 remarked first in [20] and proved later in [26]. The proofs are very similar to those of [15] and give again at least for moderate degrees c.f. with real nodes as the examples for instance in [11,21] show.

THEOREM 2. Denoting by $P^{m-i,i}$ the i-th basic orthogonal polynomial of degree m , let

$$Q_k := P^{s+1-k,k} + \sum_{i=0}^{s} a_{ik} P^{s-i,i} \quad , \quad k=0,\ldots,s+1 \quad ,$$

$$R_k := x Q_k - y Q_{k-1} \quad , \quad k=1,\ldots,s+1 \quad .$$

If $R_k \in \text{span}\{Q_0,\ldots,Q_{s+1}\}$ for $k=1,\ldots,s+1$, and if the common zeros of $Q_0,\ldots,Q_{s+1}$ are real and simple, then a cubature formula of degree $2s$ exists having the $(s+1)(s+2)/2$ common zeros of $Q_0,\ldots,Q_{s+1}$ as nodes.

THEOREM 3. Let I possess central symmetry, i.e. $I(p)=0$ for all odd $p \in P$, let

$$Q_k := P^{s+1-k,k} + \sum_{i=0}^{s-1} a_{ik} P^{s-1-i,i} \quad , \quad k=0,\ldots,s+1 \quad ,$$

$$R_k := x Q_k - y Q_{k-1} \quad , \quad k=1,\ldots,s+1 \quad .$$

If $x R_k$, $y R_k \in \text{span}\{Q_0,\ldots,Q_{s+1}\}$ for $k=1,\ldots,s+1$, and if the common zeros of $Q_0,\ldots Q_{s+1}$ are real and simple, then a cubature formula of degree $2s-1$ exists having the $(s+1)(s+2)/2 + [s/2]$ common zeros of $Q_0,\ldots,Q_{s+1}$ as nodes.

Another similar method for constructing c.f. with a small number of nodes is used in [5]. We formulate it here as a theorem although the (first) proof will be given in the next section.

THEOREM 4. Let I satisfy $I(x^i y^j) = 0$ whenever i or j is odd. Let s be odd and

$$Q_k := P^{s+1-2k,2k} + \sum_{i=0}^{[s/2]} a_{ik} P^{s-1-2i,2i} \quad , \ k=0,\ldots,\frac{s+1}{2},$$

$$R_k := y^2 Q_k - x^2 Q_{k+1} \quad , \ k=0,\ldots,\frac{s-1}{2}.$$

If $R_k \in \text{span} \{Q_0,\ldots,Q_{(s+1)/2}\}$, $k=0,\ldots,(s-1)/2$, and if the common zeros of the Q_k are real and simple, then a cubature formula of degree $2s-1$ exists having the at most $(s+1)(s+3)/2$ common zeros of the Q_k as nodes.

3. CUBATURE FORMULAE AND GROEBNER BASES

For completeness of this paper, we give a short summary of the definition and the main properties of Groebner bases needed in our context. For more details we refer to Buchberger's survey [1] or to [19]. Since the theory holds for arbitrary fields, we consider now polynomials over a field K and not necessarily $K= \mathbb{R}$.

Let the set of terms (power products)

$$T = \{\ x_1^{i_1} \ldots x_n^{i_n} \ / \ i_1 \geq 0,\ldots, \ i_n \geq 0 \ \}$$

be ordered by $<_T$, such that for all $\varphi,\varphi_1,\varphi_2 \in T$

$$1 \leq_T \varphi \quad \text{and} \quad \varphi_1 <_T \varphi_2 \implies \varphi\varphi_1 <_T \varphi\varphi_2 \ .$$

Examples for such orderings are the lexicographical ordering

$$x_1^{i_1} \ldots x_n^{i_n} <_T x_1^{j_1} \ldots x_n^{j_n} \quad :<=> B(j_n - i_n, \ldots, j_1 - i_1)$$

and the graduated lexicographical ordering

$$x_1^{i_1} \ldots x_n^{i_n} <_T x_1^{j_1} \ldots x_n^{j_n} \quad :<=> B(\Sigma(j_k - i_k), j_n - i_n, \ldots, j_1 - i_1),$$

where the Boolean function B is defined by $B(0):=$"false" and $B(0,a_2,\ldots,a_m) = B(a_2,\ldots,a_m)$ and $B(a_1,\ldots,a_m)$ is "true" for $a_1 > 0$ and "false" for $a_1 < 0$. For $n=2$ this gives, using $x=x_1$, $y=x_2$,

$$1 <_T x <_T x^2 <_T \ldots <_T y <_T xy <_T x^2 y <_T \ldots <_T y^2 <_T xy^2 <_T \ldots \qquad \text{(lex. ord.)}$$

$$1 <_T x <_T y <_T x^2 <_T xy <_T y^2 <_T \ldots \qquad \text{(grad.lex. ord.)}$$

DEFINITION 3. Let

$$f = \sum_{k=1}^{m} c_k \varphi_k \quad \text{with} \quad \varphi_k \in T \ , \quad \varphi_1 <_T \ldots <_T \varphi_m \ \text{and} \ c_k \in K-\{0\}.$$

Then $\text{Hterm}(f) := \varphi_m$ (head term of f) and $M_T(f) := c_m \varphi_m$ (maximal part of f). For $f,g \in P\backslash\{0\}$ let $H(f,g) := \mathrm{lcm}\{\mathrm{Hterm}(f),\mathrm{Hterm}(g)\}$.

DEFINITION 4. Let $F \subset P\backslash\{0\}$ be a finite set. We write $f \xrightarrow{F} g$, if $f,g \in P$ and $h \in P$, $f_i \in F$ exist, such that

$$f = g + hf_i \quad , \quad \mathrm{Hterm}(g) <_T \mathrm{Hterm}(f) \quad \text{or} \quad g = 0 \ .$$

We write $\underline{f}$ if $f \xrightarrow{F}\hspace{-1.5em}\diagup \ \ g$ for no $g \in P$. By $\xrightarrow{F}^+$ we denote the reflexive transitive closure of $\xrightarrow{F}$.

DEFINITION 5. $F = \{f_1,\ldots,f_s\} \subset P\backslash\{0\}$ is a Groebner basis (G-basis) of the ideal a generated by F , if

$$f \in a \implies f \xrightarrow{F}^+ 0 \ .$$

THEOREM 5. Let $F = \{f_1,\ldots,f_s\} \subset P\backslash\{0\}$ and a be the ideal generated by F . Then the following conditions are equivalent.

i) F is a Groebner basis of a .

ii)   For all $f \in a$ there are $g_1,\ldots,g_s \in P$, such that
$$f = \sum_{i=1}^{s} g_i f_i \quad , \quad \mathrm{Hterm}(g_i f_i) \leq_T \mathrm{Hterm}(f) \quad \text{or} \quad g_i = 0.$$

iii) For all (i,j) with $1 \leq i < j \leq s$
$$S(f_i,f_j) := \frac{H(f_i,f_j)}{M_T(f_i)} f_i - \frac{H(f_i,f_j)}{M_T(f_j)} f_j \xrightarrow{F}^+ 0.$$

iv)   Let $B_1,\ldots,B_r$ generate the module of syzygies
$$\{(g_1,\ldots,g_s) \in P^s \ / \ \overset{s}{\underset{1}{\Sigma}} \ g_i M_T(f_i) = 0 \ \}, \text{ and } B_i = (b_{i1},\ldots,b_{is}).$$
Then
$$\sum_{k=1}^{s} b_{ik} f_k \xrightarrow{F}^+ 0 \ , \quad i=1,\ldots,r \ .$$

The equivalence of the first three condition is shown in [19], and the equivalence to the fourth condition in [17]. Condition (iv) indicates a connection to syzygies. In [19] it is described how to find all syzygies when $\{f_1,\ldots,f_s\}$ is a Groebner basis.

COROLLARY 1. If $<_T$ is compatible with the partial ordering by the degrees, i.e.

$$\deg(f) < \deg(g) \implies \mathrm{Hterm}(f) <_T \mathrm{Hterm}(g) \ ,$$

then a G-basis w.r.t. $<_T$ is also an H-basis.

PROOF. This is a consequence of (ii).

COROLLARY 2. Let $<_T$ be the lexicographical ordering with $x_1 <_T \ldots <_T x_n$ and let $\{f_1,\ldots,f_s\}$ be a G-basis of a w.r.t. $<_T$. Then the elimination ideal $a \cap K[x_1,\ldots,x_k]$ has $\{f_1,\ldots,f_s\} \cap K[x_1,\ldots,x_k]$ as G-basis.

PROOF. Let $f \in a \cap K[x_1,\ldots,x_k]$. Then $Hterm(f) \leq_T Hterm(f_i)$ $\leq_T Hterm(f_i g_i)$ for all $f_i \notin K[x_1,\ldots,x_k]$ and $g_i \in P\backslash\{0\}$. Hence in the representation for f given by (ii) , only $f_i \in K[x_1,\ldots,x_k]$ are involved.

Corollary 2 is fundamental for solving systems of nonlinear equations, because it reduces the system to a series of univariate equations: By substituting any zero $(a_1,\ldots,a_k)$ of $\{f_1,\ldots,f_s\} \cap K[x_1,\ldots,x_k]$ into all $f_j \in K[x_1,\ldots,x_{k+1}]\backslash K[x_1,\ldots,x_k]$, we get univariate equations $f_j(a_1,\ldots,a_k,x_{k+1}) = 0$, $k=0,\ldots,n-1$. For this and many other applications of Groebner bases see [1].

Using condition (iii) Buchberger introduced in 1965 an algorithm for computing Groebner bases. This algorithm is slightly improved over the years but the original algorithm and its improvements are up to now the main algorithms for computing Groebner bases. Here we quote the basic version.

ALGORITHM

Input: $\{f_1,\ldots,f_r\} \subset P\backslash\{0\}$

Step 1: $G := \{(f_i,f_j) \;/\; 1 \leq i < j \leq r\};$ $F := \{f_1,\ldots,f_r\}$.

Step 2:   Take $(f_i,f_j) \in G$ s.t. $H(f_i,f_j) = \min\{H(f_k,f_l) \;/\; (f_k,f_l) \in G\}$. Remove (f_i,f_j) from G and compute $S(f_i,f_j) \xrightarrow{+}_F^* \underline{h}$.

Step 3:   If $h \neq 0$, enlarge F by $f_{r+1}:=h$, G by $\{(f_1,h),\ldots,(f_r,h)\}$ and enlarge finally r by 1.

Step 4:   If $G \neq \emptyset$ go to step 2, else go to Output.

Output: Groebner basis F .

Modifications of the algorithm cancel redundant elements in F (f_i is redundant in F , if $F\backslash\{f_i\}$ is a Groebner basis generating the same ideal) and reduce the number of pairs (f_i,f_j) to be inserted in G . It can be shown [10], that a pair (f_i,f_j) is superfluous if

   $H(f_k,f_j)$ divides properly $H(f_i,f_j)$ and $k < j$,

or if $H(f_k,f_j) = H(f_i,f_j)$ for a $k < i$,

or if $Hterm(f_k)$ divides $H(f_i,f_j)$, $k > j$, and $H(f_i,f_k) \neq H(f_i,f_j) \neq H(f_j,f_k)$.

Buchberger's algorithm can be applied for constructing c.f. This procedure is considerably simpler than the methods mentioned in the preceding section as the following example shows.

EXAMPLE. For the construction of a cubature formula of degree 9 with 17 nodes for the integral over the unit square, it is known [15], that the orthogonal polynomials

$$P_1 = 21P^{50} + 20P^{32} + 21P^{14} \;, \quad P_2 = 21P^{41} + 20P^{23} + 21P^{05}$$

and two other lin. independent orthogonal polynomials of degree 5 vanish
in the nodes. Taking

$$P_3 = BP^{50} + AP^{41} + P^{23} \quad , \quad P_4 = DP^{50} + CP^{41} + P^{32} \quad ,$$

then in the graduated lexicographical ordering

$$\text{Hterm}(P_1) = xy^4 \quad , \text{Hterm}(P_2) = y^5,$$

$$\text{Hterm}(P_3) = x^2y^3 \quad , \text{Hterm}(P_4) = x^3y^2 \quad .$$

Starting Buchberger's algorithm with input P_1, P_2, P_3, P_4 , one gets a
polynomial P_5 with head term x^5y and a polynomial P_6 with head
term x^6 . If all other polynomials h calculated in the algorithm are
0 , we have a Groebner basis $\{P_1, \ldots, P_6\}$ and the ideal a generated
by these polynomials satisfies

$$H(4,a) = \dim P_4 = 15 \quad , \quad H(k,a) = 17 \quad \text{for} \quad k \geq 5.$$

If these 17 common zeros of $P_1, \ldots, P_6$ are real and simple, then a c.f.
of degree 9 with these 17 nodes exists by theorem 1.
    The condition, that all other polynomials h in step 2 are 0 ,
leads to a series of nonlinear equations for A,B,C,D, which can be
found in principle by Computer Algebra Systems. The resulting system of
equations for computing the zeros can be solved as well by formula manipulation as indicated in the remarks following corollary 2.   *
    In fact, there are two solutions (A,B,C,D):

$$\left(\frac{31}{81}, \frac{\sqrt{43}}{162}, \frac{\sqrt{43}}{162}, -1 \right) \quad , \quad \left(\frac{31}{81}, -\frac{\sqrt{43}}{162}, -\frac{\sqrt{43}}{162}, 1 \right) \quad ,$$

which follows by the results in [15]. But a straighforward test using
the SAC 2 system on an IBM 370/165 for solving only the system in A,B,
C,D failed because of storage problems (too few cells reclaimed).
    For proving and generalizing the theorems 2, 3 and 4 the following
notation introduced in [27] is useful.

DEFINITION 6. $R_{\alpha_1}, \ldots, R_{\alpha_m}$ is called a fundamental set of degree s, if
$\{\alpha_1, \ldots, \alpha_m\} = \{\alpha \in (\mathbb{N}_0)^n \not{/} \quad |\alpha| = s \}$ and

$$R_{\alpha_i} = x^{\alpha_i} + \text{lower degree terms} \quad .$$

In the following, the ordering $<_T$ is always the graduated lexicographical one, such that all Groebner bases are automatically by corollary 1
H-bases. Therefore we formulate in the following theorem only the condition that a given polynomial set is a Groebner basis. Assuming d-or-
thogonality of the polynomials and assuming that their common zeros are
real and simple, theorem 1 gives the desired c.f. of degree d.

THEOREM 6. Let $R_{\alpha_1}, \ldots, R_{\alpha_m}$ be fundamental of degree s and let for all
$\beta \in (\mathbb{N}_0)^n$ with $|\beta| = s - 1$

$$Q_{ij}^{(\beta)} := \cdot x_i R_{\beta+e_j} - x_j R_{\beta+e_i} \, , \quad 1 \le i < j \le m \, ,$$

where e_k denotes the k-th unit vector, and F the set of all R_{α_j} and $Q_{ij}^{(\beta)}$. If for all β, γ with $|\beta| = |\gamma| = s-1$ and all $(i,j), (k,l)$ and all t

$$S(Q_{ij}^{(\beta)}, R_{\alpha_t}) \xrightarrow[F]{} {}^+ 0 \, ,$$

$$S(Q_{ij}^{(\beta)}, Q_{kl}^{(\gamma)}) \xrightarrow[F]{} {}^+ 0 \, ,$$

then F is a Groebner basis.

PROOF. We have $S(R_{\beta+e_j}, R_{\beta+e_i}) = Q_{ij}^{(\beta)} \xrightarrow[F]{} 0$ because $Q_{ij}^{(\beta)} \in F$. The other pairs $(R_{\alpha_i}, R_{\alpha_j})$ have not to be considered by the remarks following Buchberger's algorithm. The rest follows by (iii) of theorem 5.

A generalization of theorem 2 is obtained by the following

COROLLARY 3. Let $R_{\alpha_1}, \ldots, R_{\alpha_m}$ be fundamental of degree s+1 and 2s-orthogonal, i.e.

$$R_{\alpha_i} = P^{\alpha_i} + \sum_{|\beta|=s} A_\beta^{(i)} P^\beta$$

and let $F = \{ R_{\alpha_1}, \ldots, R_{\alpha_m} \}$. If for all β with $|\beta| = s$ and $1 \le i < j \le m$ the polynomial

$$Q_{ij}^{(\beta)} := x_i R_{\beta+e_j} - x_j R_{\beta+e_i}$$

belongs to span F , then F is a Groebner basis and the polynomials of F have exactly $\binom{s+n}{n}$ common zeros (counted with multiplicity).

PROOF. Because of $Q_{ij}^{(\beta)} = \Sigma \lambda_k R_{\alpha_k}$, we have

$$S(Q_{ij}^{(\beta)}, R_{\alpha_t}) = \Sigma \lambda_k S(R_{\alpha_k}, R_{\alpha_t}) \xrightarrow[F]{} {}^+ 0 \, ,$$

since $S(R_{\beta+e_\mu}, R_{\beta+e_\nu}) = Q_{\mu\nu}^{(\beta)} \xrightarrow[F]{} 0$ implying as shown in th. 6, that $S(R_{\alpha_k}, R_{\alpha_t}) \xrightarrow[F]{} {}^+ 0$. Analogously $S(Q_{ij}^{(\beta)}, Q_{kl}^{(\gamma)}) \xrightarrow[F]{} {}^+ 0$. By th. 6, F is a Groebner basis. The ideal a generated by F satisfies $a \cap P_s = (0)$, $a \cap P_{s+1} = \dim P_{s+1} - \dim P_s$. This gives $H(k,a) = \binom{k+n}{n}$ for $k \le s$ and $H(k,a) = \binom{s+n}{n}$ for $k \ge s+1$.

Theorem 3 is generalized in the following way:

COROLLARY 4. Let I be a centrally symmetric integral and let R_{α_i} , $i=1,\ldots,m$, be fundamental of degree s+1, (2s-1)-orthogonal and each R_{α_i} odd or even, i.e.

$$R_{\alpha_i} = P^{\alpha_i} + \sum_{|\beta|=s-1} A_\beta^{(i)} P^\beta .$$

Let

$$Q_{ij}^{(\beta)} := x_i R_{\beta+e_j} - x_j R_{\beta+e_i} , \qquad |\beta|= s , \ i<j ,$$

and

$$F := \{ R_{\alpha_1},\ldots,R_{\alpha_m} \} \cup \{ Q_{ij}^{(\beta)} / \ |\beta|=s , \ i<j \}.$$

If for all $\nu \in \{1,\ldots,n\}$, $|\beta|=s$, $i<j$

$$x_\nu Q_{ij}^{(\beta)} \in \text{span} \{R_{\alpha_1},\ldots,R_{\alpha_m} \}$$

and if each $Q_{ij}^{(\beta)}$ is $(2s-1)$-orthogonal, then F is a Groebner basis and the polynomials of F have at most $\binom{s+n+1}{n}$ common zeros (counted with multiplicities).

PROOF. In case $Q_{ij}^{(\beta)}$ and $Q_{kl}^{(\gamma)}$ have the same head term, then $S(Q_{ij}^{(\beta)};Q_{kl}^{(\gamma)})$ is orthogonal of degree s and member of span F. A linear combination of orthogonal polynomials is orthogonal of the same degree or 0. Hence in this case $S(Q_{ij}^{(\beta)}, Q_{kl}^{(\gamma)}) \xrightarrow[F]{+} 0$. If $Q_{ij}^{(\beta)}$ and $Q_{kl}^{(\gamma)}$ have not the same head term, then x_ν and x_μ exist, such that $S(Q_{ij}^{(\beta)},Q_{kl}^{(\gamma)}) = S(x_\nu Q_{ij}^{(\beta)},x_\mu Q_{kl}^{(\gamma)})$. Replacing $x_\nu Q_{ij}^{(\beta)}$ and $x_\mu Q_{kl}^{(\gamma)}$ by their resp. linear combinations in terms of $R_{\alpha_1},\ldots,R_{\alpha_m}$, we obtain as in the preceding proof $S(Q_{ij}^{(\beta)},Q_{kl}^{(\gamma)}) \xrightarrow[F]{+} 0$. Analogously $S(Q_{ij}^{(\beta)},R_{\alpha_t}) \xrightarrow[F]{+} 0$. The upper bound for the number of common zeros follows from

$$m = \dim a \cap P_{s+1} - \dim a \cap P_s = \dim P_{s+1} - \dim P_s$$

giving $H(k,a) = H(s+1,a) \leq \dim P_{s+1}$ for $k \geq s+1$.

Theorem 4 is generalized by the following

COROLLARY 5. Let I satisfy $I(x_1^{i_1}\cdots x_n^{i_n}) = 0$, whenever one exponent i_ν is odd. Let s be odd and $\tilde{R}_{\alpha_1},\ldots,\tilde{R}_{\alpha_m}$ be fundamental of degree $\frac{s+1}{2}$, such that $R_{\alpha_1},\ldots,R_{\alpha_m}$, defined by

$$R_{\alpha_i}(x_1,\ldots,x_n) := \tilde{R}_{\alpha_i}(x_1^2,\ldots,x_n^2) , \quad i=1,\ldots,m ,$$

are $(2s-1)$-orthogonal. If for all β with $|\beta| = \frac{s-1}{2}$ and all $i<j$

$$\tilde{Q}_{ij}^{(\beta)} := x_i \tilde{R}_{\beta+e_j} - x_j \tilde{R}_{\beta+e_i}$$

is member of $\tilde{F} := \text{span } \{\tilde{R}_{\alpha_1}, \ldots, \tilde{R}_{\alpha_m}\}$, then $F := \{R_{\alpha_1}, \ldots, R_{\alpha_m}\}$ is a Groebner basis. In the bivariate case, the R_{α_i} have exactly $(s+3)(s+1)/2$ common zeros (counted with multiplicity).

PROOF. Considering the symmetries of the integral, one easily finds, that for odd s any $(2s-1)$-orthogonal polynomial of degree $s+1$ is a polynomial in $x_1^2, \ldots, x_n^2$. By corollary 3, $\tilde{R}_{\alpha_1}, \ldots, \tilde{R}_{\alpha_m}$ is a G-basis, i.e. (iii) of theorem 5 holds for the $\tilde{R}_{\alpha_i}$. Substituting $x_1, \ldots, x_n$ by $x_1^2, \ldots, x_n^2$ we see that (iii) holds also for $R_{\alpha_1}, \ldots, R_{\alpha_m}$, i.e. they constitute a G-basis, too. Observing, that in the bivariate case the polynomials $R_{\alpha_1}, \ldots, R_{\alpha_m}$ are fundamental of degree $s+2$ and using $a \cap P_{s+1} = \text{span}\{R_{\alpha_1}, \ldots, R_{\alpha_m}\}$, we obtain by simple calculation $H(k,a)$ $= (s+1)(s+3)/2$.

4. FINAL REMARKS

The use of Groebner bases techniques simplified the ideal theoretic approach for constructing bivariate cubature formulae and allowed the extension to the multivariate case. This method requires the solution of two different systems of nonlinear equations. First, when one starts with a set of polynomials depending on parameters, one needs a system for determining the parameters, such that the common zeros of the polynomials can be used as a set of nodes for a cubature formula. The calculation of these zeros requires the solution of a second system. Both systems are hard to solve by paper and pencil methods except when the degrees of exactness and the number of variables are moderate.

Computer Algebra Systems (CAS) seem to resolve this problem. In principle, many of these systems are able to derive the systems of non-linear equations for the parameters and to solve the systems for the zeros. By experiences with different CAS, the author has the impression that the automatic derivation of the systems for the parameters causes no great problem, when the CAS is used interactively. The solution of nonlinear equations is nowadays the bottleneck. The complexity of Buchberger's algorithm depend strongly on the number of unknowns. For the moment, it allows the solving of such polynomial systems for mode-rate number of unknowns or in case of many unknowns only for low degrees of polynomial equations, as the corresponding articles in [4] show. However, the rapid development of CAS promises, that more and more com-plicated systems of nonlinear equations can be solved automatically.

One possibility to compute with the present CAS-facilities cubature formulae of higher degree or in more variables is the use of symmetries of the integral. For instance to obtain fully symmetric c.f. (for a fully symmetric integral) it is sufficient to consider only fully symmetric polynomials, i.e. polynomials in the variables $\sigma_1, \ldots, \sigma_n$

$$\sigma_1 = x_1^2 + \ldots + x_n^2, \quad \sigma_2 = x_1^2 x_2^2 + x_1^2 x_3^2 + \ldots + x_{n-1}^2 x_n^2, \ldots, \quad \sigma_n = x_1^2 \ldots x_n^2$$

as shown by Sobolev [7]. Hence the Groebner bases techniques can be applied as before, but now to $K[\sigma_1,\ldots,\sigma_n]$. The only difference is, that now

$$\deg(\sigma_1^{i_1}\cdots \sigma_n^{i_n}) = 2i_1 + 4i_2 + \ldots + 2n\, i_n$$

holds and that now the common zeros of the polynomials are always orbit sets.

REFERENCES

[1] BUCHBERGER, B.: 'Gröbner bases: an algorithmic method in polynomial ideal theory'. in: *Progress, directions and open problems in multi-dimensional systems theory* (N.K.Bose,ed.), D. Reidel Publ. Comp., pp.184 - 232, 1985.

[2] BUCHBERGER, B., COLLINS, G.E., and LOOS, R. (eds.): Computer Algebra, Symbolic and Algebraic Computation, *Computing Supplementum* $\underline{4}$, Springer Verlag 1982.

[3] BUCHBERGER, B.(ed.): EUROCAL '85, *Lecture Notes in Comp.Sci.* $\underline{203}$ and $\underline{204}$, Springer Verlag 1985.

[4] CHAR, B.W. (ed.): *Proceedings of the 1986 Symposium on Symbolic and Algebraic Computation,* ACM , 1986.

[5] COOLS, R. and HAEGEMANS, A.: 'Construction of fully symmetric cubature formulae of degree 4k-3 for fully symmetric planar regions'. Report TW 71, Dept.of Comp.Sci., Univ. Leuven, 1985.

[6] COOLS, R. and HAEGEMANS, A.: 'Optimal addition of knots to cubature formulae for planar regions', *Numerische Mathematik* $\underline{49}$, pp.269 -274 1986.

[7] COOLS, R. and HAEGEMANS, A.: 'Automatic computation of knots and weights of cubature formulae for circular symmetric planar regions'. This volume.

[8] DAVIS, P.J. and RABINOWITZ,P.: *Methods of Numerical Integration,* Academic Press 1975.

[9] ENGELS, H.: *Numerical Quadrature and Cubature,* Academic Press 1980.

[10] GEBAUER, R. and MÖLLER, H.M.: 'A variant of Buchberger's algorithm'. Submitted to *J.of Symb.Comp.* 1986.

[11] KROLL,N., LINDEN,J. and SCHMID, H.J.: 'Minimale Kubaturformeln für Integrale über dem Einheitsquadrat'. Preprint 373 des SFB72, Bonn 1980.

[12] LYNESS, J.N. and JESPERSEN, D.:'Moderate degree symmetric quadrature rules for the triangle', *J. Inst. Math. Appl.* 15, pp. 19-32, 1975.

[13] MACAULAY, F.S.: *The Algebraic Theory of Modular Systems*, Cambridge Tracts in Math. and Math. Physics no. 19, Cambridge Univ. Press, 1916.

[14] MÖLLER, H.M.:'Polynomideale und Kubaturformeln', thesis, Univ. Dortmund, 1973.

[15] MÖLLER, H.M.:'Kubaturformeln mit minimaler Knotenzahl', *Numerische Mathematik* 25, pp. 185-200, 1976.

[16] MÖLLER, H.M.:'The construction of cubature formulae and ideals of principal classes', in: *Multivariate Approximation Theory* (W. Schempp and K. Zeller, eds.) ISNM 51, pp. 249-264, Birkhäuser Verlag 1979.

[17] MÖLLER, H.M.:'A reduction strategy for the Taylor resolution', in [3], vol. 204, pp. 526-534, 1985.

[18] MÖLLER, H.M. and MORA, F.:'The computation of the Hilbert function', in: Computer Algebra, Eurocal '83(J.A. van Hulzen, ed.), *Lecture Notes in Comp. Sci.* 162, Springer Verlag 1983, pp.157-167.

[19] MÖLLER, H.M. and MORA, F.:'New constructive methods in classical ideal theory', *Journal of Algebra* 100, pp. 138-178, 1986.

[20] MORROW, C.R. and PATTERSON, T.N.L.:'Construction of algebraic cubature rules using polynomial ideal theory', *SIAM J. Numer. Anal.* 15, pp.953-976, 1978.

[21] MÜNZEL, G. and RENNER, G.:'Zur Charakterisierung und Berechnung von symmetrischen Kubaturformeln', *Computing* 31, pp. 211-230, 1983.

[22] MYSOVSKIH, I.P.: *Interpolatory cubature formulae* (in Russian), Izdat. Nauka, Moscow 1981.

[23] MYSOVSKIH, I.P.:'The approximation of multiple integrals by using interpolatory cubature formulae', in: *Quantitative Approximation* (R. DeVore and K. Scherer, eds.), Academic Press 1980, pp. 217-243.

[24] RASPUTIN, G.G.:'Cubature formulae and common zeros of orthogonal polynomials'(in Russian), *Vestnik Leningradskogo Univ.* 13, No 3, pp. 40-45, 1978.

[25] RASPUTIN, G.G.: 'On the construction of cubature formulae with preassigned nodes' (in Russian), *Metody vyčislenii* 13, pp. 122-129, 1981.

[26] SCHMID, H.J.: 'On Gaussian cubature formulae of degree 2k-1', in: *Numerische Integration* (G. Hämmerlin, ed.), ISNM 45, Birkhäuser Verlag 1979, pp. 252-263.

[27] SCHMID, H.J.: 'Interpolatorische Kubaturformeln und reelle Ideale', *Math. Z.* 170, pp. 267-282, 1980.

[28] STROUD, A.H.: *Approximate Calculation of Multiple Integrals*, Prentice Hall, Englewood Cliffs, 1971.

QUASI-RANDOM SEQUENCES FOR OPTIMIZATION AND NUMERICAL INTEGRATION

J. P. Lambert
Department of Mathematical Sciences
University of Alaska, Fairbanks
Fairbanks, Alaska 99775-1110
U.S.A.

ABSTRACT. This paper concerns certain deterministic, or 'quasi-random,' uniformly distributed sequences and their implementation in numerical practice. We consider two types of problems from numerical analysis: numerical integration, and the search for functional extrema. The sequences presented here, defined in the unit cubes of dimensions two, three, and four, are well distributed according to two pertinent measures, discrepancy and dispersion. Additionally these sequences are easily generated and are comprised only of dyadic fractions, making them amenable to efficient accurate computer utilization. Numerical examples, in two and four dimensional settings, will serve to illustrate the usefulness of computational schemes based on these sequences.

1. INTRODUCTION

For $\underline{x}_1, \underline{x}_2, \cdots, \underline{x}_N$ a set of N points in the s-dimensional unit cube $I^s = [0,1]^s$, the notions of discrepancy and dispersion are defined as follows:

Let $\mathcal{E} = \{[0,\xi_1) \times \cdots \times [0,\xi_s) : \underline{\xi} = (\xi_1, \ldots, \xi_s) \in I^s\}$. Then the discrepancy is:

$$D_N^* = \sup_{E \in \mathcal{E}} |A(E;N)/N - \mu(E)| ,$$

where $A(E;N)$ is the number of indices m for which $\underline{x}_m \in E$ and $\mu(E)$ is the s-dimensional volume (Lebesgue measure) of E. The dispersion is:

$$d_N = \sup_{\underline{\xi} \in I^s} \min_{1 \leqslant m \leqslant N} d(\underline{\xi}, \underline{x}_m) ,$$

where $d(\cdot, \cdot)$ is a metric, which we will take to be standard Euclidean distance. For an infinite sequence $\underline{x}_1, \underline{x}_2, \cdots$, D_N^* and d_N are understood to be the discrepancy and dispersion of the first N terms. Cf.

P. Keast and G. Fairweather (eds.), Numerical Integration, 193–203.
© 1987 by D. Reidel Publishing Company.

[4,7] for results and references on discrepancy (and a great deal more); for work on dispersion, cf. [6,8,9,10,13].

Discrepancy and dispersion are measures of irregularity of distribution. More specifically, discrepancy can be viewed as a measure of uniformity, and dispersion as a measure of denseness, of the initial terms of a sequence. An infinite sequence is uniformly distributed in I^S if and only if $\lim_{N\to\infty} D_N^* = 0$ (cf. [4, Ch. 2, §1]) and is dense in I^S if and only if $\lim_{N\to\infty} d_N = 0$ (cf. [9, p. 1164]). Dispersion and discrepancy are related by an inequality (which follows from [8, Theorem 3] and [7,(3.1)]):

$$d_N \leqslant 2^{(s+1)/2} \sqrt{D_N^*} \; .$$

Sequences with uniformly low discrepancy are often useful as sources of nodes for numerical quadrature. Consider the (quasi-) Monte Carlo approximation for the integral of a function f of s real variables over I^S:

$$\int_{I^S} f(\underline{\xi}) \, d\underline{\xi} \approx \frac{1}{N} \sum_{m=1}^{N} f(\underline{x}_m) \; . \tag{1}$$

If the nodes $\underline{x}_1, \ldots, \underline{x}_N$ have discrepancy D_N^* and if f is of bounded variation in the sense of Hardy and Krause [7, p. 967], then the approximation error in (1) is bounded by the product of D_N^* with the variation of f. This result, known as the Koksma-Hlawka inequality (cf. [4] or [7]), suggests the importance of low-discrepancy sequences.

Of course if the nodes $\underline{x}_1, \ldots, \underline{x}_N$ are presumed 'random' points from a uniform distribution in I^S, then (1) is the classical statistical Monte Carlo approximation. Typically, in practice, Lehmer's linear congruential pseudo-random number generator (cf., e.g., [7]) is used to supply a 'pseudo-random' number sequence in I, and successive terms of this sequence are taken, as needed, to be coordinates for the $\underline{x}_m$'s .

Quasi-random sequences with low discrepancy frequently provide greater accuracy than do such pseudo-random sequences. (Sequences well distributed in the sense of having low discrepancy or low dispersion are loosely termed 'quasi-random.')

In numerical practice it is often desirable to work with an infinite sequence so that the number of terms N may be increased at will, without loss of data, to provide greater accuracy. It is also desirable that sequences be dyadic (coordinates of terms are dyadic fractions), and be easily and rapidly generated, making them amenable to accurate and efficient computer implementation.

A class of infinite sequences with uniformly low discrepancy was introduced by Halton [2]. For s = 1, 2, ... , the Halton sequence in I^S is $\{(\phi_{p_1}(m-1), \phi_{p_2}(m-1), \ldots, \phi_{p_s}(m-1))\}_{m=1}^{\infty}$, where ϕ_p is the base p radical inverse function (cf. $[7, p. 978]$), and $p_1, p_2, \ldots, p_s$ are relatively prime. We will denote by $\phi^{(s)}$ the s-dimensional Halton sequence for which $p_1, p_2, \ldots, p_s$ are the first s primes. An algorithm for generating terms of the Halton sequence is available in $[3]$. The Halton sequence is infinite and has particularly low discrepancy, $D_N^* = O(\ln^s N / N)$ $[2]$. In the one dimensional case, $\phi^{(1)} = \{\phi_2(m-1)\}_{m=1}^\infty$ is known as the van der Corput sequence, an especially important dyadic sequence with discrepancy $O(\ln N / N)$, which is asymptotically best possible for an infinite sequence in the unit interval $[7, p. 972]$. For $s > 1$ the Halton sequence is not dyadic.

Faure $[1]$ has constructed sequences in I^s which have uniformly smaller discrepancies than the Halton sequence (although with the same asymptotic order); these are non-dyadic for $s > 2$. A sequence in I^2 with $D_N^* = O(\ln N / N)$ was formulated by K. F. Roth, and a sequence in I^s with $D_N^* = O(\ln^{s-1} N / N)$ was introduced by J. M. Hammersley, but neither of these sequences is infinite (they require N to be prescribed), and Hammersley's is not dyadic (cf. $[7, pp. 977,978]$). I. M. Sobol $[14]$ has developed a general theory of sequences having certain uniformity properties. His (dyadic) one-dimensional $'P_\tau$-nets' and higher dimensional $'LP_\tau$-sequences' are discussed by Niederreiter in $[7]$. In our experience these sequences require considerable computational effort to produce.

Sequences with uniformly low dispersion are useful for quasi-random search methods for estimating the extreme values of functions. This application was developed by Niederreiter $[8,9,10]$ and has been the subject of some numerical experimentation (e.g., $[11,12]$). The idea, a deterministic analog of random search, is as follows: Let f be a bounded real-valued function defined on a bounded subset E of $\mathbb{R}^s$ (s $\geqslant 1$) and let $\underline{x}_1, \ldots, \underline{x}_N$ be points in E. Then use $M_N = \max_{1 \leqslant m \leqslant N} f(\underline{x}_m)$ to approximate $M = \sup_{\underline{\xi} \in E} f(\underline{\xi})$. Then $[8, \text{Theorem 1}]$ $M - M_N \leqslant \omega(d_N)$, where $\omega(\cdot)$ is the modulus of continuity of f on E (we will assume that $E = I^s$); that is,

$$M - M_N \leqslant \sup_{\underline{\xi}_1, \underline{\xi}_2 \in E} |f(\underline{\xi}_1) - f(\underline{\xi}_2)|.$$
$$d(\underline{\xi}_1, \underline{\xi}_2) \leqslant d_N$$

This inequality suggests the usefulness of points $\underline{x}_1, \ldots, \underline{x}_N$ having small dispersion. Little seems to be known about the dispersion of particular sequences (except of course in so far as may be inferred from discrepancy and the inequality relating dispersion with discrepancy, noted earlier; also, cf. $[6]$ and $[13]$).

2. THE SEQUENCES $\underline{\eta}^{(2)}$, $\underline{\eta}^{(3)}$, AND $\underline{\eta}^{(4)}$

We define a sequence $\underline{\eta}^{(2)}$ in I^2 as follows. Let the first term be $(0,0)$. For $m = 2,3, \ldots$, determine the m+1 st term (x_{m+1}, y_{m+1}) from the m th term (x_m, y_m) recursively: Suppose r is the smallest positive integer such that $x_m < 1-2^{-r}$ or $y_m > 2^{-r}$. Then, if $x_m < 1-2^{-r}$, let $x_{m+1} = x_m + 3 \cdot 2^{-r} - 1$ and $y_{m+1} = y_m + \gamma \cdot 2^{-r}$, where $\gamma = 1$ if $y_m < 2^{-r}$ and $\gamma = -1$ if $y_m \geq 2^{-r}$. Otherwise let $x_{m+1} = x_m + 2^{-r} - 1$ and $y_{m+1} = y_m$. The sequence $\underline{\eta}^{(2)}$ begins with the points $(0,0)$, $(1/2,1/2)$, $(0,1/2)$, $(1/2,0)$, $(1/4,1/4)$, $(3/4,3/4)$, $\ldots$ The first 32 terms are depicted in Figure 1. If k is odd, the first 2^k terms fill out a 'diagonal' lattice (as in Figure 1); if k is even, the first 2^k terms fill out a regular $2^{k/2}$ by $2^{k/2}$ square lattice (consider, e.g., the first 16 terms in Figure 1). We were originally led to this sequence by a more intuitive geometric construction [5] based on a certain two-dimensional 'splitting and stacking' technique, a two-dimensional version of a scheme due to von Neumann and Kakutani for constructing ergodic measure preserving transformations on the unit interval. The sequence $\underline{\eta}^{(2)}$ is clearly infinite and dyadic, and it is very readily generated -- a computer code is provided in the Appendix. $\underline{\eta}^{(2)}$ has discrepancy and dispersion both of the order of $N^{-1/2}$, as has been pointed out in [5] and [6].

The sort of geometric perspective which initially led to the construction of $\underline{\eta}^{(2)}$ has also been applied to three and four dimensional settings, resulting in infinite dyadic sequences $\underline{\eta}^{(3)}$ in I^3, and $\underline{\eta}^{(4)}$ in I^4. Algorithms (computer codes) for generating these sequences are provided in the Appendix. For $s = 2, 3, 4$ and $j = 1,2, \ldots$, the first 2^{js}

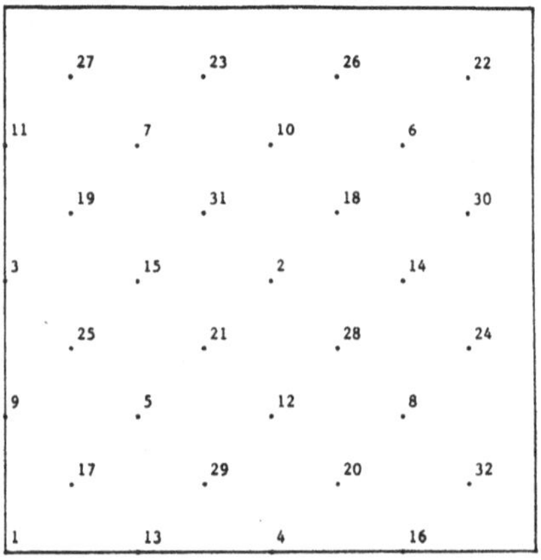

Figure 1. The first 32 terms of $\underline{\eta}^{(2)}$ in I^2.

terms of $\underline{n}^{(s)}$ fill out a regular "cubic" $2^j \times \cdots \times 2^j$ lattice in a certain prescribed order. Although we have not worked out the details, it seems very likely that for $s=3$ and $s=4$ it is again true that the discrepancy and dispersion of $\underline{n}^{(s)}$ are of order $N^{-1/s}$. It is known [8, Theorem 2] that this asymptotic order for dispersion, $d_N = O(N^{-1/s})$, is best (lowest) possible. In the case of discrepancy however, the asymptotic order $O(N^{-1/s})$ is not ideal and in particular is not as good as that for the Halton sequence which, as noted above, has $D_N = O(\ln^2 N/N)$.

3. NUMERICAL RESULTS

We illustrate the implementation of two of these new sequences, $\underline{n}^{(2)}$ and $\underline{n}^{(4)}$ in numerical practice. To begin with, we use the sequence $\underline{n}^{(2)}$ to estimate some integrals, allowing for comparisons with results from using the two-dimensional Halton sequence $\phi^{(2)}$ and also from using classical Monte Carlo. For the case of classical Monte Carlo we will use a standard pseudo-random number generator to compute a two-dimensional (pseudo-)random point sequence, which we will denote $\underline{\rho}^{(2)}$. As sample 'test' functions we use the following:

$$f_1(x,y) = 396900\, x^4(1-x)^4 y^4(1-y)^4, \qquad f_2(x,y) = e^{-(1-x)^2 - 2y^2},$$

$$f_3(x,y) = \begin{cases} \dfrac{280}{19}(y-x)^3, & y \geq x \\ \dfrac{280}{19}(y-x)^6, & y < x \end{cases}, \qquad f_4(x,y) = \dfrac{2y-x}{y^2+1}.$$

We estimate the integrals $I_n = \int_0^1 \int_0^1 f_n(x,y)\, dy\, dx$ for $n = 1, 2, 3, 4$.

Actual values are: $I_1 = 1$, $I_2 = 0.44670838$, $I_3 = 1$, $I_4 = 0.30044081$ ($= \ln 2 - \pi/8$). There is some precedent for using these sample integrals in testing and comparing numerical quadrature methods. For instance I_1, I_2, and I_3 have been used for that purpose by Zaremba [15]. For $n = 1, 2, 3$, and 4, we let $J_n(\eta)$, $J_n(\phi)$, and $J_n(\rho)$ denote the estimates for I_n obtained from using respectively the quasi-random sequences $\underline{n}^{(2)}$ and $\phi^{(2)}$, and the pseudo-random sequence $\underline{\rho}^{(2)}$. By '% error' we will mean $100(J_n(\cdot) - I_n)/I_n$, where '$\cdot$' is to be replaced by η, ϕ, or ρ as appropriate. Results of computations are printed out in Tables 1 through 4. (These calculations were carried out on a Columbia Data Products Personal Computer, Model 1600-VP/110.) In each case $N = 60, 300, 1200, 3600,$ and 7200 nodes (quasi- or pseudo-random points) were used. The tables

Table 1. Comparison of I_1 estimates $J_1(\eta)$, $J_1(\phi)$, $J_1(\rho)$.

N	$J_1(\eta)$	% error	$J_1(\phi)$	% error	$J_1(\rho)$	% error
60	0.98516	-1.48425	0.95339	-4.66088	1.27060	27.05986
300	1.00368	0.36835	0.99719	-0.28105	1.06631	6.63119
1200	1.00008	0.00779	0.99956	-0.04451	1.04583	4.58291
3600	1.00003	0.00327	1.00013	0.01263	0.98766	-1.23400
7200	1.00000	0.00005	0.99968	-0.03207	0.99681	-0.31954

Table 2. Comparison of I_2 estimates $J_2(\eta)$, $J_2(\phi)$, $J_2(\rho)$.

N	$J_2(\eta)$	% error	$J_2(\phi)$	% error	$J_2(\rho)$	% error
60	0.44638	-0.07347	0.45418	1.67328	0.41721	-6.60349
300	0.44972	0.67420	0.44801	0.29235	0.41650	-6.76342
1200	0.44850	0.40098	0.44729	0.13069	0.43553	-2.50194
3600	0.44763	0.20619	0.44960	0.04302	0.44312	-0.80433
7200	0.44681	0.02164	0.44680	0.02045	0.44594	-0.17284

Table 3. Comparison of I_3 estimates $J_3(\eta)$, $J_3(\phi)$, $J_3(\rho)$.

N	$J_3(\eta)$	% error	$J_3(\phi)$	% error	$J_3(\rho)$	% error
60	0.91426	-8.57411	0.92442	-7.55756	0.86342	-13.65799
300	0.99915	-0.08494	0.99967	-0.03298	0.81260	-18.74034
1200	1.00297	0.29739	0.99964	-0.03637	0.96492	-3.50823
3600	1.00111	0.11106	1.00008	0.00806	1.02735	2.73482
7200	1.00191	0.19084	0.99971	-0.02877	1.01707	1.70670

Table 4. Comparison of I_4 estimates $J_4(\eta)$, $J_4(\phi)$, $J_4(\rho)$.

N	$J_4(\eta)$	% error	$J_4(\phi)$	% error	$J_4(\rho)$	% error
60	0.29994	-0.16847	0.28762	-4.26851	0.37576	25.06688
300	0.29465	-1.92852	0.29797	-8.25343	0.35464	18.03870
1200	0.29749	-0.98543	0.29945	-0.33118	0.32704	8.84987
3600	0.29900	-0.48334	0.30010	-0.11687	0.30985	3.12982
7200	0.30049	0.01486	0.30028	-0.05482	0.30445	1.33243

indicate that $\underline{\eta}^{(2)}$ is more accurate than is $\underline{\phi}^{(2)}$ in estimating I_1, less accurate in estimating I_3, and of a similar level of accuracy in estimating I_2 and I_4; in each instance, computations involving $\underline{\eta}^{(2)}$ were much faster than those involving $\underline{\phi}^{(2)}$. Classical Monte Carlo estimates using $\underline{\rho}^{(2)}$ are considerably less accurate in each case than are the others.

In light of the Koksma-Hlawka inequality to the effect that, for a given function with the appropriate bounded variation, the approximation error is $O(D_N^*)$, it may seem surprising that the new sequence $\underline{\eta}^{(2)}$ does, comparatively, so well in these numerical experiments; it does as well on average, for a given number of terms, as does the Halton sequence $\underline{\phi}^{(2)}$ despite the fact that the Halton sequence has discrepancy of asymptotically smaller order (recall that the respective orders of magnitude of discrepancy for $\underline{\eta}^{(2)}$ and $\underline{\phi}^{(2)}$ are $N^{-1/2}$ and $\ln^2 N/N$). The following two comments may be of interest in this regard: (i) For the number of terms considered ($\leqslant 7200$) there is not much difference in the respective discrepancies (in fact note that $\ln^2 N/N$ is actually larger than $N^{-1/2}$ for $5 \leqslant N \leqslant 5503$). (ii) The Koksma-Hlawka inequality can only provide an upper bound on the approximation error, without any further indication as to the quality of approximation. It is interesting to observe in the tables that the absolute percentage error in most cases does not decrease monotonically with increasing N when either of the quasi-random sequences $\underline{\eta}^{(2)}$ or $\underline{\phi}^{(2)}$ is used (the absolute percentage error does monotonically decrease when the pseudo-random sequence $\underline{\rho}^{(2)}$ is used, although remaining much larger in each instance than it is for the other sequences).

It might also be worth remarking that, for the case of classical Monte Carlo, theoretically (presuming random uniformly distributed nodes in I^2), with high probability the integration error is $O(N^{-1/2})$ -- the same as we obtain in using $\underline{\eta}^{(2)}$ -- and yet, as has been noted, the pseudo-random sequence $\underline{\rho}^{(2)}$ does far worse than $\underline{\eta}^{(2)}$ at integrating the test functions.

As a second set of numerical examples we use the sequence $\underline{\eta}^{(4)}$ and, for comparison, the Halton sequence $\underline{\phi}^{(4)}$ in estimating maxima of functions of four variables on I^4. The sample functions considered in Examples 1 through 4 have previously been used for a comparative study of quasi-random search methods by Niederreiter and Peart [12] (see also

[11]). In these examples, M is the actual value of the maximum. For

(a) N=10,000, (b) N=50,000, and (c) N=100,000, first using the sequence

$\underline{\phi}^{(4)}$ and then the sequence $\underline{n}^{(4)}$, the value $M_N = \max_{1 \leqslant m \leqslant N} f(x,y,z,w)$ is

computed; ℓ is the value of m at which the maximum occurs; and $x_\ell, y_\ell,$

z_ℓ, w_ℓ are the coordinates of the ℓ th term of the sequence. Computations

involving $\underline{n}^{(4)}$ were substantially faster than were those involving $\underline{\phi}^{(4)}$,

and a comparison of M_N values in the examples seems to indicate that

overall the sequence $\underline{n}^{(4)}$ is significantly more effective for the

purpose of estimating maxima of these functions. (Calculations for these

examples were carried out on a VAX 11-750 computer.)

EXAMPLE 1: $f(x,y,z,w) = e^{xyzw} \sin(x+y+z+w)$

$M = 1.02620 = f(.409888, .409888, .409888, .409888)$

using $\phi^{(4)}$

	M_N	ℓ	x_ℓ	y_ℓ	z_ℓ	w_ℓ
(a)	1.02565	3502	.459717	.414876	.404864	.351163
(b)	1.02597	16102	.404236	.395773	.436813	.386863
(c)	1.02597	16102	.404236	.395773	.436813	.386863

using $\eta^{(4)}$

	M_N	ℓ	x_ℓ	y_ℓ	z_ℓ	w_ℓ
(a)	1.02521	819	.375	.5	.375	.375
(b)	1.02577	16657	.375	.4375	.375	.4375
(c)	1.02610	65809	.40625	.40625	.40625	.40625

EXAMPLE 2: $f(x,y,z,w) = \dfrac{100\, x\, y\, z\, w}{(1+x\,y\,z)^2\, e^w}$

$M = 9.19699 = f(1, 1, 1, 1)$

using $\phi^{(4)}$

	M_N	ℓ	x_ℓ	y_ℓ	z_ℓ	w_ℓ
(a)	9.07946	5039	.960083	.937205	.889024	.970964
(b)	9.14326	40319	.995010	.922302	.940634	.970148
(c)	9.17054	75599	.948799	.966423	.990707	.966477

using $\eta^{(4)}$

	M_N	ℓ	x_ℓ	y_ℓ	z_ℓ	w_ℓ
(a)	9.09278	4370	.9375	.9375	.9375	.9375
(b)	9.09278	4370	.9375	.9375	.9375	.9375
(c)	9.17158	69906	.96875	.96875	.96875	.96875

EXAMPLE 3: $f(x,y,z,w) = -|2x+y-z+w-5/12|-|2y-3z|-|x+4y+6z-4|-|x+5w-5/4|$

$M = 0 = f(0, 1/2, 1/3, 1/4)$

using $\phi^{(4)}$

	M_N	ℓ	x_ℓ	y_ℓ	z_ℓ	w_ℓ
(a)	-.484781	2416	.056885	.514556	.335360	.183733
(b,c)	-.332418	15016	.083435	.467498	.321536	.219432

using $\eta^{(4)}$

	M_N	ℓ	x_ℓ	y_ℓ	z_ℓ	w_ℓ
(a)	-.416667	2755	0	.5	.375	.25
(b,c)	-.208333	41155	0	.5	.3125	.25

EXAMPLE 4: $f(x,y,z,w) = 10^4\, x\, y\, z\, w\, e^{-x-2y-3z-4w}$

$M = 7.63152 = f(1, 1/2, 1/3, 1/4)$

using $\phi^{(4)}$

	M_N	ℓ	x_ℓ	y_ℓ	z_ℓ	w_ℓ
(a)	7.50826	3991	.913818	.525072	.353984	.217469
(b)	7.59520	16591	.949249	.509628	.347533	.233177
(c)	7.61082	82111	.988319	.479370	.313987	.251536

using $\eta^{(4)}$

	M_N	ℓ	x_ℓ	y_ℓ	z_ℓ	w_ℓ
(a)	7.51227	1342	.875	.5	.375	.25
(b)	7.60047	22334	.9375	.5	.3125	.25
(c)	7.60047	22334	.9375	.5	.3125	.25

APPENDIX. Programs, in VAX Basic, for deriving and printing terms of

(i) $\underline{\eta}^{(2)}$,   (ii) $\underline{\eta}^{(3)}$,   (iii) $\underline{\eta}^{(4)}$.

```
(i)   100   REM    ETA2
      110   INPUT "# OF TERMS ";N%
      120   X=0\ Y=0
      130   FOR M%=1% TO N%
      140   PRINT X;Y
      150   T=1
      160   T=T/2\ U=1-T
      170   IF X>=U AND Y<T THEN GOTO 160 ELSE X=X-U
      180   IF X>=0 THEN GOTO 200 ELSE X=X+2*T
      190   IF Y<T THEN Y=Y+T ELSE Y=Y-T
      200   NEXT M%

(ii)  100   REM    ETA3
      110   INPUT "# OF TERMS ";N%
      120   X=0\ Y=0\ Z=0
      130   FOR M%=1% TO N%
      140   PRINT X;Y;Z
      150   T=1
      160   T=T/2\ U=1-T
      170   IF X>=U AND Y>=U AND Z<T GOTO 160
      180   X=X-U\ Y=Y-U\ Z=Z-T \ S=2*T
      190   IF X<0 THEN X=X+S ELSE GOTO 230
      200   IF Y<0 THEN Y=Y+S
      210   IF Z<0 THEN Z=Z+S
      220   GOTO 260
      230   IF Y<0 AND Z<0 THEN Y=Y+T\ Z=Z+S\ GOTO 260
      240   IF Y<0 THEN Y=Y+S\ Z=Z+T \ GOTO 260
      250   Y=Y+T
      260   NEXT M%

(iii) 100   REM    ETA4
      110   INPUT "# OF TERMS ";N%
      120   X=0\ Y=0\ Z=0\ W=0
      130   FOR M%=1% TO N%
      140   PRINT X;Y;Z;W
      150   T=1
      160   T=T/2\ U=1-T
      170   IF X>=U AND Y>=U AND Z>=U AND W<T GOTO 160
      180   X=X-U \ Y=Y-U\ Z=Z-U\ W=W-T\ S=2*T
      190   IF X<0 THEN X=X+S ELSE GOTO 240
      200   IF Y<0 THEN Y=Y+S
      210   IF Z<0 THEN Z=Z+S
      220   IF W<0 THEN W=W+S
      230   GOTO 310
      240   IF Y<0 AND Z<0 AND W<0 THEN Y=Y+S\ Z=Z+S\ W=W+T\ GOTO 310
      250   IF Y<0 AND Z<0 THEN Y=Y+T \ Z=Z+S\ GOTO 310
      260   IF Y<0 AND W<0 THEN Y=Y+S\ Z=Z+T \ W=W+S\ GOTO 310
      270   IF Y<0 THEN Y=Y+S\ W=W+T\ GOTO 310
      280   IF Z<0 AND W<0 THEN Z=Z+T\ W=W+S\ GOTO 310
      290   IF Z<0 THEN Z=Z+S\ W=W+T\ GOTO 310
      300   Y=Y+T
      310   NEXT M%
```

REFERENCES

1. H. Faure, Discrépance de suites associées à un système de numération (en dimension s), Acta Arithmetica 41 (1982), 337-351.

2. J. H. Halton, On the efficiency of certain quasi-random sequences of points in evaluating multi-dimensional integrals, Numer. Math. 2 (1960), 84-90.

3. J. H. Halton and G. B. Smith, Algorithm 247: radical-inverse quasi-random point sequence, Comm. ACM 7 (1964), 701-702.

4. L. Kuipers and H. Niederreiter, Uniform Distribution of Sequences, Wiley, New York, 1974.

5. J. P. Lambert, On the development of infinite low-discrepancy sequences for quasi-Monte Carlo implementation, Technical Report, University of Alaska, Fairbanks, March 1986.

6. J. P. Lambert, A sequence well dispersed in the unit square, preprint (1986).

7. H. Niederreiter, Quasi-Monte Carlo methods and pseudo-random numbers, Bull. Amer. Math. Soc. 84 (1978), 957-1041.

8. H. Niederreiter, A quasi-Monte Carlo method for the approximate computation of the extreme values of a function, Studies in Pure Mathematics (To the Memory of Paul Turán), Birkhäuser, Basel, 1983, 523-529.

9. H. Niederreiter, On a measure of denseness for sequences, Colloq. Math. Soc. Janos Bolyai, Topics in Classical Number Theory, (Budapest, 1981), North-Holland, Amsterdam, 1984, 1163-1208.

10. H. Niederreiter, Quasi-Monte Carlo methods for global optimization, Proc. 4th Pannonian Symp. on Math. Stat., (Bad Tatzmannsdorf, Austria, 1983), Akadémiai Kiadó, Budapest, 1986, 251-267.

11. H. Niederreiter and K. McCurley, Optimization of functions by quasi-random search methods, Computing 22 (1979), 119-123.

12. H. Niederreiter and P. Peart, A comparative study of quasi-Monte Carlo methods for optimization of functions of several variables, Caribbean J. Math. 1 (1982), 27-44.

13. P. Peart, The dispersion of the Hammersley sequence in the unit square, Monatsh. Math. 94 (1982), 249-261.

14. I. M. Sobol', Multidimensional quadrature formulas and Haar functions, Izdat. 'Nauka,' Moscow, 1969. (Russian)

15. S. K. Zaremba, A quasi-Monte Carlo method for computing double and other multiple integrals, Aequationes Math. 4 (1970), 11-22.

NON-FORTUITOUS, NON-PRODUCT, NON-FULLY SYMMETRIC CUBATURE STRUCTURES

Francis Mantel
351 N. Stanley Avenue
Los Angeles, CA 90036
U.S.A.

ABSTRACT

We consider cubatures invariant under the order four subgroup of the fully symmetric group over regions invariant under right angle rotations. The moment fitting system (MFS) includes only equations for monomials invariant under this subgroup. The linear consistency conditions for the nonlinear system of equations obtained are calculated for all parameter limited subspaces of the linear span of the equations in the MFS. Using the linear consistency conditions we determine all possible structures of cubatures possessing the symmetry of this subgroup.

P. Keast and G. Fairweather (eds.), Numerical Integration, 205.
© *1987 by D. Reidel Publishing Company.*

A PARALLELIZATION OF ADAPTIVE INTEGRATION METHODS

Elise de Doncker and John A. Kapenga
Western Michigan University
Computer Science Department
Kalamazoo, MI 49008
U.S.A.

ABSTRACT. A meta-algorithm for adaptive integration is specified. It includes a class of adaptive integrators which apply extrapolation to deal with boundary singularities. Methods adhering to the meta-algorithm are parallelized in a straightforward way with use of a set of macros layered over the Argonne macro package.

1. INTRODUCTION

On a high enough level the meta-algorithm description by Rice [24] remains valid as a representation of its parallel analogue investigated in [23,25,26]. The model fits on a Multiple Instruction Multiple Data architecture with shared memory.

The integral computation involves an adaptive partitioning of the interval and henceforth the manipulation of an *interval collection*. Intervals are taken from the collection and "processed". In Rice's approach, the subprogram for *processing an interval* (subdividing it and estimating the integrals and errors over the parts) runs on each of a set of n physical processors. Two other processors are assigned the tasks of *interval collection management* (sending intervals to the interval processors and putting the resulting subintervals with their integral and error estimates back in the collection) and *algorithm control* respectively.

Let the integral to be evaluated be

$$If = \int_0^1 f(x)\, dx\ .$$

Under certain conditions, Rice [23] obtains convergence and speedup results. Apart from related assumptions on the integrand and on the local error estimates, there are conditions on the times characterizing the interval collection management and interval processing in parallel. For example, in intervals which do not contain "singularities" of the integrand, the first two assumptions involve the existence of a constant p

P. Keast and G. Fairweather (eds.), Numerical Integration, 207–218.
© *1987 by D. Reidel Publishing Company.*

such that $f^{(p)}(x)$ is continuous and the error estimate is of order $O(h^{p+1})$ where h is the partition length. The timing conditions lead to an upper bound for the *cycle time* T_c which comprises *delivery* of an interval to an (idle) interval processor, processing, *return* of the results to the manager and the controller and *insertion* of the returned intervals into the data structure.

The interval collection consists of two disjoint sets (*boxes*) for distinguishing between active and discarded intervals. The convergence and speedup results are derived on the basis of the *proportional partition* algorithm [24]: an interval J belongs to the *active box* if its error estimate exceeds $\varepsilon|J|$ for a prescribed ε and where $|J|$ represents the length of J; otherwise it is in the *discard box*.

Under the assumed conditions the following two main results are derived [23]:

1.

$$|If - Q_N f| = O(N^{-p}) \qquad \text{as } N \to \infty$$

where N is the number of function evaluations used and $Q_N f$ is the integral approximation;

2. if $\dot N > n^2$ there is a constant K (independent of n and T_c) such that

$$T_N f \leq KT_c \frac{N}{n}$$

with $T_N f$ the time needed to compute the approximation $Q_N f$.

Basically, the latter result is obtained by deriving a bound for the length of the longest path d in the tree of intervals produced by the subdivision strategy. In this context, T_c corresponds to the time required for processing a node. If n or more nodes were active all the time, then the total processing time for the assumed model would not exceed DT_c/n where D is the total number of nodes. In order to take account of situations where some processors may be idle some of the time, the bound $(D/n + d)T_c$ is adopted. It emerges that algorithms which generate shorter but wider interval trees (wide enough to guarantee n active intervals all the time) tend to have a better speedup. Note that a long, narrow path in the tree is generally an indication of an integrand problem spot, for example a strong isolated singularity.

Rice concludes that, if the cycle time is a constant independent of n, there is a constant C such that

$$T_N f \leq C \frac{N}{n}$$

for $N > n^2$. This covers the situation where no interaction is required between the processes. The practical ramifications of this result are unclear since, for all commercially available systems at present, T_c will vary (even linearly) in n. However, more practical results could be obtained using a technique not unlike Rice's in conjunction with an accurate model for the performance of the interval manager, to produce realistic bounds on the delivery, return and insertion components.

In the present paper, unlike Rice, we refer to *processes* rather than processors, leaving the mapping of processes onto processors unspecified. This mapping may be effectively carried out in several ways on various architectures but will not be discussed here. We shall parallelize a meta-algorithm for adaptive integration by considering n parallel processes which perform *interval subdivision steps* as outlined in Figure 1.

> *Subdivision step:*
> Select the interval to be processed next;
> process the interval, possibly generating
>   more active intervals;
> update the global results;
> if termination conditions occur then stop;
> add any new intervals to the interval collection;

Figure 1. Interval subdivision step.

One of the processes (the *master*) also controls the algorithm and is responsible for creating the others (*slaves*). All the processes access and maintain the interval collection (as opposed to Rice's single interval collection manager). This approach of parallelizing algorithms which involve the manipulation of a global work pool is advocated by Lusk and Overbeek [15,16]. It was also used by Overbeek to parallelize a simple local adaptive integrator [19].

In section 2 we shall outline the meta-algorithm we intend to parallelize. It was designed to include the algorithm QAGS of the subroutine package QUADPACK [22]. In this algorithm the interval partitioning is organized into stages. Apart from allowing for extrapolations in between the stages, this feature is connected to a widening of the interval partitioning tree in a logical way. This has a favorable effect on the speedup because it helps to ensure "useful" work for the n processes especially when dealing with integrand singularities.

We will discuss the parallelization in section 3. A set of simple macros were layered over the Argonne macros, hiding the synchronization efforts. As a result, algorithms belonging to the considered meta-algorithm can be parallelized efficiently using the macros, thereby retaining most of the serial code. Performance modeling can be done on the basis of the serial implementation at hand, as presented by Kapenga [12] and applied to task partitioning algorithms in [4].

2. THE META-ALGORITHM

First consider the meta-algorithm given by Figure 2 below.

>Initialize;
>While the global acceptance criterion is not satisfied do
>   Subdivision step; Enddo;

Figure 2. General adaptive algorithm.

This contains local and global adaptive algorithms [18,24,27].
In a *local adaptive* algorithm,

- newly obtained intervals are subject to a local acceptance
  criterion and classified thereupon as active or to be discarded;
- for normal termination the set of active intervals must be
  empty.

The two-box algorithm considered by Rice [23] is local adaptive. In
this case the global acceptance condition in the description above requires an empty active box; i.e. the while clause comes down to "While
the active box is not empty".
In a basic *global adaptive* strategy,

- all intervals remain active throughout the computations;
- the intervals are ranked with respect to their desirabili-
  ty of partitioning;
- the most desirable interval is selected at each step;
- the criterion for normal termination depends on the estimated error for the entire collection.

For example, the algorithms AIND [21] and QAG in QUADPACK [22] belong
to this category. Here the intended "While the user-prescribed tolerance is not met" is translated into "While the global error estimate is
larger than the required accuracy". The global error estimate is obtained by summing the local error estimates over the interval collection.

In the case of an isolated singularity these algorithms will subdivide around this location and try to single it out until its presence influences the global quadrature error estimate to within the requested accuracy. Note that this process, which results automatically
from the adaptive strategies, is inherently sequential.

The algorithm [22,2] at the basis of QAGS in QUADPACK, and of
TRIEX [5,6], TRIAN [3] and CUBEX [7] for multidimensional integration,
has the ability to handle interesting classes of singular integrand
behaviour at the boundaries of integration or on the subdivision grid.
A sophisticated strategy is applied to mimic the generation of a sequence of 2^{ℓ}-copies of a Gauss-Kronrod pair of rules in 1D and extrapolating to the limit with Wynn's ε-algorithm [29,30]. An approximate
$2^{\ell}-$ copy is formed during *stage* ℓ of the algorithm (for $\ell=0,1,\ldots$).
The exact copy would involve the subdivision of the original interval
into 2^{ℓ} equal parts and application of the scaled quadrature rule to

each part. The approximate copy will contain some intervals of the
smallest size (at least two, of size $2^{-\ell}$ times the length of the original
interval) and a set of bigger intervals where the integration error has
to be made small enough to let the integration result over the approximate copy replace the result over the full copy as the ℓ-th element of
the sequence to which the extrapolation has to be applied. The bigger
intervals are those which can still be subdivided in the course of the
current stage; they belong to the *active box*. The "small" intervals
will have to await subdivision. Until the end of this stage they remain
in the *passive box*. At all times intervals can only be selected from
the active set. The two intervals resulting from a subdivision can
either go back into the active box or go into the passive box depending
on their size. At the end of a stage (after the extrapolation), the
active and the passive intervals are merged and all are made active
again to begin the next stage. The algorithm is represented in Figure 3.

> Initialize;
> While the global acceptance criterion is not satisfied do
> *Stage:*
> While the error sum over the active box is too big do
> Subdivision step (on the active set); Enddo;
> Extrapolate;
> If termination conditions occur then stop;
> Merge the passive set with the active set;
> Enddo;

Figure 3. Meta-algorithm to be parallelized.

In the inner While condition, the sum of the error estimates over the
active intervals is compared to a fraction of the user-requested accuracy.(The test is made relative with respect to the integral result
obtained so far if the user requested a relative accuracy.) In this way
an attempt is made to ensure a couple of guard digits in the input to
the extrapolation.
     The use of the ε-algorithm for integration was demonstrated by
Kahaner [11] in a non-adaptive way. The idea of using it with (in some
other sense) approximate quadrature rule copies was incorporated into
a local adaptive strategy by Genz [9]. The extrapolation is justified
for several types of asymptotic error functional expansions [9] (a study was also done in [28]), most importantly the forms in inverse powers of the mesh ratio m, possibly multiplied with log(m) to an integer power, and with constant coefficients. This includes the expansion
valid for the error of a (1D) $m=2^{\ell}$-copy quadrature rule approximation
when the integrand has algebraic and/or logarithmic end-point singularities [17].

3. PARALLELIZATION

The part to be executed in parallel is the code of the subdivision
steps. It is embedded in a "work module" as suggested by Lusk and Overbeek [15,16]. The work module is executed by the slave processes as
soon as they are created and it is also run by the master process. This
feature allows   some debugging   on a serial machine, where the program runs with just one (the master) process [15,16]. The body of the
loop which constitutes the work module is shown in Figure 4.

> *getwork:* select the interval to be processed next;
> *getspace* for the subintervals to be obtained in
>   the subdivision;
> Process the interval;
> *update* the global results;
> If program termination conditions occur then
>   *progend;* Stop;
> Endif;
> *addwork:* add the new intervals to the interval collection;
> If stage end conditions occur then *probend;*

Figure 4. Parallel version of Subdivision step.

In this description, *getwork, getspace, update, progend, addwork*
and *probend* represent macro invocations. The macros *progend* and *probend*
used to flag the end of the program or end of a stage to the running
processes, belong to the Argonne macro package [15,16]. The other macros are layered over the Argonne package, for the implementation of
the synchronization primitives needed. The *getwork* macro uses the Argonne *askfor* to request a new interval from the active box in the interval collection. When the stage end is flagged, a request from the
master process for a new interval causes it to return from the work
routine and resume the control of the algorithm. In the present implementation, the processes are then all held at a barrier, until a new
stage begins. Otherwise, when an interval is available, it gets processed. The global results, such as the sum of the local integral
approximations, the error estimate and flags are accessed and updated
via the macro *update.* Through *addwork* the data structure on the interval collection is built up and maintained.
    After the necessary initializations and the creation of the slave
processes (using the Argonne *create* macro), the sequencing of the stages is governed by the outer While loop of Figure 3. This is coded into
the control part of the program, executed by the master process. Within
the While loop each stage is initiated by a call to the work module.
If termination conditions are flagged upon its return, *progend* is invoked and the computations are ended. This is either a "normal" termination with the sum of the local integral contributions as the result,
or an abnormal ending with an error flag issued, for example after the
detection of roundoff error by one of the processes. In the latter case

either the sum of the local integral contributions or an extrapolated
result may be selected to be returned to the user as the integral
approximation. Usually the result with the smallest error estimate is
selected; exceptions are made in some flagged error conditions.
If the computations do not have to be ended upon the return of the work
routine, an extrapolation is performed with the current sum of the local integral contributions as input to the extrapolation procedure.
This yields a new extrapolation result, with its error estimate. If the
error estimate does not exceed the user's requested accuracy, the computations are ended (after *progend* is invoked). Otherwise, the algorithm proceeds by merging the contents of the active and of the
passive box (so that all intervals become active), and goes to the next
stage. An outline of the control part of the algorithm is given in Figure 5.

> Initialize;
> *create* the slave processes;
> While the global acceptance criterion is not satisfied do
>   *Stage*:
>   Call the work module;
>   If termination conditions exist then stop;
>   Extrapolate;
>   If termination conditions occur then
>   *progend*; Stop;
>   Endif;
>   *merge* the passive set with the active set;
> Enddo;

Figure 5. Control part of the algorithm.

For the synchronization of the running processes we have to consider their accessing and updating the global data. These operations give
raise to critical sections in the code. The execution of the critical
sections is embedded in *monitors* [10,15,16]. On their use the high level
synchronization primitives implemented by the Argonne macro package are
based. The following definition is given: "a conceptual abstraction composed of: the data shared between the processes, the operations that
represent critical sections associated with the shared data, and the code
required to initialize the data structures".
Apart from the states of the control variables, the shared data in our
application are constituted by the interval collection (subdivided into a passive and an active set) and the global variables specific to
the algorithm (such as the global integral result, error estimate and
error flags).
The principal operations performed on the global data via the work
routine are the selection of a new interval and insertion of the subintervals, and furthermore the updates of the global results and flags.
These all involve intra-stage updates. The merge of the active and
passive sets and the updates of the results from the extrapolation

are done in between the stages, and via the control part (although they
do not have to be).

Note that the global data in the interval collection, and the
operations performed on it, are quite distinct from the "user's" or algorithm specific data. While the latter are accessed and updated via
the user defined macro *update* , or via the control part, the management of the interval collection can be handled in a more general framework, in a problem independent manner.   This feature supports an attempt
to "mechanize" the parallelization of the integrators belonging to the
meta-algorithm (not unlike the Argonne attempt to relieve the user from
various synchronization efforts). The macro *getwork* passes to the Argonne *askfor* a piece of code to select an element from the interval
collection. We had to make a choice as to the data structure on the
collection. Two interleaved linked lists were used, representing the
data in the active and in the passive boxes respectively. Macro *getspace*
then requests available nodes and *addwork* links in new information.

4. CONCLUDING REMARKS

We implemented a preliminary version of an algorithm adhering to the
meta-algorithm in Fortran, under Unix, on a Denelcor Hep. At the present
time, after the Hep disappeared, it is fortunate that the program including the macros, can be ported together with the Argonne macros (for
example, to a Cray XM-P). The portability of the macro layers is an
advantage added to those of making the parallelization easier for the
user while retaining the possibility to let him insert code into the
synchronization tools on lower levels.

In the conversion of the code we found that one of the biggest
tasks was the replacement of the data structure in the existing serial
code to a linked representation. The former did not allow insertion of
an interval without shifting part of an array over one location; we
wanted to improve the efficiency of this operation. From that point on
we were able to keep almost all the serial code in our conversion to
the parallel program. The biggest change here was breaking the (top-
level) integration routine up into a control part and a work routine.
This took a couple of days worth of work and we expect that no bigger
effort would have to go into the parallelization of any member of the
meta-algorithm or even any adaptive integrator, using the sets of macros available or slight modifications thereof. Note that, in this respect, changing to a linked stack representation would involve only a
minor modification of the macros *addwork* and *getspace*, and would be
appropriate for use in a local adaptive algorithm. Modifications to
other data structures (trees, heaps), which might be more efficient in
some applications, are straightforward as well. Note also that, with
the present approach using the additional layer of macros, the data
structure involved is invisible to the application program.

A preliminary version of the parallel integrator run with different numbers (n) of processes on the Hep revealed that the least execution time was needed with 9 processes and the obtained speedup,

$$S^{(n)} = \frac{T^{(1)}}{T^{(n)}} \,,$$

was around 6 for $n = 9$. Several improvements could be suggested to increase $\max_{n=1,2,\ldots} S^{(n)}$ for the parallel implementation, for example
- by allowing $n-1$ processes to continue the subdivision work while one process is doing the extrapolation;
- by allowing several processes to insert new intervals into the lists at the same time.

In [12,4] it is proposed to predict, based on properties of the serial program and a few machine characteristics, if parallelizing in a given manner will let the speedup reach the "effectve parallelism" (n_e) of the target parallel machine. Using 1 PEM (8 stage instruction pipeline) on a Hep suggests a value of 8 for n_e, but other hardware characteristics explain higher possible speedups. On a machine with P parallel processors one usually assumes $S^{(P)} < P$. However, it is shown in [20] that "superlinear speedups" $(S^{(P)} \geqslant P)$ can be obtained by the way the hardware operates. Contrary views on this subject [8] seem to be due to differences in concepts.

A final remark concerns the eventual port of the meta-algorithm to MIMD machines without shared memory, like the hypercube. On such a system we have to deal with a distributed task pool, i.e. several nodes all have to store part of the interval collection. For example, a hierarchy of nodes can be considered where the leaf nodes are responsible for the work involving the interval subdivisions. Berntsen [1] mapped a global adaptive integrator (QAGE [22], without extrapolation) onto the hypercube using a method of this type. Fixed parts of the original interval are assigned to the work nodes in the beginning of the algorithm, where their corresponding sets of subintervals will be managed during the computations. The results of each local integration are sent up into the hierarchy, in order to update the global results and test the stop criteria at each step.

In view of the high overhead of the message sending it seems appropriate to let the work nodes send the results of their local integrations less often. Moreover, in order to avoid long waits for nodes which happen to have difficult parts of the interval, the work nodes should communicate with one or more load balancing nodes. A general load balancing system is under study [13]. Note the load balancing scheme applied by Lemme and Rice [14] to the interval queues in their map of (local) adaptive integration methods onto MIMD machines.

In conclusion, as opposed to the monitors used on MIMD machines with shared data (and critical operations), one needs a distributed interval collection management system on machines without global memory. This system would again be invisible to the application program, as it can be hidden under the macros.

ACKNOWLEDGEMENT

The authors acknowledge the support of P. Messina, E. L. Lusk and R. A. Overbeek, and furthermore of Bob Reschly, Eric van de Velde and Jarle Berntsen.

REFERENCES

[1]   J. Berntsen, *Private communication*.

[2]   E. de Doncker, An Adaptive Extrapolation Algorithm for Automatic Integration. *SIGNUM Newsletter 13* (1978), 12-18.

[3]   E. de Doncker-Kapenga and D.K. Kahaner, Adaptive Integration over a Triangulated Region, *Unpublished*.

[4]   J.A. Kapenga and E. de Doncker, A Parallelization of Adaptive Task Partitioning Algorithms, *Submitted for publication*.

[5]   E. de Doncker and I. Robinson, An Algorithm for Automatic Integration over a Triangle using Nonlinear Extrapolation. *ACM Trans. Math. Softw. 10*, 1 (1984), 1-16.

[6]   E. de Doncker and I. Robinson, Algorithm 612 - TRIEX: Integration over a TRIangle using Nonlinear EXtrapolation. *ACM Trans. Math. Softw. 10*,1 (1984), 17-22.

[7]   E. de Doncker and I. Robinson, Integration over an N-Cube, *Unpublished*.

[8]   V. Faber, O.M. Lubeck and A.B. White, Jr., Superlinear Speedup of an Efficient Sequential Algorithm is not possible. *Parallel Computing 3* (1986), 259-260.

[9]   A.C. Genz, The Approximate Calculation of Multidimensional Integrals using Extrapolation Methods. *Ph.D. Thesis, Univ. of Kent at Canterbury* (1975).

[10]   C.A.R. Hoare, Monitors: An Operating System Structuring Concept. *CACM 17* (1974), 549-557.

[11]   D.K. Kahaner, Numerical Quadrature by the ε-Algorithm. *Math. Comp. 26* (1972), 689-693.

[12]   J.A. Kapenga, High Level Parallelization supported by Macro Implemented Monitors. *Unpublished*.

[13]   J.A. Kapenga, A Method for Distributive Priority Queue Management. *Unpublished*.

[14]   J.M. Lemme and J.R. Rice, Speedup in Parallel Algorithms for
       Adaptive Quadrature. *Report CSD-TR 192, Math. Sciences, Purdue
       Univ.* (1976).

[15]   E.L. Lusk and R.A. Overbeek, Implementation of Monitors with
       Macros: A Programming Aid for the HEP and other Parallel Pro-
       cessors. *Report MCS ANL-83-97, Argonne National Lab.* (1983).

[16]   E.L. Lusk and R.A. Overbeek, Use of Monitors in FORTRAN: A
       Tutorial on the Barrier, Self-Scheduling DO-loop, and Askfor
       Monitors. *Report MCS ANL-84-51, Argonne National Lab.* (1984).

[17]   J.N. Lyness and B.W. Ninham, Numerical Quadrature and Asymptotic
       Expansions. *Math. Comp. 21* (1967), 162-178.

[18]   M.A. Malcolm and R.B. Simpson, Local vs. Global Strategies for
       Adaptive Quadrature. *ACM Trans. Math. Softw. 1* (1975), 129-146.

[19]   R.A. Overbeek, *Private communication.*

[20]   D. Parkinson, Parallel Efficiency can be greater than Unity.
       Parallel Computing 3 (1986), 261-262.

[21]   R. Piessens, An Algorithm for Automatic Intergation. *Angewandte
       Informatik 9* (1973), 399-401.

[22]   R. Piessens, E. de Doncker-Kapenga, C.W. Überhuber and D.K.
       Kahaner, *QUADPACK - A Subroutine Package for Automatic Integra-
       tion.* Springer Series in Computational Mathematics 1 (1983),
       Springer-Verlag, N.Y.

[23]   J.R. Rice, Parallel Algorithms for Adaptive Quadrature -
       Convergence. *Proc. IFIP Congress 1974,* North-Holland, Amsterdam
       (1974), 600-604.

[24]   J.R. Rice, A Metalgorithm for Adaptive Quadrature. *JACM 22*
       (1975), 61-82.

[25]   J.R. Rice, Parallel Algorithms for Adaptive Quadrature II -
       Metalgorithm Correctness. *Acta Informatica 5* (1976), 273-285.

[26]   J.R. Rice, Parallel Algorithms for Adaptive Quadrature III -
       Program Correctness. *TOMS 2, 1* (1976), 1-30.

[27]   H.D. Shapiro, Increasing Robustness in Global Adaptive Quadra-
       ture through Interval Selection Heuristics. *TOMS 10, 2* (1984),
       117-139.

[28]   C.W. Überhuber and G. Zöchling, Quadratur singulärer Funktionen
       mittels Extrapolation. *Bericht Nr. 40/79, Institut fur Numerische
       Mathematik, TU Wien* (1972).

[29] P. Wynn, On a Device for Computing the $e_m(S_n)$ Transformation. *Mathematical Tables and Aids to Computing* 10 (1956), 91-96.

[30] P. Wynn, Acceleration Techniques in Numerical Analysis, with Particular Reference to Problems in One Independent Variable. *Proc. IFIP Congress 1962*, North-Holland, Amsterdam (1963), 149-156.

THE NUMERICAL EVALUATION OF MULTIPLE
INTEGRALS ON PARALLEL COMPUTERS

Alan Genz
Computer Science Department
Washington State University
Pullman, WA 99164-1210
U. S. A.

1. INTRODUCTION

Many problems in the applied sciences require the numerical calculation of integrals. Sometimes a problem requires the calculation of a single integral in one or more dimensions, but more often the calculation of a large number of integrals is required, with the integration region fixed for all of the integrals. Two examples of this second class of integration problem occur a) in quantum chemistry when the Self Consistent Field method [14] is used for the study of molecules and large numbers of similar two and three dimensional electron orbital integrals are required, and b) with the finite element method for the numerical solution of partial differential equations [1], where the Galerkin formulation of the method requires the computation of large numbers of one, two or three dimensional integrals before an approximate solution to the partial differential equation can be found. For many problems, the numerical integration occupies a significant proportion of the total processing time. All of the commonly used integration methods (two good general references are [2] and [16]) use a linear combination of the integrand function values to estimate the integrals, so this numerical integration time is usually dominated by the repeated calculation of the integrand values. However, some adaptive algorithms also spend a significant proportion of the computation time with further processing of the integrand values in order to obtain a good subdivision of the integration region.

There has not been much published research work in the area of parallel numerical calculation of integrals, and the work that has appeared so far has mostly considered the computation of single integrals. A series of papers by Rice [9-12] consider algorithms for single one dimensional integrals on MIMD parallel computers. A report by Simpson and Yazici [15] discussed the implementation of an extrapolation method for integration over a triangle on a CDC STAR-100. The present author has studied [5] the implementation of a globally adaptive algorithm for single multiple integrals on an SIMD computer.

The main purpose of this paper is to give an overview of the ways parallelism could be introduced in software for numerical multiple integration. We will consider the problem of the numerical evaluation of a single integral using a single integration rule or an automatic algorithm based on an integration rule. We also consider the problem of the numerical evaluation of a group of related integrals. Current software for multiple integration is designed to take as input a function or vector of functions to be integrated, a description of the integration region, an error tolerance and a limit on the total work to be allowed. When the software

P. Keast and G. Fairweather (eds.), Numerical Integration, 219–229.

is used, it returns an estimate for the required integral(s) and some information about whether the desired accuracy was achieved in the allowed time. We would like to know to what extent this simple interface can be preserved for software that runs efficiently on a variety of parallel computers.

We begin with a general description of some currently available parallel computers. This is followed by a general discussion of parallelization of the evaluation of a single integration rule and a generic adaptive algorithm. Finally we describe in detail the implementation of a globally adaptive algorithm on several parallel machines and report of some test results that were obtained.

2. PARALLEL COMPUTERS AND LANGUAGE SUPPORT

For the purpose of discussing the parallel algorithms in this paper we will consider four general types of parallel computers: SIMD, shared memory MIMD, distributed memory MIMD and vector computers. The SIMD/MIMD classification is due to Flynn [4]. A typical SIMD machine is the ICL DAP with 1024-4096 one bit processors arranged in a two dimensional array. For a given instruction a selected set of active processors all execute that instruction on their own data. The primary programming language is FORTRAN with array extensions similar to those in the proposed ANSI F8x standard [17]. A typical shared memory MIMD machine has 2-20 processors which can work independently using data from a common memory area. Examples are the Sequent Balance and Alliant FX computers. Large granularity parallelism is most easily introduced in standard FORTRAN codes using compiler directives to isolate loops where memory conflicts do not occur. The Alliant FORTRAN language also includes FORTRAN 8x extensions and the compiler does automatic vectorization of simple inner loops. Examples of distributed memory machines are the Intel iPSC and NCUBE machines. The machines have 16-1024 nodes connected in a hypercubic network with each node also directly connected to a host processor. Each processor has local memory. Standard FORTRAN with the addition of message passing subroutines is used on both machines. A typical program consists of a host program along with separate programs for the nodes. The programs run independently and communicate using the message passing subroutines. A vector computer uses pipelining to efficiently execute simple instructions on vectors of operands. Good examples are the CRAY-1S and CYBER-205 computers. Standard FORTRAN is used. The newer CRAY machines can have 2-4 vector processors and the programmer will eventually have control over the independent use of the processors. Further details about the hardware for the parallel machines briefly described here and for other parallel machines can be found in the report by Dongarra and Duff [3].

3. SINGLE RULE METHODS

In this section we discuss the parallel implementation of single rule methods. These methods use a single integration rule of the form

$$R(f) = \sum_{i=1}^{N} w_i f(x_i),$$

to provide an estimate for a single integral or a group of integrals. Associated with this calculation there is sometimes a similar calculation for an error estimate. For the single integral case we have N independent computations for the function values followed by the final sum for R(f). In many cases function values for different x values will each require the same amount of time to compute, but this does not always occur. For example, if f is defined by a series, some x values could require more terms of the series (and more time) for convergence.

For parallel algorithms we can decompose the computation of R(f) into three steps:

i) prepare vectors $W = (w_1, w_2, ..., w_N)$ and $X = (x_1, x_2, \cdots, x_N)$,

ii) compute the vector $F = (f(x_1), f(x_2), \cdots, f(x_N))$ and

iii) compute $R(f) = W \cdot F$.

For a serial algorithm the steps are normally carried out together in a single loop which includes a function call for the integrand, but different parallel machines may often require very different code structures. If step i) is separate from step ii), then N(n+1) extra memory locations are required.

We first discuss SIMD machines. We begin with step ii), where we would expect to achieve a high degree of parallelism because step ii) is an SIMD step consisting of the single instruction "compute f for different X values". As long as N ≈ kP, when k is an integer for a P processor machine, step ii) is efficient for an SIMD machine where each processor can compute k function values. If f requires different computation times for different x values, then the work of the processors that already have accurate f values can be discontinued while the remaining processors finish. This is easy to organize on a machine like the ICL DAP, where a vector valued function for the integrand f can be defined directly using the DAP FORTRAN and masking vectors can be used to control the activity of specific processors.

On commercial SIMD computers P is usually large, with typical values in the thousands. For many integration problems, however, the most appropriate integration rule might have an N value that is small relative to P or not close to an integer multiple. The programmer could then be faced with the choice of using this rule and computing the integral without good processor utilization or using a more efficiently implementable rule (e.g. Monte-Carlo) that is not as accurate. It will not always be clear which of these choices is consistent with the overall aim of fast accurate calculation. This is often more of a problem for the smaller values of the integral dimension n.

Step i) will often be much more difficult to parallelize. This is due to the complicated structure of the set {x_i} for many integration rules. If the rule is a nonproduct polynomial rule it can often be written as a sum of product rules. The points and weights used by product rules can be determined from some smaller sets using an indexing function, and step i) could be combined with step ii), but in some cases significantly increasing the computation time. Most number-theoretic rules and Monte-Carlo rules have a regular structure for {x_i} that will allow straightforward parallelization of step i). Storage of the X and W vectors at compile time in appropriate data structures is a possible means of reducing the execution time at step i) but this strategy would often significantly increase the length of the code. In any case, the amount of time needed for step i) will not depend on the integrand, so even when fairly extensive calculations are needed for step i) there will always be integrands that are difficult enough to make the computations for step ii) dominate the total time. These problems are the ones most likely to be considered for parallel computers.

Step iii) would usually be implemented as a cascade sum [7]. A procedure for this would usually be provided as part of a good parallel extended FORTRAN and would typically require O(log(N)) (with a small constant) time, if N is not significantly larger than P. For many integration rules steps i) and ii) could be efficiently parallelized and it might appear that step iii) could dominate the computation. The constant associated with the O(log(N)) term is usually so small that this will not happen.

On MIMD machines a similar discussion applies, except that P is much smaller for the shared memory machines. The smaller P means that efficiently matching N with an integer multiple of P will not usually be a problem. Step i) would normally include some partioning of X and W. For distributed memory machines step i) could also involve some distribution time, unless the points and weights required by a particular processor were prepared by the same processor. For a distributed memory machine the communication time for step iii), where all the results are combined, could be significant.

The efficient use of a vector machine will depend on whether or not the steps can be written as loops that vectorize. If the function evaluations in step ii) can be decomposed into simple combinations of the basic intrinsic FORTRAN functions, then it might be possible for step ii) to be programmed as a sequence of vectorizable loops and run efficiently on a CRAY-like machine. For more complicated integrands good vectorization could be difficult to achieve.

We next consider the problem of rule evaluation for a large group of integrals in the form

$$R(f_j)=\sum_{i=1}^{N} w_i f_j(\mathbf{x}_i), \quad j = 1, 2, ..., M.$$

Here, the most straightforward way to utilize parallel machines is to parallelize over the j index. As long as M is close to an integer multiple of P, good parallelism can be achieved with very little code modification on SIMD machines and shared memory MIMD machines. For vector machines significant speedups could be achieved if the evaluation of the functions f_j is vectorizable. On distributed memory MIMD machines this approach would have a host or master processor compute the rule sum and node processors computing the integrand values. In this case the communication times could be significant unless the integrands f_j were relatively time consuming to compute.

4. ADAPTIVE METHODS

In this section we are concerned with integration methods where the integrand evaluation points are not fixed at the start of a calculation, but are chosen dynamically, according to the behavior of the integrand. Serial algorithms for these methods can often be much more efficient than nonadaptive algorithms, because computation time is concentrated in subregions of the total integration region where the integral is more difficult. Adaptive algorithms proceed in stages, each stage using information from earlier stages to determine where to best spend integrand evaluation time. The main components of an adaptive algorithm are a basic integration rule (which also provides an error estimate), a subdivision strategy and a data structure for the subregions. A typical serial adaptive algorithm has a loop that contains four steps:

    i) determine a new subdivision of the integration region,
    ii) apply the basic rule to any new subregions,
    iii) combine new results from step ii) with previous results and
    iv) check for convergence.

The work involved for step ii) usually takes the most time because all of the integrand evaluations occur in step ii).

A serial adaptive algorithm is easily parallelized if the basic rule has an efficient parallel implementation as a single rule method. Then the modification to the software is straightforward and only involves the replacement of the serial basic rule by its parallel form. The effect of this simple change could be that steps i) and iii) now dominate the time taken by the algorithms and then the parallelization of these steps would also have to be considered. If step ii) involves a large number of subregions (P or more) then the basic rule applications could be carried out in parallel.

If the calculation of a large group of integrals is required then a parallel loop over the index for the integrands is most easily introduced inside step ii). The decisions for the optimal subdivision that the adaptive algorithm makes then have to be based on some overall measure of the local difficulty of the integrands in particular subregions. This approach to parallelization will be efficient if the integrands are similar. This small granularity parallelization should be most efficient on SIMD, shared memory MIMD and vector machines (when the calculation of the integrands vectorizes). The granularity can be increased by interchanging the integrand and rule evaluation loops in step ii). With this loop organization near maximum parallelism should be achievable on the widest range of architectures, when the integrands are similar. When the integrands are not similar then the loop over the integrand index could be moved completely outside the loop for the adaptive algorithm. This approach allows for the largest granularity parallelism and should be time efficient for SIMD and both types of MIMD machines as long as P is not significantly larger than M, but it is unlikely to work efficiently on a vector machine.

The general discussion so far has indicated that there should be some parallel machines that will allow the efficient implementation of any particular multiple integration algorithm, although a particular algorithm will not necessarily be efficiently implementable on all parallel machines. Ideally, we would like to take a good serial algorithm and see if the coded version of such an algorithm could be organized so that it would also run efficiently on a variety of parallel machines. In the next sections we consider an attempt to do this with an adaptive algorithm that has been used in the NAG [8] library for several years.

5. PARALLELIZATION OF A GLOBALLY ADAPTIVE ALGORITHM

The algorithm that we discuss is an algorithm originally described by van Dooren and de Ridder [13] and modified by Genz and Malik [6]. This is a globally adaptive algorithm that is efficient for a wide range of integrals with dimensions from 2-8. If we follow the outline for a generic adaptive algorithm given in the previous section we can describe the essential features of this algorithm. Step i) chooses from a list (initially with one element, the whole integration region) of subregions the one that has the largest estimated integration error. Two new subregions are obtained from this subregion by dividing the subregion in half along the axis where the integrand has the largest fourth difference. Step ii) applies a degree seven basic rule to the two new subregions, using differences with results from an imbedded fifth degree rule for error estimation in each subregion. Step iii) incorporates the new subregion dimensions, integral estimates and error estimates in the list, which is organized as an error keyed heap data structure. Step iv) updates the global error and integral estimates and checks for convergence.

We will first discuss in some detail parallelization of this algorithm for the case of a single integral. The degree seven basic rule used by the algorithm uses $2^n + 2n^2 + 2n + 1$ integrand values when $1 < n < 10$. In order to attempt to make the algorithm as portable as possible over a range of parallel machines we did not attempt introduce parallelism at the basic rule level. This would have involved extensive reorganization of the way in which the rule calculation was done, and, for efficient parallelization, the reorganization would be different for each machine and possibly each n. Instead of introducing parallelism at the basic rule level we introduced it at the next level up, the subregion level. This involved a change at step i) where the $P/2$ subregions with the largest errors were taken off the list of subregions. Each of the $P/2$ were divided into two parts and the P processors were then each given one subregion for application of the basic rule in step ii), where a loop of length P was introduced. The algorithm was initialized by subdividing the whole integration region into $P/2$ equal parts. This was done by halving along successive coordinate axes until there were enough subregions. A loop of length P also had to be introduced at step iii) where the results from step iii) were incorporated into the heap.

On a serial machine this form of the algorithm might run less efficiently on some problems where the errors associated with subregions in the subregion list had large differences in magnitude. In this case time could be wasted by the serial algorithm working on subregions (among the top $P/2$ in the current list) where the error was already small. When $P > 2$

processors are available, however, there is usually no reason to keep some idle. The work the idle ones could be doing will almost always provide some reduction in the total error, and unless the communication time is relatively high, this error reduction will contribute to the overall efficiency of the algorithm.

The implementation of this modified globally adaptive algorithm on the two shared memory MIMD machines that were available for tests (an Alliant FX/8, with P = 8 and a Sequent Balance 8000, with P = 12) was fairly straightforward. For the Alliant machine all that was necessary was to introduce a compiler directive immediately before the the step ii) loop and compile the program with the appropriate compiler option. For the Sequent machine a similar modification was made but the code did not run correctly. This was found to be due to the use of the integrand name as a parameter in the basic rule subroutine call. Elimination of this parameter allowed the code to run correctly, but this had the disadvantage of having to use a fixed name for the integrand. The Sequent compiler that was used was an alpha test version; future versions will hopefully have this compiler fault eliminated. The length P loop in step iii) cannot be as easily parallelized because there would be memory contention problems as different processors try to update the heap. We did not try to parallelize this section of the code but introduced timing code in order to see how important the non-parallel work in step iii) was relative to the parallel work in step ii).

In order to produce parallel code for the modified algorithm for the INTEL iPSC and NCUBE machines separate programs had to be written for the host and node processors. Node programs were written to do the basic rule work. These programs waited for a subregion from the host, evaluated the basic rule on that subregion and sent the results back to the host. The host program contained most of the original code for the modified algorithm. The primary differences were that the basic rule calls in step ii) were replaced by calls to message passing subroutines that sent messages to the appropriate node programs, and the loop in step iii) required calls to message receiving subroutines for results from the node programs.

The modified algorithm is unlikely to run any faster on a vector machine because with vectorization, parallelism is achieved only at a low granularity level. In order to provide an example of possible vectorization for large scale calculations with an adaptive algorithm we include some some test results at the end of the next section. These results were obtained using the original adaptive algorithm modified to take as an input parameter a subroutine for a vector of integrands, all of which were to be integrated over the same region. Each function call in the original algorithm was replaced by a call to the subroutine. Other appropriate minor changes were made and the sup norm was used to determine the error for each subregion.

6. TEST RESULTS

In order to test the efficiency of the modified adaptive algorithm we used a single test integrand

$$f(x_1, x_2, \cdots, x_n) = 1/(x_1 + x_2 + \cdots + x_n).$$

The integration region was the unit hypercube, with n = 2-10. One reason for choosing this integrand is the presence of a peak at the origin. With this feature we can compare the efficiency of adaptivity of any parallel algorithm with that of the serial algorithm. For tests on the Sequent and Alliant machines that had a fixed number of processors the parameter P in the subroutine was varied in order to see if the speedups that were achieved were consistent with the actual number of processors. Two parts of the program were timed: step ii) (the parallel part) and the rest (the nonparallel part). For each test the subroutine was terminated when the total number of basic rule subroutine calls reached 100. In Tables 1 and 2 below we give the times obtained for the tests on the Sequent and Alliant machines. The first row of times is for the subroutine run with P = 2, but compiled without the parallel code generation flag set. These are times obtained using only one processor. For each (P,n) pair the times are given for the parallel part of the program. At the end of each table average times are given for the nonparallel parts of the program.

Table 1: Sequent Times in Seconds									
P: n	2	3	4	5	6	7	8	9	10
Serial	1.5	2.3	3.4	5.3	8.5	13.5	23.9	40.8	76.1
2	1.4	2.0	3.1	4.4	5.8	8.7	13.4	22.5	40.0
4	1.3	1.6	1.9	2.3	3.1	4.6	7.1	11.5	20.3
6	.96	1.1	1.4	1.7	2.3	3.1	4.9	7.8	13.7
8	.94	1.2	1.5	2.0	2.8	4.1	6.4	10.1	19.4
10	.86	.98	1.2	1.6	2.3	3.2	5.0	8.2	14.4
12	.75	.95	1.2	1.6	2.1	3.1	4.5	7.8	13.3
Non ‖	.6	.6	.7	.7	.8	.8	.9	1.0	1.1

Table 2: Alliant Times in Seconds									
P: n	2	3	4	5	6	7	8	9	10
Serial1	.39	.53	.82	1.31	2.10	3.48	6.00	10.9	20.5
Serial2	.21	.28	.37	.54	.89	1.61	2.51	4.89	8.54
2	.25	.27	.34	.47	.70	1.08	1.81	3.11	5.77
4	.14	.16	.20	.26	.37	.57	.91	1.57	2.89
6	.10	.12	.14	.18	.25	.38	.60	1.04	1.89
8	.09	.10	.12	.16	.21	.31	.49	.84	1.51
10	.09	.11	.13	.17	.25	.38	.62	1.08	1.99
12	.08	.10	.12	.16	.23	.35	.56	.97	1.79
Non ‖	.07	.07	.08	.05	.07	.07	.06	.06	.07

If we consider the Sequent times first we can see that near optimal speedups were obtained for P in the range 2-6, but there is no significant speedup with P > 6. During the time the test runs were made the job status was monitored, and it was found that no more than 6 processors were ever allocated to the job. The lack of increase in the speedup for 6 < P < 13 if probably due to either a fault in the compiler or the way the system was configured. The Alliant times include 2 sets of serial times. The Serial1 times were obtained using no optimization in the compilation and the Serial2 times were obtained with the optimization flags set. The compilation with optimization parallelizes some of the inner loops and probably did this in portions of the basic rule. This compiler parallelization was more effective than the programmed parallelization for P = 2 and n = 2,3. For P < 9 Alliant times show good to super-optimal speedups, depending on how the optimization is considered. For P > 8 there is some degradation in the times, and this might be expected because there were only 8 processors.

In the previous section we discussed why the modified adaptive algorithm might not be as efficient when the total computation done was compared with the error for a specific calculation. We compared the errors for all of the calculations that were done and found that there were no significant differences between the serial and parallel errors for n > 2. For n = 2, however, there were significant differences, and we summarize these in the first half of Table 3. The n = 2 case is the case where the integrand is most singular and therefore adaptivity is more important. The results for n = 2 indicate that the speedups from parallelization do not compensate for the larger errors which come from the reduced adaptivity. When these results were presented at the workshop Philip Rabinowitz suggested that the adaptivity could be improved if subregions with large errors were subdivided into more than two pieces. In order to test this, a simple modification was made to the parallel algorithm. When the subregions were removed from the top of the heap, any subregion with an error estimate that was more than four times the error estimate for the next subregion on the heap was divided into four parts instead of two and was allocated four processors instead of two. At step i) subregions were removed from the heap in this way until at most P processors were allocated. This strategy was not expected to give an optimal processor allocation, but was expected to improve the adaptivity for the parallel algorithm. The second set of results in Table 3 support this hypothesis.

Table 3: Errors x 100000 for n = 2						
P	1	2	4	6	8	10
Estim.	.98	.98	.98	1.1	4.4	31
Actual	.01	.01	.02	.06	.45	3.3
Estim.	.98	.98	1.0	1.2	1.3	1.9
Actual	.01	.02	.02	.02	.02	.02

In Tables 4 and 5 following we give corresponding times for the Intel and NCUBE machines. The Intel machine that was used could be configured with from 1 to 32 processors and the NCUBE from 1 to 64. Both of these machines had timing software for the node processors but not the host. The times given are for the time spent by the basic rule. We also include average times for each P for message passing. These times include times receiving, sending and waiting for messages by the nodes. The Intel times show that for large P and small to moderate size n these times dominated the basic rule times and there was no speedup. These times also were dominant on the NCUBE for n < 5. However, the test integrand was very easy to compute, and these tests indicate that for problems where the test integrands are, for example, 100 times more difficult to compute, the results would show good overall speedups for wide ranges of n and P.

Table 4: Intel Times in Seconds										
P: n	2	3	4	5	6	7	8	9	10	Mess.
4	.61	1.0	1.9	3.2	5.4	9.4	16.9	31.2	60.0	13
8	.29	.48	.88	1.52	2.55	4.42	7.85	14.6	28.0	20
16	.12	.24	.40	.68	1.16	1.99	3.54	6.53	12.5	35
32	.05	.10	.18	.30	.51	.89	1.59	2.94	5.62	66

Table 5: NCUBE Times in Seconds										
P: n	2	3	4	5	6	7	8	9	10	Mess.
4	.31	.55	.96	1.62	2.73	4.70	8.32	15.3	29.2	1.6
8	.15	.27	.47	.80	1.35	2.31	4.08	7.51	14.3	1.3
16	.07	.12	.21	.36	.62	1.60	1.87	3.45	6.57	1.1
32	.03	.06	.10	.16	.28	.48	.85	1.57	2.99	1.2
64	.01	.02	.04	.07	.11	.19	.34	.63	1.20	.96

Because the hypercube times were mostly dominated by communication, a different modification of the adaptive algorithm was considered for the Intel machine. This version, which we call the "ring algorithm", assumed the node processors were connected in a ring.

Using Gray codes it is easy to map a ring of processers onto a hypercube in such a way that the two neighbors for a ring processor are also neighbors on the hypercube. The host program initially divided the integration region into P equal subregions and sent one to each processor; it then waited for the final results. The node processors used the original adaptive algorithm on their specific subregions until the required amount of work was completed. In order to balance the work between nodes, each node was required to stop frequently and send its current error to the next node along the ring. If at any stage a node found that its current error was more than twice the error from its predecessor on the ring it sent the top subregion on its heap to the predecessor. The times given for the ring algorithm are given in Table 6. Here we can see that this algorithm had substantially reduced communication times compared to the original modified algorithm and good speedups were achieved for larger n values. Unfortunately, the actual and estimated errors for the n = 2 case were much higher for this algorithm than than other hypercube algorithms. This was because the difficult subregions were not subdivided and distributed around the ring rapidly enough. This algorithm could be modified to send pieces of very difficult subregions to more than one processor, but this would increase communication. It would also require more complicated message passing protocols. It is clear from these results that the ring algorithm should achieve good processor utilization without excessive communication costs when the integrands are difficult to evaluate and any difficult featuress that the integrands posses are distributed throughout the integration region.

Table 6: Ring Algorithm Intel Times in Seconds										
P: n	2	3	4	5	6	7	8	9	10	Mess.
4	1.0	1.7	2.7	4.0	6.4	10.7	19.6	33.0	62.4	.79
8	.49	.72	1.08	1.72	2.89	4.92	8.64	15.9	30.3	1.2
16	1.36	1.00	1.81	2.66	1.89	2.68	4.27	7.52	13.9	4.1

None of the modifications described so far would provide any significant speedups on a vector computer. In order to demonstrate speedups possible with this type of architecture, we used a version of the globally adaptive algorithm which has a subroutine parameter instead of the usual integrand function parameter. The subroutine takes as input a point x and returns a vector of integrand values for this point. This vector version of the algorithm returns vectors of integral and error estimates, for M integrals over the same region. For step i) in this vector algorithm the subregion with the largest error is divided, but the subregion error is determined by taking the maximum of the estimated errors for all of the integrals over that subregion. This subroutine will be in Mark 12 of the NAG [8] library.

The integrands used for the vector tests were

$$f_j(\mathbf{x}) = log\,(\sum_{i=1}^{4} x_i)sin\,(j + \sum_{i=1}^{4} ix_i), \ \text{for } j = 1, ..., M.$$

The integration region was the unit four-cube. The serial results in Table 7 are total times obtained running the original adaptive algorithm M times and the vector results were obtained running the vector version of the adaptive algorithm. For the serial algorithm the routine was stopped for each j when 1000 integrand calls had been done, and 1000 integrand subroutine calls were allowed for the vector algorithm. These results show that when the integrand calculation can be vectorized significant speedups can be achieved.

Table 7: Cray Times in Seconds									
M	10	20	30	40	50	60	70	80	90	100
Serial	.53	1.05	1.59	2.13	2.65	3.17	3.70	4.23	4.76	5.30
Vector	.085	.098	.112	.126	.142	.156	.191	.204	.217	.230

7. CONCLUDING REMARKS

After a general discussion of the parallelization of different methods for the calculation of multiple integrals we considered the parallel implementation of a globally adaptive algorithm. Minor changes to the original algorithm were shown to provide good speedups on two shared memory MIMD machines without changing the user interface. For distributed memory MIMD machines good speedups are probably possible for difficult or numerous integrands but significant changes to the user interface are necessary on presently available hardware because of the necessity of having separate programs running on node and host processors. We did not consider an implementation of the globally adaptive algorithm for a single integral that would run efficiently on a vector machine. The results given for a version of the algorithm designed for estimating a vector of integrals show that a single integral version could run efficiently, for some integrals at least, on a vector machine. Such a single integral subroutine would require the user to supply a integrand subroutine parameter. The integrand subroutine would take as input a set of integration points and return a set of integrand values. This integrand subroutine would be called by the basic rule subroutine. This modification should also work well on shared memory MIMD machines and SIMD machines like the ICL DAP, but is unlikely to work efficiently on distributed memory MIMD machines except for very difficult integrands.

8. ACKNOWLEDGMENTS

The author wishes to thank the people at Argonne National Laboratory for advice and access to the machines in the Advanced Computing Research Facility, where the majority of the tests reported here were carried out during a summer faculty research appointment. Particular thanks go to James Lyness, Jack Dongarra, Danny Sorenson, Gene Rackow and Rick Stevens. The author also wishes to thank David Kahaner for helpful discussions about some of this work and help in accessing an NCUBE machine at the National Bureau of Standards. The timings were obtained using a CRAY machine at Boeing Computer Services, using an NSF supercomputer time allocation grant.

9. REFERENCES

[1] J. E. Akin, *Application and Implementation of the Finite Element Method*, Academic Press, New York, 1982.

[2] P. J. Davis and P. Rabinowitz, *Methods of Numerical Integration*, Academic Press, New York, 1984.

[3] J. Dongarra and I. Duff, Advanced Computer Architectures, Argonne National Laboratory MCS Technical Memorandum No. 57 (1985).

[4] M. Flynn, Some Computer Organizations and Their Effectiveness, IEEE Trans. Comput. C-21(1972), pp. 948-60.

[5] A. C. Genz, Parallel Methods for the Numerical Calculation of Multiple Integrals, Comp. Phys. Comm., 26(1982), pp. 349-352.

[6] A. C. Genz and A. A. Malik, An Adaptive Algorithm for Numerical Integration over an N-Dimensional Rectangular Region, J. Comp. Appl. Math., 6(1980), pp. 295-302.

[7]   R. Hockney and C. Jesshope, *Parallel Computers*, Adam Hilger, Bristol, U.K., 1981, pp. 178-192.

[8]   Numerical Algorithms Group Limited, Mayfield House, 256 Banbury Road, Oxford OX2 7DE, United Kingdom.

[9]   J. R. Rice, Parallel Algorithms for Adaptive Quadrature, Convergence, *Proc. IFIP Congress '74*, North Holland, New York, 1974, pp. 600-604.

[10] J. R. Rice, Parallel Algorithms for Adaptive Quadrature II, Metalgorithm Correctness, Acta Informatica 5 (1975), pp. 273-285.

[11] J. R. Rice, Parallel Algorithms for Adaptive Quadrature III, Program Correctness, ACM TOMS 2 (1976), pp. 1-30.

[12] J. R. Rice and J. M. Lemme, Speedup in Parallel Algorithms for Adaptive Quadrature, Purdue University Computer Science Technical Report CSD-TR 192 (1976).

[13] P. van Dooren and L. de Ridder An Adaptive Algorithm for Numerical Integration over an N-Dimensional Rectangular Region, J. Comp. Appl. Math. 2(1976), pp. 207-217.

[14] H. E. Schaefer III (Ed.), *Methods of Electronic Structure Theory*, Plenum Press, New York, 1977.

[15] R. B. Simpson and A. Yazici, An Organization of the Extrapolation of Multidimensional Quadrature for Vector Processing, University of Waterloo Computer Science Research Report CS-78-37 (1978).

[16] A. H. Stroud, *Approximate Calculation of Multiple Integrals*, Prentice-Hall, New Jersey, 1971.

[17] J. L. Wagener, Status of Work Toward Revision of Programming Language FORTRAN, ACM SIGNUM Newsletter, 19 (1984), Number 3.

VECTORISATION OF ONE DIMENSIONAL QUADRATURE CODES

I. Gladwell
NAG Central Office   Department of Mathematics
Mayfield House   University of Manchester
256 Banbury Road   Manchester M13 9PL
Oxford OX2 7DE   U.K
UK.

ABSTRACT. We investigate certain vectorisation aspects of one-dimensional quadrature codes. For purposes of illustration we use the NAG code D01AKF [2] (corresponding to the QUADPACK code QAG-6, [1]). This is an adaptive integration code using 61-point Gauss Kronrod quadrature with a 31-point rule for error estimation. We aim to use little or no more arithmetic than in D01AKF and to obtain an improvement on scalar machines as well as on vector processors. We close with an outline algorithm suitable for multiprocessors which has a similar philosophy.

1. FIRST OBSERVATIONS

The only vectorisable loops in D01AKF which are significant in cost are the ones at the innermost level. In the case of D01AKF these correspond to the point at which the integration rule and its error estimate are evaluated. Here a number of inner products roughly of the form

$$\sum_{i=1}^{N} w_i \, f(x_i) \quad \text{and} \quad \sum_{i=1}^{N/2} \hat{w}_i \, f(x_i)$$

are evaluated; N=61 in the case of D01AKF.

To achieve the minimal level of vectorisation we must move the call of the user's function outside the loop in which we compute the inner product. With that change we have a piece of code which could be replaced by calls to SDOT (a Basic Linear Algebra Subroutine (BLAS)). The cost is working space to keep the function values.

In general, on scalar machines, there is a potential gain to be made by inserting a call to SDOT as long as there is an optimised (manufacturer's or library) version available.

P. Keast and G. Fairweather (eds.), Numerical Integration, 231–238.
© *1987 by D. Reidel Publishing Company.*

On vector machines the code should be left in line because

i. if a call to a subroutine is made then, after vectorisation, the overhead of the linkage is likely to dominate,

ii. insertion of the call can prevent the compiler or preprocessor recognising level 2 vector constructs that are extended BLAS (for example a matrix times vector operation).

The SDOT operation is relatively slow on most vector machines. We would prefer to use a linked triad type of operation if that were possible.

2. THE USER'S FUNCTION AND VECTORISATION

Of course, the change in the code to take the call to the user's function outside the innermost loop is only half the story. The user is still expected to evaluate the function at each integration point. If we ask him to evaluate the integrand at all integration points for the current rule in a subroutine, the code to call a user's function F

```
        DO 10 I = 1,N
     10 FVAL(I) = F(X(I))
```

becomes a code to call the user's subroutine F

```
        CALL F(N,X,FVAL)
```

where N is the largest of the values associated with the rule.
The user may then vectorise internally to the subroutine F.
This way we save on linkage (1 subroutine call instead of N function calls) for all machines. For example, in Table 1 we show the timings on 100 test problems of the form

$$\int_0^1 e^{-x} \cos wx \, dx$$

for varying w.

	ICL 2988	Cray 1
FUNCTION	13.11	.316
SUBROUTINE	12.43	.258
Saving	5%	18%

Table 1 Scalar Timings

The only cost incurred is the requirement for additional workspace. About cN locations are required (c=2 for D01AKF).

With vectorisation enabled on the Cray 1 the timings
corresponding to the SUBROUTINE results in Table 1 are given in
Table 2. The gain in the second column of Table 2 is sufficient
alone to justify the modification. To judge whether it would be
adequate for the purposes of vectorisation just to move the FUNCTION
evaluation outside the quadrature rule evaluation loop but leave the
interface essentially unchanged we have also run the tests on the
Cray 1 with only the integrator vectorised (but leaving the
integrand evaluation in subroutine form). The results are given in
column 3 of Table 2. Clearly it is crucially important to permit
the user to simultaneously evaluate the integrand even in our simple
case.

Scalar	Full Vectorisation	Only Integrator Vectorised
.258	.060	.244

Table 2   Comparison of Timings on Cray 1.

3. ADDITIONAL POSSIBILITIES FOR VECTORISATION

D01AKF and all the other similar QUADPACK derived codes have
essentially the same strategy. Starting initially with the whole
interval, at each stage they take the current subinterval with
largest error estimate, bisect it and calculate the integral and
error estimate for each half interval. Next the current overall
error estimate is adjusted to take account of the estimates on the
new subintervals. As long as the overall error estimate is still
too large, the interval with largest error estimate is found and the
process is restarted.
Clearly at any stage normally there will be a number of
subintervals which we know we must bisect at some later stage. If
$e_1 < e_2 < \ldots < e_n$ are the error estimates and

$$\sum_{i=1}^{r-1} e_i < TOL, \quad \sum_{i=1}^{r} e_i \geq TOL$$

then if we adopt the basic algorithmic strategy of D01AKF, at some
stage we must bisect intervals $r, r+1, \ldots, n$. D01AKF works on the
interval corresponding to e_n. If we concatenate the lists of
integration points for all the intervals corresponding to e_r,
$e_{r+1}, \ldots, e_n$, then clearly we will have $(n-r+1)*N$ integrand
evaluations to be computed in one subroutine call and so we can
produce another saving on linkage and a reduction in start up times
for the user's vector computation. Then we may rank all the newly
computed intervals in increasing order of size of error estimates in
one computation by a call to NAG routine M01DAF instead of to
D01AJX(QSORT) which takes error estimates for just two new
subintervals.

We need an additional workspace of the order of cN(n−r+1) where
c is as defined earlier. This process requires an amount of
workspace which is dependent on values n and r, determined during
the integration. If this proves a problem we need not perform all
n−r+1 integrations simultaneously.

Can we improve the linear algebra? We have to compute the
rules

$$s_J = \sum_{I=1}^{N} w_I\ f(x_{J,I}) \quad J = r,r+1,\ldots,n$$

where $f(x_{J,I})$ are stored in locations $F((J-1)*N+I)$, $I = 1,2,\ldots,N$,
$J = 1,2,\ldots,n-r+1$, after the evaluation of the integrand at the
concatenated set of integration points. Hence we could use the code

```
    DO 20 J = 1,n−r+1
    S(J) = 0.0
    DO 10 I = 1,N
    S(J) = S(J) + W(I)*FVAL((J−1)*N+I)
 10 CONTINUE
 20 CONTINUE
```

to evaluate the rules, performing (n−r+1) inner products; that is
calculating $\underline{s} = F\ \underline{w}$ where $F = (F_{J,I})$ is rectangular. This is an
extended BLAS operation and can be coded in the alternative form

```
    DO 10 J = 1,n−r+1
    S(J) = 0.0
 10 CONTINUE
    DO 20 I = 1,N
    DO 15 J = 1,n−r+1
    S(J) = S(J) + W(I)*FVAL((J−1)*N+I)
 15 CONTINUE
 20 CONTINUE
```

involving linked triads. Similar codes can be used for computing
the error estimates. What are the problems with this approach?
First normally n−r+1 << N and so we get a short vector operation at
the innermost level. Second there is a stride of N (slower on the
Cyber 205 and probably on other memory to memory machines). Third,
and most important, we have assumed the set of intervals have been
ordered so that those intervals labelled r,r+1,...,n are stored
consecutively in the list. D01AJX (and M01DAF) only rank the error
estimate and so if we use only these ranking routines we would have
to replace S(J) in the above codes by S(IRANK(J)) with all the
vectorisation problems associated with indirect addressing (e.g.
there will be a requirement for implicit or explicit "gathers" on
most machines and we will be unable to use extended BLAS code from
elsewhere because it assumes direct addressing).

An alternative, necessary only for this purpose, is to reorder the four lists (left and right subinterval end points, computed integrals and error estimates) in line with the ranking. We have tested this last possibility by inserting calls to the appropriate NAG routines (M01DAF and M01EAF) in place of the call to D01AJX. In the same calculation as in Table 2 the sorting (which does not vectorise) costs 0.020 seconds on the Cray 1, that is one third of the total cost of the vectorised codes, and there is relatively little sorting for these simple examples. Given that the total cost of the linear algebra (on the Cray 1) is only 0.014 seconds no amount of code optimisation can make this approach pay off.

   If we accept that we cannot make a sufficient gain by reordering the integration intervals as ranked by their error estimates and then using an appropriate extended BLAS, we are left with the question of whether simply evaluating as many integrand values as possible in one call will pay off. (There will always be at least 2N integrand values which could be evaluated because of the bisection algorithm). Here some care must be exercised. Our example may not be typical and we should test independently. Accordingly we evaluated the integrand a total $61*2^{JMAX}$ times but in $JMAX+1=9$ different ways. In the Jth experiment, we evaluated $61*2^{JMAX-J}$ integrand values, in 2^J subroutine calls, J=0,1,...,9. These experiments were intended to embrace all the possibilities ranging from evaluating just the integrand values for a single rule in the subroutine to evaluating the integrand values for 2^{JMAX} rules in one subroutine call. (This number of integrand values in one call is never likely to be realised in one-dimensional quadrature.) As J decreases, that is the number of integrand values evaluated in one subroutine call increases, the cost decreases in vectorised mode on the Cray 1. However this decrease is only about 10% between J = JMAX and J = 0. Even in the non-vectorised case on the Cray 1 we observe a 3% variation. In the vectorised example considered in Table 2, the cost of integrand evaluations is about 25% of the total. The saving of up to 10% of this cost does not seem worthwhile for the disruption to the code that would be involved in reorganising it from its current form of one rule evaluation per subroutine call.

4. AN ALTERNATIVE DESIGN FOR VECTORISATION

   An alternative to the approaches outlined above is to attempt to obtain the speed-up available on a vector processor by treating several integrands simultaneously. This is a feature that can be useful in some applications where the integrands are sufficiently alike in behaviour that much the same choice of rules made by the adaptive integrator for one integrand will be suitable for them all. In such a code the choice of subintervals is made on a worst case basis (see for example ADAPTV(D01EAF) by A. Genz) and so each integral uses at least as many function evaluations as it would if it were computed alone. It is hoped that savings will be made in two ways:

i.   by permitting calculation of the whole set of integrands in a
     fast manner (that is on a vector processor);

ii. by taking advantage of common terms in the integrands a
     considerable saving can be made in some cases, even on scalar
     machines.

     There would need to be several integrals to be evaluated before
(i) alone makes a significant impact, though the saving in linkage
and other integrator overhead is not inconsiderable.

5. MULTI-VECTOR PROCESSOR IMPLEMENTATION

     The call to the subroutine to evaluate the integrand values for
a single rule plus the calculation of the rule and its error
estimate is one possible level of granularity for a multiprocessor
implementation. In this case we should take advantage of our
earlier observation that we can identify (n-r+1) integrals to be
computed simultaneously. The key idea of our approach is to
postpone re-ranking of the error estimates (an expensive operation)
until all the refinements known to be necessary have been completed.
Each stage of the algorithm, except the last, finishes with a
ranking of tentatively accepted error estimates during which time
further computing of unprocessed integrals normally will proceed.
We assume that we reserve one processor for control purposes and
that we keep the code to evaluate the integral and error estimate
and the code to rank the list of error estimates etc. on each of
the processors.
     At the start of a stage, we have a newly ranked list of error
estimates of length n, we have calculated r and we have s integrals
either under calculation or computed whilst the list is being
ranked. Initially r=0, n=0 and s is the initial subdivision of the
interval. We send out as many integrals for computation as there
are processors available and each time one returns we send out
either another integral or a ranking depending on the current state
of affairs. When an integral and its error estimate are returned,
the error estimate, ee, is checked as follows.

i.   When $r \neq 0$, ee is checked against e_r. If it is larger than e_r
     then it is too large to be included in the list so the
     corresponding interval is bisected and the integrals over the
     two resulting subintervals are put at the bottom of the stack
     for evaluation.

ii. If the ee can be added to

     $$\sum \text{(tentatively accepted error estimates)} \quad + \quad \sum_{}^{r-1} e_i$$

$$i=1$$

and TOL still not be exceeded then ee is tentatively accepted in the set to be ranked but no further action need be taken. However, if ee > e_{r-1} then we can redefine e_r = ee and set r = r+1.

iii. When r≠0, if adding ee to

$$\sum \text{(tentatively accepted error estimates)} \quad + \sum_{i=1}^{r-1} e_i$$

would exceed TOL then the integrals over the bisected subintervals corresponding to whichever of ee and e_{r-1} is larger are sent to the bottom of the stack to be evaluated (possibly together with all the unprocessed intervals with error estimates greater than or equal to this one). The other estimate is tentatively accepted. The remainder (that is $e_1, e_2, \ldots, e_{r-2},$ the tentatively accepted estimates and those unaccepted estimates whose intervals were not bisected and sent to the stack for evaluation) are ranked. When this ranking is complete a new stage starts.

Notes

1. This algorithm is designed to avoid ranking as far as possible but at the same time does not "waste" processing power by evaluating rules which are not needed, except possibly initially when the processors may otherwise be idle.

2. The process will terminate in the control mode when all the intervals have been accepted. Then the controller does the book-keeping, that is it evaluates the integral.

3. The process to be performed at most nodes is vectorisable hence lending itself to that class of multiprocessors with vector processing power at the nodes (Cray X-MP, Cray 2, some Intel Hypercubes, Alliant etc.).

4. So as to ensure that communication time relative to processor time is not a bottleneck we need to make sure that large enough computations are performed on each processor and large enough packets of information are communicated between processors. Hence, we may send several integrals for evaluation to a processor simultaneously. We would then expect the results of these computations to be communicated simultaneously

ACKNOWLEDGEMENTS

The author wishes to acknowledge the helpful comments and criticism of this manuscript by W.H.Enright and N.J.Higham.

REFERENCES

[1] R.Piessens, E.de Doncker Kapenga, C.W.Uberhuber and D.K.Kahaner,
    (1980) QUADPACK - A Subroutine Package for Automatic Integration,
    Springer-Verlag, Berlin.

[2] The NAG Fortran Library Manual, Mark 12, (1986) Numerical
    Algorithms Group Ltd., 256 Banbury Road, Oxford.

PERFORMANCE OF SELF-VALIDATING ADAPTIVE QUADRATURE [1]

George F. Corliss
Department of Mathematics, Statistics, and Computer Science
Marquette University
Milwaukee, WI 53233 USA

ABSTRACT. SVALAQ is a suite of programs for self-validated quadrature. For the usual quadrature problem on a finite interval,

$$If = \int_A^B f(x)\, dx,$$

we compute an interval $[c, d]$ in which If is guaranteed to lie. The inclusion $If \in [c, d]$ is automatically validated by the computer program provided only that f can be evaluated at every point of $[A, B]$. This paper addresses two issues related to the performance of self-validated quadrature:

- How much does validation cost? and

- How accurate can it be?

The answers are that the cost in CPU time for validation typically varies from a factor of 3 - 5 for very stringent accuracy requests to 3 - 15 for modest accuracy requests compared to the CPU time required by the general purpose routine QAGS from QUADPACK. Accuracies of a few units in the last place (ULP) can be achieved, while for modest accuracy requests, validation assures that the request has been met without costly excess accuracy.

SVALAQ runs in any IBM System 370 environment using ACRITH. The algorithms could be implemented in any environment which supports interval calculations and an accurate scalar product, such as Pascal-SC.

1. INTRODUCTION.

Consider the usual quadrature problem

---

[1] Supported in part by IBM Deutschland GmbH.

P. Keast and G. Fairweather (eds.), Numerical Integration, 239–259.
© 1987 by D. Reidel Publishing Company.

$$If = \int_A^B f(x)\, dx \qquad (1)$$

consisting of

- Integrand - A Fortran-like expression for the integrand f as a finite sequence of $+$, $-$, $*$, $/$, $\sqrt{\ }$, $\exp$, $\ldots$.

- Limits - Finite limits of integration A and B (perhaps intervals).

- Tolerances - α and ρ, tolerances for the desired absolute and relative errors.

- Evaluations - A limit M on the number of function evaluations allowed.

We wish to compute an interval $J = [c, d]$ satisfying the goals:

- Inclusion - $If \in [c, d]$.

- Accuracy - $w(J) = d - c \leq \max\{\alpha,\ \rho \cdot |If|\}$.

- Bounded cost - At most M function evaluations are used.

The algorithms for self-validating, adaptive quadrature (SVALAQ) described in [2] and summarized here nearly meet these goals by using information about the continuous set of values of the integrand. In contrast, standard methods use only a finite set of values of f and can be fooled badly.

SVALAQ is a general-purpose quadrature suite; it does not yet include special handling of singularities or of weight functions. In order to estimate the cost of validation, we compare its performance to the general-purpose routine QAGS from QUADPACK [10].

2. VALIDATION.

QUADPACK [10] is a well-known state-of-the-art package for numerical quadrature. It computes an estimate for If and an estimate for the absolute error.

SVALAQ [2] computes an interval $J = [c, d]$ in which the solution lies. The midpoint $(c + d)/2$ has a guaranteed absolute error of at most $(d - c)/2$, and the geometric mean $\sqrt{c \cdot d}$ has a guaranteed relative error of at most $1 - \sqrt{c/d}$ (for $0 < c$).

Validation is achieved by arranging the computations in such a manner that the computer verifies that the hypotheses of an appropriate mathematical theorem hold. Then the conclusions of the theorem must also hold. In the context of quadrature, this is done by using interval mathematics to bound the rule and the truncation error of a standard quadrature rule as described by Moore [7]. INTE [4,5], was an earlier implementation of self-validated quadrature, although that term was not used there.

Gaussian Quadrature Theorem. If $f \in C^{(2n)}[-1, 1]$ and let x_i and w_i be the Gaussian nodes and weights, respectively, then

$$\int_{-1}^{1} f(x)\, dx = \sum_{i=1}^{n} w_i f(x_i) + c_n f^{(2n)}(\xi), \tag{2}$$

where c_n is the error coefficient, and $\xi \in [-1, 1]$.

The validated inclusion of Problem (1) is achieved by verifying the hypotheses of the Gaussian Quadrature Theorem, and by performing all computations in interval arithmetic to capture the rule and the truncation error. It is important to emphasize that equality holds in Equation (2). That observation is usually of no practical value because the derivatives of f are allegedly hard to compute, and because the location of ξ is unknown. However, as Moore [8,9], Rall [11], Kagiwada, et. al [6], and others have shown, it is quite practical to compute $f^{(2n)}$ by automatic differentiation. Automatic differentiation is neither symbolic nor numeric; it can compute high derivatives efficiently and accurately by using recurrence relations which embody the rules of calculus.

The rule, $\sum w_i f(x_i)$, is captured by performing all arithmetic operations in the evaluation of f and in the accumulation of the scalar product using interval arithmetic. The truncation error is captured by observing that $f^{(2n)}(\xi) \in f^{(2n)}[-1, 1]$, which is evaluated using the same recurrence relations (using interval arithmetic) which are used to evaluate derivatives at a point.

The calculations outlined in the preceding paragraph make heavy use of ACRITH, an IBM product [1] which runs in any IBM System 370 environment. For example, much of the development work for SVALAQ was done on a AT/370 Personal Computer. We use ACRITH's support for interval calculations throughout the program. The accurate scalar product is used to minimize the accumulation of round-off in the recurrence relations for the derivatives and in summing the rule. The same algorithms could be implemented in Pascal-SC or in any other environment which supports interval arithmetic.

In practice, we use a slightly weaker version of the Gaussian Quadrature Theorem which requires only that $f^{(2n)}$ be bounded. The computation of $f^{(2n)}[-1, 1]$ is arranged in such a way that the boundedness of $f^{(2n)}$ on $[-1, 1]$ is validated. For example, suppose that $f(x) = x^{10/3}$ so that $f \in C^3[-1, 1]$, but $f^{(4)}$ is not bounded in $[-1, 1]$. Then an attempt to compute $f^{(p)}[-1, 1]$ for $p \geq 4$ results in a notification that f''' is the highest derivative which can be computed. This accomplishes two things:

- validate the boundedness of derivatives which are computed, and

• signal the quadrature routine that is must use a low order method.

SVALAQ also provides validated quadrature by Newton-Cotes formulas or by the integration of Taylor polynomials based on theorems corresponding to the Gaussian Quadrature Theorem. The performance of Newton-Cotes and Taylor polynomial quadrature is nearly always inferior to Gaussian quadrature in this setting, so these methods are not considered further in this paper.

The preceding discussion shows how the guaranteed inclusion of If is achieved. However, for nearly all problems, the accuracy (the width of the interval J) exceeds the requested tolerance, and SVALAQ uses a global subinterval adaptation strategy based on guaranteed bounds for the integral on each subinterval.

Unless f depends on interval-valued parameters, the width of the integral on each subinterval is usually only slightly wider than the width of the truncation error. Hence the subinterval adaptation proceeds based only on the width of the truncation error (with no integrals being computed) until the user requested tolerance is probably satisfied. At that point, integration is performed on all pending subintervals. If necessary, subinterval adaptation continues with the integrals on subsequent subintervals being computed. This "wait to integrate" strategy is made possible by the use of guaranteed bounds for the truncation errors, and it typically saves about 35% in execution time.

3. TEST ENVIRONMENT.

The purpose of the tests reported here is to give a preliminary indication of the cost of validation relative to a standard (not validated) method. The tests were conducted on an IBM 4341 under VM at the University of Wisconsin - Madison. Computations were done in double precision (14 hexadecimal digits) in Fortran using the IBM product ACRITH, release 3. Self-validating, adaptive Gaussian quadrature is compared with the general purpose integrator DQAGS from QUADPACK.

The test problems used are problems 2, 3, 7, 8, 10, and 12 from the QUADPACK testing reported in [10]. Problems 1, 9, and 11 were omitted because they involve singularities which SVALAQ cannot handle. For several of the test problems, other QUADPACK routines perform better than QAGS, but we choose to use QAGS for comparison in each case because it, like SVALAQ, is a general purpose routine.

The user interface for SVALAQ is designed to allow tuser to enter the integrand *at run time*. This limits its usefulness to integrands which can be expressed as one line Fortran expressions. The expression for the integrand is parsed into a code list. Whenever a function value or a derivative value is needed, the code list is interpreted and the appropriate subroutine is called to perform the arithmetic operation on the (interval-valued Taylor series) operands. This mechanism is also used to evaluate f for QAGS. If the interpretation is replaced by a subroutine written specifically for one integrand, the performance of both SVALAQ and QAGS is improved by about 10%.

4. NUMERICAL RESULTS.

Each test problem was run requesting both absolute and relative accuracy tolerances of 10^{-4} and 10^{-9}. SVALAQ was limited to 2000 and 4000 function evaluations for the tolerances of 10^{-4} and 10^{-9}, respectively.

The "function evaluations" used by SVALAQ requires explanation since that routine requires the evaluation of high derivatives. There is no completely satisfactory way to count the evaluations of derivatives. The cost of evaluating $f^{(p)}$ using automatic differentiation is $\alpha p^2 + \beta p + \gamma$, where α, β, and γ depend on f. For some integrands, evaluations of derivatives are inexpensive relative to evaluation of the function itself, while for other integrands, derivatives are very costly. SVALAQ uses

$$\text{Cost of evaluating } f^{(p)} = \frac{\text{CPU time required to evaluate } f^{(p)}}{\text{CPU time required to evaluate } f},$$

so that its "number of function evaluations" is roughly proportional to CPU time.

The results are summarized by pairs of graphs showing the performance of the two methods at each tolerance as a function of the parameter α.

In each graph, an integer near a point is the abnormal error code (if any) returned by a routine:

QUADPACK QAGS ERROR CODES

IER.GT.0 Abnormal termination of the routine. The estimates for integral and error are less reliable. It is assumed that the requested accuracy has not been achieved.

$= 5$ The integral is probably divergent, or slowly convergent. It must be noted that divergence can occur with any other value of IER.

SVALAQ ERROR CODES

0 Normal end.

1 Tolerance relaxed - relative noise in function evaluation.

2 Tolerance relaxed - excessive number of function evaluations.

5 Unable to evaluate the integrand.

QUADPACK problem 2:

$$\int_0^1 \frac{4^{-\alpha}\, dx}{(x - \pi/4)^2 + 16^{-\alpha}}$$

QUADPACK problem 3:

$$\int_0^1 \cos(2^\alpha \sin(x))\, dx$$

QUADPACK problem 7:

$$\int_0^1 |x - 1/3|^\alpha\, dx$$

QUADPACK problem 8:

$$\int_0^1 |x - \pi/4|^\alpha\, dx$$

QUADPACK problem 10:

$$\int_0^1 \sin^\alpha(x)\, dx$$

QUADPACK problem 12:

$$\int_0^1 \exp(20(x - 1)) \cdot \sin(2^\alpha x)\, dx$$

In order to provide validation, SVALAQ must be able to evaluate f on the entire interval $[A, B]$. That is why Figures 7 and 7 show that SVALAQ was unable to evaluate the integrands for $\alpha < 0$. When SVALAQ was able to evaluate the integrand, it was 100% reliable in the sense that it correctly warned of a relaxed tolerance in every case when the tolerance was not met, and that it did not warn of a relaxed tolerance in any case where the tolerance actually *was* met.

These graphs show that the cost of validation is quite variable, ranging from a factor of 0.28 to 33.5 for a tolerance of 10^{-4} and from 0.19 to 81.1 for a tolerance of 10^{-9}. Half of the values are in the range from 3.67 to 8.76 for 10^{-4} and from 6.50 to 15.5 for 10^{-9}. This suggests that the cost of validation is comparable to the cost of a few extra runs of QAGS with slightly perturbed data.

To address the issue of accuracy, we ran 8 test problems from [3] requesting absolute and relative tolerances of 0.0. QAGS was not designed for such tolerances, and it usually had trouble telling when to quit. SVALAQ could detect when no further reduction in the width of the bounds for If, so it halted more promptly. These tests were run in single precision with a limit of only 500 function evaluations.

Accuracy test 1:

$$\int_0^1 \frac{dx}{1+x^2}$$

Accuracy test 2:

$$\int_0^1 x^{3/2}\,dx$$

Accuracy test 3: (Does the power operator behave the same as $\sqrt{}$?)

$$\int_0^1 x \cdot \sqrt{x}\,dx$$

Accuracy test 4:

$$\int_0^1 \frac{4\,dx}{1+256(x-3/8)^2}$$

Accuracy test 5:

$$\int_0^1 \frac{dx}{1+0.5\sin(10\pi x)}$$

Accuracy test 6:

$$\int_0^1 f(x)\,dx \qquad f(x) = \begin{cases} e^x & 0 \le x \le 1/2 \\ e^{1-x} & 1/2 \le x \le 1 \end{cases}$$

Accuracy test 7:

$$\int_0^1 \frac{dx}{\sqrt{x(1-x)}}$$

Accuracy test 8:

$$\int_0^1 \frac{dx}{1+e^x}$$

We remark that the characteristic function, CHAR, is provided so that simple piece-wise functions like that in Accuracy test 6 above can be entered as one line Fortran expressions:

```
CHAR (0.0, 0.5) * EXP (X) + CHAR (0.5, 1.0) * EXP (1 - X).
```

| | SVALAQ | | QAGS | | |
Test	Function Evaluations	Units in Last Place	Function Evaluations	Ratio of CPU Times	Comments
1	204	16	609	0.7	outside
2	377	14	819	1.9	outside
3	383	14	861	1.9	outside
4	> 500	194	777	1.8	error 2
5	> 500	> 2000	1113	1.1	error 2
6	280	5	1155	2.0	outside
7					error 5
8	247	10	525	0.9	outside

Table 1. Performance of SVALAQ and QAGS on Dixon's test problems.

For each problem, SVALAQ computes an interval $[c, d]$ in which If is validated to lie. "Units in Last Place" (ULP) is a measure of $d - c$. For example in Problem 1,

$$[c, d] = [0.78539\underline{764}, 0.78539\underline{861}]_{10} = [0.C90F\underline{D2}, 0.C90F\underline{E2}]_{16},$$

which differ by 16 ULP's.

The ratios of CPU times are much smaller than those shown in the accompanying graphs primarily because at such stringent tolerances, QAGS had difficulty determining when to quit. For a wider range of test problems, including those used for modest tolerance testing reported above, factors of 3 - 5 are more typical.

For Problems 1, 2, 3, 6, and 8, SVALAQ returned an error code = 1, indicating that the noise in function evaluation prevented the program from satisfying the requested accuracy. In each of these cases, the estimate for If computed by QAGS was *outside* the interval in which SVALAQ guaranteed If to lie. For example in Problem 1,

$$0.78539\underline{658} \notin [0.78539\underline{764}, 0.78539\underline{861}].$$

In each case, the distance of the estimate from QAGS from the interval from SVALAQ was comparable to the width of the interval. In 460 test problems we have run, we have observed no cases in which QAGS failed badly.

ACKNOWLEDGMENT.

The author wishes to thank Professor L. B. Rall for many helpful discussions.

REFERENCES.

[1] ACRITH High Accuracy Subroutine Library: General Information Manual. IBM publications, GC33-6163-02, 1986.

[2] G. F. Corliss and L. B. Rall. Adaptive, Self-Validating Numerical Quadrature. To appear in *SIAM Journal on Scientific and Statistical Computation*. Also available as *MRC Technical Summary Report No. 2815*, University of Wisconsin-Madison, 1985.

[3] Valerie A. Dixon. Numerical Quadrature. A Survey of the Available Algorithms. *Software for Numerical Mathematics*, ed. by D. J. Evans. Academic Press, London, 1974.

[4] Julia H. Gray and L. B. Rall. A computational system for numerical integration with rigorous error estimation. *Proceedings of the 1974 Army Numerical Analysis Conference*, pp. 341-355. U. S. Army Research Office, Research Triangle Park, N. C., 1974.

[5] Julia H. Gray and L. B. Rall. INTE: A UNIVAC 1108/1110 program for numerical integration with rigorous error estimation. *MRC Technical Summary Report No. 1428*, University of Wisconsin-Madison, 1975.

[6] Harriet Kagiwada, Robert Kalaba, Nima Rasakhoo, and Karl Spingarn. *Numerical Derivatives and Nonlinear Analysis*. Plenum Press, New York, 1986.

[7] R. E. Moore. The automatic analysis and control of error in digital computation based on the use of interval numbers, pp. 61-130 in *Error in Digital Computation, Vol. 1*, ed. by L. B. Rall. Wiley, New York, 1965.

[8] R. E. Moore. *Interval Analysis*. Prentice-Hall, Englewood Cliffs, N. J., 1966.

[9] R. E. Moore. *Techniques and Applications of Interval Analysis*. SIAM Studies in Applied Mathematics, 2, Society for Industrial and Applied Mathematics, Philadelphia, 1979.

[10] R. Piessens, E. de Doncker-Kapenga, C. W. Überhuber, and D. K. Kahaner. *QUADPACK: A Subroutine Package for Automatic Integration*. Springer Series in Computational Mathematics, No. 1. Springer-Verlag, New York, 1983.

[11] L. B. Rall. *Automatic Differentiation: Techniques and Applications*. Lecture Notes in Computer Science, No. 120. Springer-Verlag, New York, 1981.

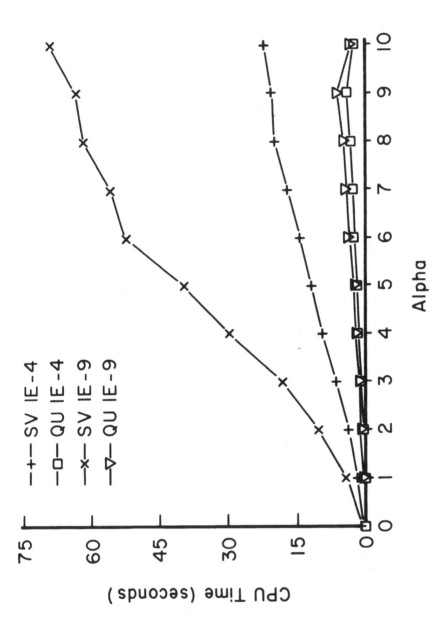

Figure I. QUADPACK test problem 2.
Measured CPU times.

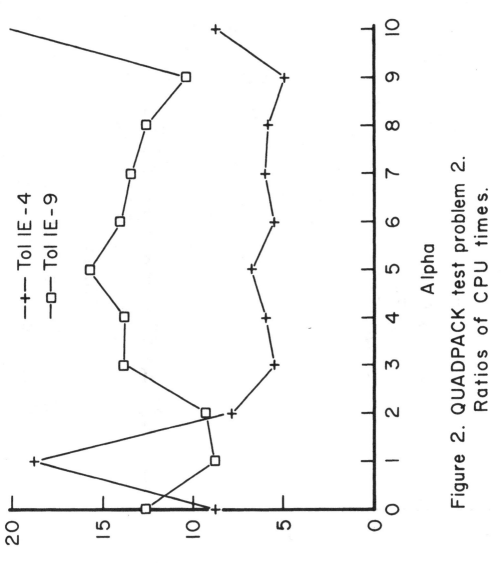

Figure 2. QUADPACK test problem 2. Ratios of CPU times.

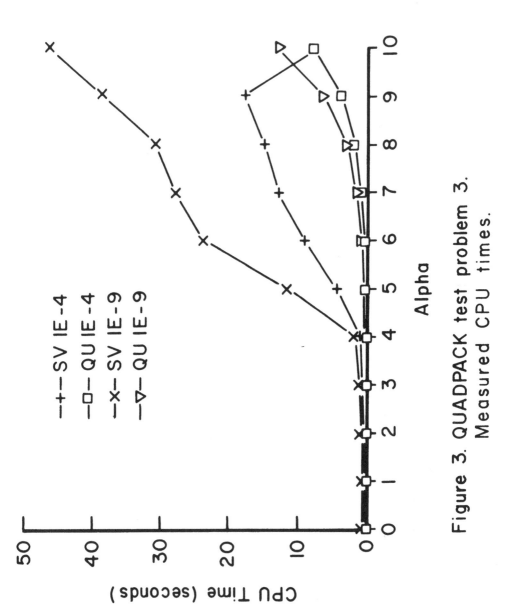

Figure 3. QUADPACK test problem 3.
Measured CPU times.

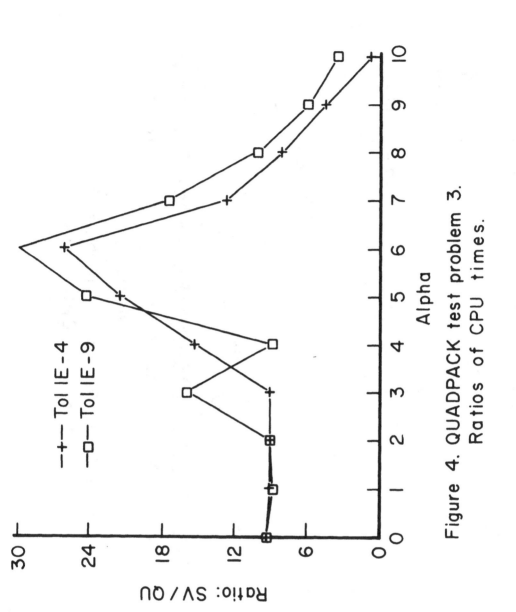

Figure 4. QUADPACK test problem 3.
Ratios of CPU times.

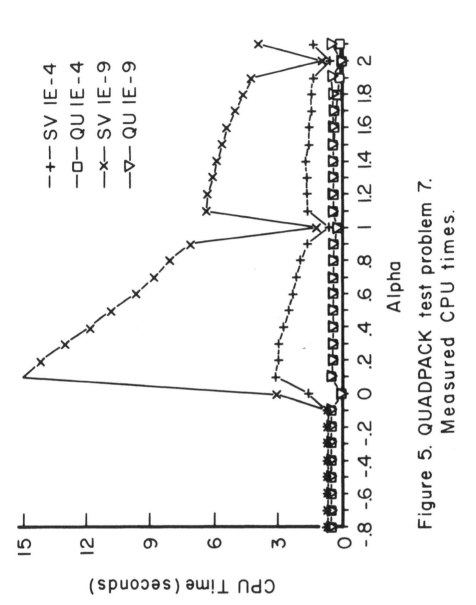

Figure 5. QUADPACK test problem 7.
Measured CPU times.

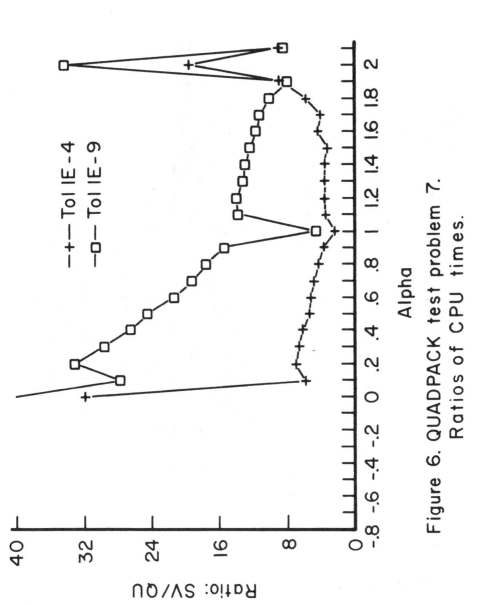

Figure 6. QUADPACK test problem 7.
Ratios of CPU times.

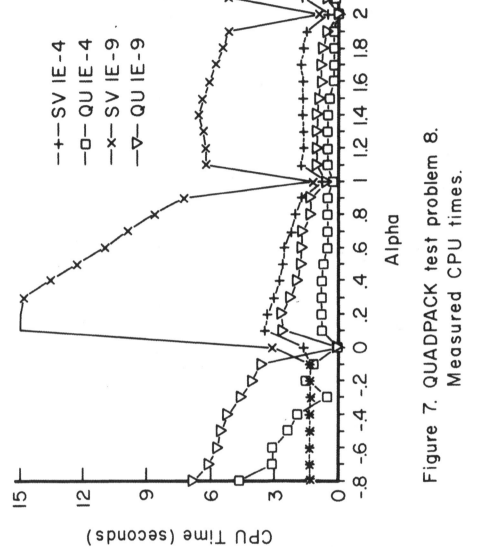

Figure 7. QUADPACK test problem 8. Measured CPU times.

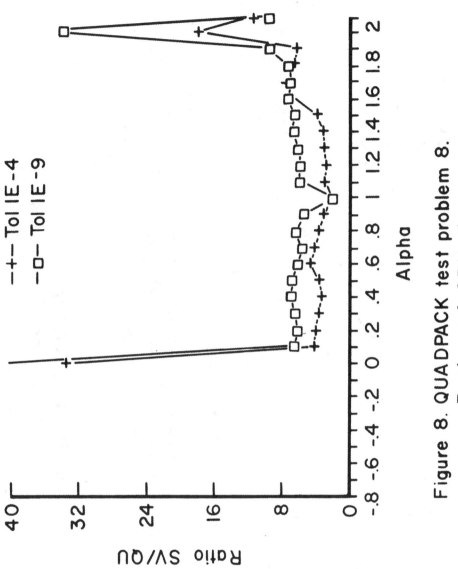

Figure 8. QUADPACK test problem 8.
Ratios of CPU times.

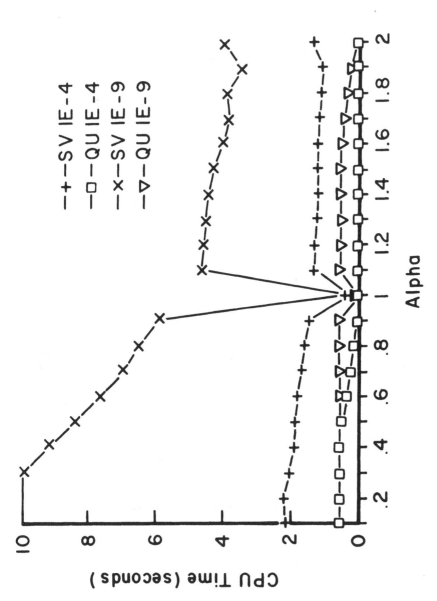

Figure 9. QUADPACK test problem IO.
Measured CPU times.

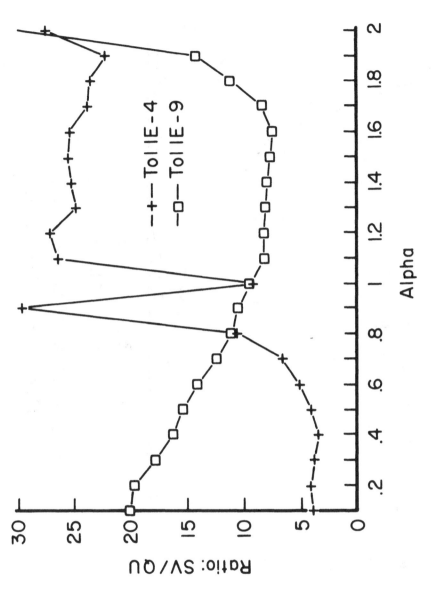

Figure 10. QUADPACK test problem 10.
Ratios of CPU times.

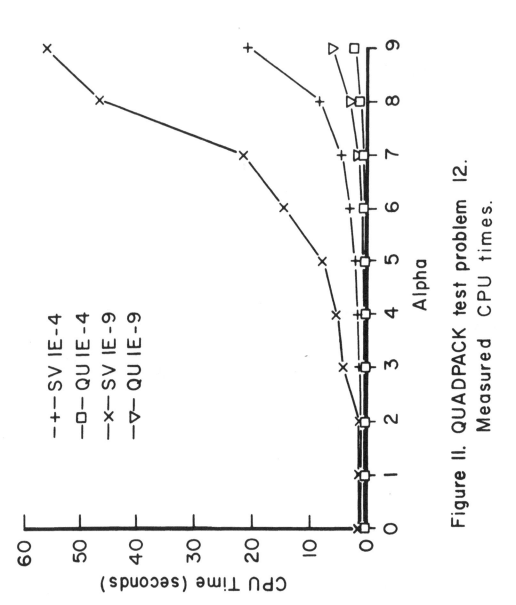

Figure II. QUADPACK test problem 12.
Measured CPU times.

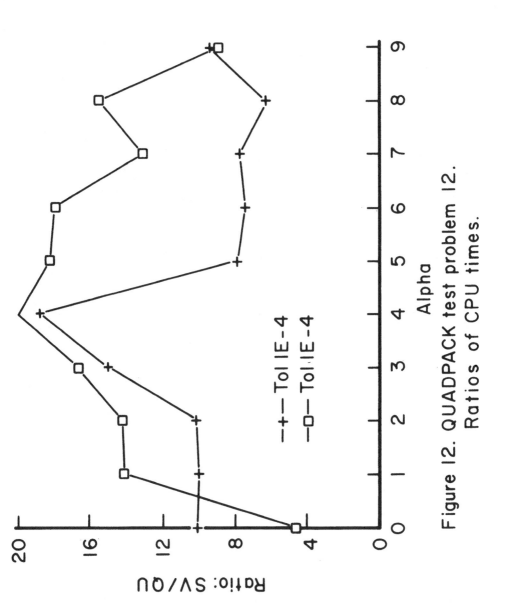

Figure 12. QUADPACK test problem 12.
Ratios of CPU times.

USING WEIGHT FUNCTIONS IN SELF-VALIDATING QUADRATURE

Gary S. Krenz
Department of Mathematics, Statistics, and Computer Science
Marquette University
Milwaukee, WI 53233 USA

ABSTRACT. We describe how weight functions are implemented in SVALAQ, Self-Validating Adaptive Quadrature, to provide a self-validating computation of the definite integral

$$If = \int_A^B f(x)\, dx.$$

SVALAQ computes an interval $[c, d]$ in which If is guaranteed to lie. The inclusion is validated by SVALAQ, provided that $f(x)$ can be evaluated on all of $[A, B]$ or if $f(x)$ has certain singularities.

Rules for evaluating definite integrals using weight functions in a self-validating environment were also developed. These rules, based on Taylor's polynomials, cover some of the difficulties associated with certain algebraic, logarithmic and sinusoidal weight functions in validated quadrature.

1. INTRODUCTION.

The attempt to create an easy interface between users and a self-validating adaptive quadrature package, SVALAQ, resulted in the development of a parser which recognizes selected weight functions and the use of several special quadrature rules for computing validated definite integrals.

The goal of a self-validating environment is to generate an interval which is guaranteed to contain the value of the posed problem. In self-validating quadrature, this means that an interval $[c, d]$ is computed for which,

$$If = \int_A^B f(x)\, dx \in [c, d],$$

where the

- Integrand - A Fortran-like expression for the integrand f as a finite sequence of $+$, $-$, $*$, $/$, $\sqrt{\ }$, $\exp$, $\ldots$.

261

P. Keast and G. Fairweather (eds.), Numerical Integration, 261–268.

- Limits - Finite limits of integration A and B (perhaps intervals).

- Tolerances - α and ρ, tolerances for the desired absolute and relative errors.

Thus, we wish to compute an interval $J = [c, d]$ satisfying the goals:

- Inclusion - $If \in [c, d]$.

- Accuracy - $w(J) = d - c \leq \max\{\alpha,\ \rho \cdot |If|\}$.

In order to verify the integral is actually contained in $[c, d]$, the computer verifies that the hypotheses of mathematical theorems hold.

Arithmetic tools employed in a self-validating environment generally include a large accumulator, a scalar product, defect correction, directed roundings, and interval arithmetic [3].

In our numerical results, ACRITH replaces the usual computer arithmetic with a validating computer arithmetic [1]. ACRITH, an IBM product, provides the large accumulator and the scalar product necessary to reduce the number of roundings that occur in the arithmetic operations. In addition, ACRITH provides the necessary validated standard function evaluations.

For the quadrature we are describing [2], all function, derivative, rule and error computations are done using interval techniques. Note that validated derivatives are easily computed by automatic differentiation, as shown by Rall [5].

We occasionally make liberal use of intersections. Unlike the usual numerical computations that yield only estimates, validated computations may be intersected to obtain a new containment of the solution. This new containment may be better than either of the solutions themselves, since the intersected solution may have a smaller width than either solution.

2. WEIGHT FUNCTIONS.

For singular integrals or integrals involving highly oscillatory functions, it is often difficult to capture the truncation error for the quadrature rule. This difficulty is exhibited by wide containments of the inclusion or the inability to compute the inclusion with guarantees.

Example:
Consider

$$\int_a^b f(x)\, dx = \sum_{i=1}^n w_i f(x_i) + c_n h \frac{f^{(p)}(\xi) h^p}{p!}$$

where $h = (b - a)$ and $a < \xi < b$. In the case where, say, $f(x) = \sqrt{x}, a = 0$ and $b = 1$ a simple guaranteed containment of the error term

$$c_n h \frac{f^{(p)}(\xi) h^p}{p!}$$

will be an unbounded interval for any $p > 0$.

A different type of difficulty is encountered when trying to compute $\sin x / \sqrt{x}$ on $[0, \epsilon]$ for $\epsilon > 0$. Rather than computing

$$S = \{\frac{\sin x}{\sqrt{x}} : x \in [0, \epsilon]\}$$

as one would like, it is common to compute instead the unified extension

$$\frac{\sin[0, \epsilon]}{[0, \epsilon]}$$

This computation leads to a solution where S is contained in an unbounded interval. Obviously, the containment for this function is far too wide. Thus, a computation similar to this one would completely corrupt an error inclusion.

Highly oscillatory functions produce another type of difficulty.

Example:

Consider $R = \{\sin(\omega x) : x \in [\alpha, \beta]\}$ where ω is quite large. Unless $[\alpha, \beta]$ is sufficiently small with respect to ω, we expect $R = [-1, 1]$. Again, this containment is too wide for our purposes. Note that using higher derivatives of $\sin(\omega x)$ only compounds the problem by introducing larger and larger powers of ω.

One way to minimize the effect of such difficulties in the truncation error is to use a weight function in the quadrature computations. Typically, the width of the inclusion is directly related to the width of the interval which captures the truncation error. Although the quadrature rule usually has a nontrivial width, by decreasing the inclusions for the error term, we make the greatest gain in accuracy.

3. THE PARSER.

The parser provides an easy interface between the user and the quadrature package. The user can enter the integrand as a character string. This character string is a Fortran-like expression for the integrand which may include a finite sequence of operations such as $+, -, *, /, \sqrt{}, \ldots$, as well as parameters which may be unknown at parse time.

The addition of a parser deviates from the customary usage of user supplied function subprograms. The parser frees the user from worrying about the coding of the interval computations and the details involved in automatic differentiation. Automatic differentiation is essential since it is unreasonable to require the user to formulate the first twenty derivatives of his integrand.

An additional benefit is obtained by analyzing the output of the parser. By comparing stored templates against parse tree substructures, it is possible to automatically recognize weight functions in the integrand. That is, in the process of generating code for the user's integrand, we can scan the integrand to locate predefined weight functions. If a weight function is recognized, an appropriate special quadrature rule is selected. Otherwise, standard Gaussian quadrature is used.

During parsing, parameters in the user's integration problem (upper and lower limits of integration and any parameters within the integrand itself) have not been specified. Thus, it is extremely difficult to decide which weight function in an integrand should be selected if two or more substructures match the stored templates. We resolve the weight function dilemma by choosing the leftmost weight function. This is consistent with the customary way of writing integrands containing weight functions.

4. QUADRATURE RULES.

Assume that $a \leq b, \forall a \in A$ and $b \in B$. We first state the rectangle rule.

Case 1. If $w(x)$ does not change sign on $[A, B]$, then

$$\int_A^B w(x)g(x)\, dx \in \{W(B) - W(A)\} \cdot g[A, B]$$

where $W(x) = \int w(x)\, dx, W(X) = \{W(x) : x \in X\}$ for $X = A, B$ and $g[A, B] = \{g(\xi) : \xi \in [a, b], a \in A, b \in B\}$.

Case 2. If $w(x)$ changes sign on $[A, B]$, then

$$\int_A^B w(x)g(x)\, dx = \int_A^B \{w_+(x) - w_-(x)\} \cdot g(x)\, dx$$

In this case, the form of the rectangle rule depends on the coordinates where the weight function changes sign.

In validated quadrature, the rectangle rule is essential for handling those functions $g(x)$ for which no derivatives can be computed (e.g., handling a step function at the step). Although the interval containment is likely to be quite wide, the resulting interval is guaranteed to contain the true solution.

We now examine higher order rules based on Taylor's polynomials. First, let $c = AR + (BL - AR)/2$ denote the midpoint of $[A, B]$, where $A = [AL, AR]$ and $B = [BL, BR]$. Provided g is sufficiently smooth (which can be checked using automatic differentiation), we may expand $g(x)$ as a Taylor's polynomial at c with a remainder term, and arrive at

$$\int_A^B w(x)g(x)\,dx = \int_A^B w(x)\{\sum_{k=0}^{n-1} \frac{g^{(k)}(c)}{k!}(x-c)^k + \text{remainder term}\}\,dx$$

$$= \sum_{k=0}^{n-1} \frac{g^{(k)}(c)}{k!} \int_A^B (x-c)^k w(x)\,dx + \text{error term}.$$

The derivatives of $g(x)$ are computed using automatic differentiation, thus making Taylor's polynomials easy to generate. We also note that in automatic differentiation, the next derivative $g^{(n)}(c)$ or $g^{(n+1)}[A,B]$ does not require reevaluation of any previous derivatives or function values.

We illustrate the technique by applying it to algebraic weights, logarithmic weights and sinusoidal weights.

Algebraic weight functions:

$$w(x) = (x-s)^p,$$

where s and p are real intervals, possibly degenerate. The associated quadrature rule and truncation error are

Rule:

$$\sum_{k=0}^{n-1} \frac{g^{(k)}(c)}{k!} \sum_{j=0}^k \binom{k}{j} (s-c)^{k-j}\{\text{anti}_{p+j}(B-s) - \text{anti}_{p+j}(A-s)\}$$

Truncation Error:

$$\frac{g^{(n)}[A,B]H^n}{n!}\{\text{anti}_p(B-s) - \text{anti}_p([c,B]-s)\} +$$

$$(-1)^n \frac{g^{(n)}[A,B]H^n}{n!}\{\text{anti}_p([A,c]-s) - \text{anti}_p(A-s)\},$$

where

$$\text{anti}_{p+j}(x) = \begin{cases} \ln|x| & \text{if } p+j+1 = \{0\} \\ \dfrac{x^{p+j+1}}{p+j+1} & \text{if } 0 \notin p+j+1 \\ (-\infty, \infty) & \text{otherwise} \end{cases}$$

and $H = [\min(c-AR, BL-c), \max(c-AL, BR-c)]$.

Logarithmic weight functions:

$$w(x) = \ln(|x-s|)$$

where, as before, s can be a real interval.

Rule:

$$\sum_{k=0}^{n-1} \frac{g^{(k)}(c)}{k!} \sum_{j=0}^k \binom{k}{j} (s-c)^{k-j}\{L_{j+1}(B-s) - L_{j+1}(A-s)\}$$

Truncation Error:

$$\frac{g^{(n)}[A,B]H^n}{n!}\{L_1(B-s)-L_1([c,B]-s)\}+$$
$$(-1)^n\frac{g^{(n)}[A,B]H^n}{n!}\{L_1([A,c]-s)-L_1(A-s)\},$$

where $L_{k+1}(x) \approx \int x^k \ln|x|\,dx$. Actually, $L_{k+1}(x)$ computes the moments of the logarithm with L'Hopital's rule built in (otherwise, we would be confronted with the same difficulty we encountered with $\frac{\sin x}{\sqrt{x}}$).

Sinusoidal weight functions:

$$w(x) = \sin(\omega x),$$

where ω can be a real interval. Let $S_k = \int_A^B (x-c)^k \sin(\omega x)\,dx$.
Rule:

$$\sum_{k=0}^{n-1}\frac{g^{(k)}(c)}{k!}S_k$$

Truncation Error:

$$\frac{g^{(n)}[A,B]H^{n+1}}{2(n!)}[-1,1]$$

Similarly for cosine, that is, replace S_k by $C_k = \int_A^B (x-c)^k \cos(\omega x)\,dx$.

The rectangle and higher order rules for algebraic weight functions were implemented in the self-validating adaptive quadrature package, SVALAQ. The following numerical examples illustrate the type of results we have been able to obtain using single precision on an IBM/AT/370. Note, in Examples 1, 2 and 4, we have underlined the digits where the upper and lower bounds of the computed inclusion disagree. However, due to the nature of Example 3, highlighting upper and lower bound disagreement would be inappropriate.

Example 1. Using $w(x) = x^{-\frac{3}{4}}$, we obtain

$$\int_0^1 x^{-\frac{3}{4}}\sqrt{x}\,dx \in [1.5554\underline{C}, 1.5555\underline{C}]_{16}$$

The hexadecimal interval $[1.5554\underline{C}, 1.5555\underline{C}]_{16}$ roughly corresponds to the decimal interval
$[1.33332\underline{4}, 1.33334\underline{0}]_{10}$.

Example 2. Using $w(x) = x^{-\frac{1}{2}}$, we obtain

$$\int_0^1 x^{-\frac{1}{2}}\sin x\,dx \in [0.9EDB7\underline{8}, 0.9EDB7\underline{F}]_{16}$$

Example 3. Using $w(x) = x^{[-\frac{33}{64}, -\frac{31}{64}]}$, we obtain

$$\int_0^1 x^{[-\frac{33}{64}, -\frac{31}{64}]} \, dx \in [1.F07C1, 2.10843]_{16}$$

The above (roughly $[1.939393, 2.064517]_{10}$) compares favorably with the exact answer $[64/33, 64/31]$.

Example 4. Using $w(x) = x^{-\frac{1}{4}}$, we obtain

$$\int_0^1 x^{-\frac{1}{4}}\left(1 - 2\chi_{[\frac{1}{16}, 1]}(x)\right) \, dx \in [-1.0000\underline{7}, -0.FFFF\underline{8C}]_{16},$$

where $\chi_{[\frac{1}{16}, 1]}(x)$ denotes the characteristic function on the interval $[\frac{1}{16}, 1]$. Of course, the inclusion (approximately $[-1.00000\underline{7}, -0.999993\underline{1}]_{10}$) contains the exact answer -1.

In all four examples, the absolute and relative error tolerances were set to zero.

5. DISCUSSION.

In most quadrature schemes which use order or subinterval adaptation, the adaptation is based on some estimate of the error. In validating quadrature, however, subinterval adaptation is based on the widths of the intervals which contain the value of the integral on the corresponding subintervals. In particular, the current implementation of the self-validating quadrature for algebraic weights uses a subinterval adaptation which bisects the subinterval that has the largest inclusion of the integral.

Similarly, order adaptation is based on a combination of the inclusion width and the cost of computing the additional higher derivatives. As long as the additional cost of generating the higher order derivatives is justified by the proportional narrowing of the integral's inclusion, we continue increasing the order of the rule.

An interesting feature of the above validating approach is its ability to use interval values both in the integrand and in the parameters. For example, 0.1 is not representable in standard floating point notation on a binary machine. Thus, in conventional computations, one would not be able to integrate $x^{0.1}$, but instead integrate something which is (generally) acceptably close. However, the use of intervals need not be restricted to handling the representation errors in a machine. Instead, the intervals could be used to represent any uncertainty in data.

Finally, in the particular cases of the algebraic and logarithmic weight functions, the location of the singularities need not be known a priori. Traditional treatment of singularities assumes that the singularity is at one endpoint or perhaps at both endpoints of the interval of integration [4]. This traditional assumption is not necessary using these rules.

REFERENCES.

[1] ACRITH High Accuracy Subroutine Library: General Information Manual. IBM publications, GC33-6163-02, 1986.

[2] G. F. Corliss and L. B. Rall. Adaptive, Self-Validating Numerical Quadrature. To appear in *SIAM Journal on Scientific and Statistical Computation.* Also available as *MRC Technical Summary Report No. 2815,* University of Wisconsin-Madison, 1985.

[3] U. W. Kulisch and W. L. Miranker. The Arithmetic of the Digital Computer: A New Approach, *SIAM Review,* 28(1986), pp. 1-40.

[4] R. Piessens, E. de Doncker-Kapenga, C. W. Überhuber, and D. K. Kahaner. *QUADPACK: A Subroutine Package for Automatic Integration.* Springer Series in Computational Mathematics, No. 1. Springer-Verlag, New York, 1983.

[5] L. B. Rall. *Automatic Differentiation: Techniques and Applications.* Lecture Notes in Computer Science, No. 120. Springer-Verlag, New York, 1981.

ON THE CONSTRUCTION OF A PRACTICAL ERMAKOV-ZOLOTUKHIN MULTIPLE
INTEGRATOR

T.N.L. Patterson
Department of Applied Mathematics & Theoretical Physics
The Queen's University of Belfast
Belfast, BT7 1NN
N. Ireland

ABSTRACT: Ermakov and Zolotukhin [1] offered the prospect of reducing
to almost any prescribed degree the variance of the Monte-Carlo
procedure for multiple integration and interpolation. However, the
original difficulties were replaced by an almost intractable
statistical sampling problem.
    The work of Bogues, Morrow and Patterson [2] indicated how the
sampling problems could be considerably reduced by an iterative
approach and details of a computational algorithm were   sketched out
which allowed the principles to be established.
    This paper describes work that has continued on the iterative
sampling technique to design and implement an efficient sampling
algorithm for product regions.

1. INTRODUCTION

The elegant work of Ermakov and Zolotukhin [1] offered in principle
the prospect of reducing in a controlled fashion the variance of the
Monte-Carlo procedure for multidimensional integration and
interpolation. However the original difficulties of accurate integral
evaluation and reliable error estimation were replaced by a statistical
sampling problem of enormous magnitude. An attempt was made by Bogues,
Morrow and Patterson [2] to make the problem   tractable. They showed
how an iterative approach could considerably reduce the complexity of
sampling and sketched out a computational procedure which   allowed the
principles to be confirmed without at the time excessive concern for
practicality. This paper describes work which has continued on the
iterative technique to implement a   more efficient and practical
algorithm for generating the nodes and weights of Ermakov-Zolotukhin
type cubatures for product regions.

P. Keast and G. Fairweather (eds.), Numerical Integration, 269–290.

2. BASIC GENERAL THEORY

The principle of the Ermakov-Zolotukhin technique is expressed in the
following theorem.

Theorem 1. (Ermakov & Zolotukhin): Let the function $f(x)$ in some
region of k-dimensional space, Ω_k , be approximated by a linear
combination of functions $\varphi_1(x),\ldots,\varphi_n(x)$ orthonormal with
respect to the weight $p(x)$.

Let the estimators a_i be the solution of the system of equations,

$$\sum_{i=1}^{n} a_i \varphi_i(x_j) = f(x_j) , \qquad 1 \leq j \leq n \qquad (1)$$

and the k-dimensional vector points x_j be distributed with density,

$$\Phi(x_1,\ldots,x_n) = p(x_1)\ldots p(x_n) F(x_1,\ldots,x_n) \qquad (2)$$

where,

$$n! \ F(x_1,\ldots,x_n) = [\det | \varphi_1(x_i)\ldots\varphi_n(x_i)|]^2$$

$$= \begin{vmatrix} \varphi_1(x_1) & \cdots & \varphi_n(x_1) \\ \cdot & & \cdot \\ \cdot & & \cdot \\ \varphi_1(x_n) & \cdots & \varphi_n(x_n) \end{vmatrix}^2 . \qquad (3)$$

Then for all $1 \leq m \leq n$,

$$\text{Mean } a_m = A_m = \int_{\Omega_k} p(x)f(x)\varphi_m(x)dx \qquad (4)$$

$$\text{Variance } a_m = \text{var}(a_m) = \int_{\Omega_k} p(x)f^2(x)dx - \sum_{i=1}^{n} A_i^2. \qquad (5)$$

The difficulties of the technique arise from the requirement to sample
from a multidimensional vector distribution. One sample point for the
estimators comprises a set of n k-dimensional vector points sampled
from the distribution (2).

Bogues, Morrow and Patterson [2] showed how the complexity of
sampling could be reduced by one "level" (iteratively sampling of only
one k-dimensional vector) by taking advantage of the observation that
the distribution (2) is invariant to any permutation of its vector
points. The following result (a consequence of Theorem 1 of Bogues,
Morrow and Patterson [2]) forms the basis of the practical algorithm
discussed in this paper.

Theorem 2: Let the vectors $x_1,\ldots,x_n$ be chosen from an arbitrary
distribution $\Phi_0(x_1,\ldots,x_n)$ and let each of the set of vectors,

$$X_s = (x_{ns+1} , \ldots , x_{ns+n}) , \quad s = 1,2,\ldots \tag{6}$$

be successively generated by sampling x_h from the partial densities,

$$D(x_h) = F(x_{h-n+1} , \ldots , x_h) \ p(x_h) \ / \int_{\Omega_k} F(x_{h-n+1}), \ldots, x_h) p(x_h) dx_h \tag{7}$$

where F is defined by (2), for $h=ns+1,\ldots,ns+n$. Then the distribution of X_s, $\Phi_s(X_s)$, satisfies,

$$\lim_{s \to \infty} \Phi_s(X_s)/\Phi(X_s) = 1 . \tag{8}$$

Briefly, Theorem 2 indicates that the distribution $\Phi_s(X)$ converges to $\Phi(X)$. Providing the rate of convergence is satisfactory it is thus permissible to sample iteratively from distribution (7). Note that each X_s comprises n k-dimensional vectors which become the sample point for (1).

Taking advantage of the orthonormality of $\varphi_i(x)$ with respect to $p(x)$, distribution (7) can be expressed in a more convenient computational form. Thus,

Theorem 3: Define the n x n matrix A_h as,

$$A_h = \begin{bmatrix} \varphi_1(x_{h-n+1}) & \cdots & \varphi_n(x_{h-n+1}) \\ \vdots & & \vdots \\ \varphi_1(x_h) & \cdots & \varphi_n(x_h) \end{bmatrix} \tag{9}$$

and let

$$\det (A_h) = \sum_{i=1}^{n} g_i \ \varphi_i(x_h) \tag{10}$$

be the expansion of the determinant of A_h with respect to its last row. Then,

$$D(x_h) = p(x_h) \ [\sum_{i=1}^{n} w_i \ \varphi_i(x_h)]^2 \tag{11}$$

where,

$$w_i = g_i / (\sum_{i=1}^{n} g_i^2)^{1/2} . \tag{12}$$

3. SOLUTION OF THE ITERATED LINEAR SYSTEMS

It is clear from Theorem 3 that generating the sequence of n vectors x_h, $h=ns+1,\ldots,ns+n$ involves solving a corresponding sequence of related linear systems given by (9). The matrix A_{h+1}

is generated by evaluating the nth row of A_h at the latest x_h and then
displacing all rows of A_h upwards by one row. The first row of A_h
is discarded. An efficient sampling scheme which takes advantage of
this structuring will now be outlined.

To simplify the discussion consider the problem of generating the
sequence of vectors,

$$x_{n+1}, \cdots, x_{2n} \tag{13}$$

from the known sequence,

$$x_1, \cdots, x_n . \tag{14}$$

The matrix (9) initially takes the form,

$$A_n = \begin{bmatrix} \varphi_1(x_1) & \cdots & \varphi_n(x_1) \\ \cdot & & \cdot \\ \cdot & & \cdot \\ \varphi_1(x_n) & \cdots & \varphi_n(x_n) \end{bmatrix} = \begin{bmatrix} s_1^T \\ \cdot \\ \cdot \\ s_n^T \end{bmatrix} , \tag{15}$$

where the vector s_i is defined as,

$$s_i = (\varphi_1(x_i), \cdots, \varphi_n(x_i)) . \tag{16}$$

We begin by performing a QR decomposition,

$$A_n Q_0 = R_0^T \tag{17}$$

where Q_0 is an orthogonal matrix and R_0 is upper triangular. This will
be possible if A_n is non-singular which is ensured by the fact that
$x_1, \ldots, x_n$ will have been previously sampled from a density proportional
to the square of the determinant of A_n. The very existence of these
vectors guarantees that this determinant cannot be zero.

To sample x_{n+1} using (11) we must use the related matrix,

$$A_{n+1} = \begin{bmatrix} s_2^T \\ \cdot \\ s_n^T \\ s^T \end{bmatrix} \tag{18}$$

formed by shifting up the rows of A_n. The last row, s^T, is an
arbitrary "place-marker" which plays no part in the computations and
is replaced by s_{n+1} once x_{n+1} has been determined. Corresponding
to (17) we would like to compute the QR decomposition,

$$A_{n+1} \ Q_1 = R_1^T \tag{19}$$

since this would imply,

$$A_{n+1} \ (Q_1 e_n) = R_1^T e_n \quad \alpha \quad e_n \quad . \tag{20}$$

From the Cramer's rule solution of (20), it follows that the vector g, whose elements are the coefficients g_i of (10), is proportional to the nth column of Q_1. Hence $D(x_{n+1})$ given by (11) is immediately calculable and can be used to sample x_{n+1}. The details of the sampling procedure will be discussed later. Once x_{n+1} has been computed, the new last row of A_{n+1} can be incorporated by replacing the last row of R_1^T by $s_{n+1}^T Q_1$. Thus corresponding to (18) we have,

$$A_{n+1} \ Q_1 = \begin{bmatrix} s_2^T Q_1 \\ \cdot \\ s_n^T Q_1 \\ s_{n+1}^T Q_1 \end{bmatrix} = R_1^T \quad . \tag{21}$$

The decomposition (19) can be accomplished efficiently by updating Q_0 and R_0^T as follows. Using (17) we may write,

$$A_{n+1} \ Q_0 = \begin{bmatrix} s_2^T Q_0 \\ \cdot \\ s_n^T Q_0 \\ s^T Q_0 \end{bmatrix} = K_0^T \tag{22}$$

where K_0^T is formed by shifting up the rows of R_0^T. The last row of K_0^T is arbitrary and clearly this matrix has lower Hessenberg form. The QR structure can be restored by applying a sequence of Givens' rotations to Q_0 and K_0^T zeroising the off-diagonal elements of K_0^T. Denoting the combined Givens' rotations by G_1 we have,

$$A_{n+1} \ Q_0 \ G_1 = K_0^T \ G_1 \tag{23}$$

and comparing with (19) may identify,

$$Q_1 = Q_0 \ G_1 \quad ; \quad R_1^T = K_0^T \ G_1 \quad . \tag{24}$$

The updating process can be repeated in a similar fashion on the matrices $A_{n+2}, \ldots, A_{2n}$, to yield the remaining sample vectors $x_{n+2}, \ldots, x_{2n}$.

The procedure described in this section substantially reduces the computational labour. In principle, the QR decomposition (an n^3 process) could be carried out for each of the n vectors, $x_{n+1}, \ldots, x_{2n}$, leading to an overall n^4 process. By updating the transformations (n^2 process) as outlined the overall operation remains an n^3 process. Although the updating could be continued into the next cycle of sample vectors, to avoid any accumulation of errors, it is best to begin each new cycle with a full QR decomposition.

4. APPLICATION TO PRODUCT REGIONS

To make further progress in developing a practical algorithm the vector sampling problem must be addressed. While the discussions of previous sections apply to the general domain of integration it is unlikely that a general algorithm could be developed to deal effectively with all regions. We thus restrict consideration for the remainder of this paper to product regions.

Consider the product region,

$$\Omega_k = [a,b]^k \tag{25}$$

with weight,

$$p(t) \equiv p(t_1) \ldots p(t_k) \tag{26}$$

where $t_1, \ldots, t_k$ are the components of t.

The orthonormal basis functions of Theorem 1 can be written as,

$$\varphi_i(t) = Z_{v(i,1)}(t_1) \ldots Z_{v(i,k)}(t_k), \quad 1 \leq i \leq n \tag{27}$$

where the $v(i,j)$ denotes the order of the one dimensional orthonormal polynomials, $Z_i(t_j)$, which satisfy,

$$\int_a^b Z_i(s) Z_j(s) p(s) ds = \delta_{i,j} \quad . \tag{28}$$

$\delta_{i,j}$ is the Kronecker δ. Note that by definition the order is one greater than the polynomial degree.

If the basis is limited to all monomials up to degree d then $v(i,j)$ can assume all values in the range,

$$1 \leq v(i,j) \leq d+1 \tag{29}$$

corresponding to the orthonormal polynomials,

$$Z_1(x), \ldots, Z_{d+1}(x) \quad , \tag{30}$$

but subject to the constraint,

$$\sum_{j=1}^{k} v(i,j) \leq d+k \ , \ \text{for all } 1 \leq i \leq n \ . \tag{31}$$

If all possible monomials are included then the total number of basis functions is given by,

$$n = \binom{k + d}{k} \tag{32}$$

and all polynomials of monomial degree not exceeding d will be integrated exactly.

Let us now consider the problem of sampling the vector x_{n+1} given $x_2, \ldots, x_n$ from the partial density defined by (10), viz.,

$$D(r) = p(r) \ [\sum_{i=1}^{n} w_i \ \varphi_i \ (r) \ ]^2 \tag{33}$$

where r is the vector variable to be sampled with components,

$$r^T = (r_1, \ \ldots \ , r_k) \tag{34}$$

Suppose that components $r_1^*, \ldots, r_{m-1}^*$ have been already determined and we wish to sample component r_m. Substituting (27) into (33) and using the orthonormality property (28) leads after some analysis to the following result.

Theorem 4: Let,

$$q_m(s) = \sum_{i=1}^{n} \ \sum_{j=1}^{n} Z_{v(i,m)}(s) Z_{v(j,m)}(s) C_{i,j,m} \ (r_1^*, \ldots, r_{m-1}^*) \tag{35}$$

where,

$$C_{i,j,m} = w_i \ w_j \ \prod_{h=1}^{m-1} [Z_{v(i,h)}(r_h^*) \ Z_{v(j,h)}(r_h^*)] \ \prod_{h=m+1}^{k} \delta_{v(i,h),v(j,h)} \tag{36}$$

and w_i is defined in (12).

Let,

$$Q_m(s) = p(s) \ q_m(s)/q_{m-1}(r_{m-1}^*) \ , \ q_0(s) = 1 \ . \tag{37}$$

Then the sample value for component r_m is determined as the solution of the equation,

$$P(r_m^*) = \int_a^{r_m^*} Q_m(s) \, ds - \sigma_m = 0 \tag{38}$$

where σ_m has random value, uniformly distributed in $[0,1]$.

The entire vector $\mathbf{x}_{n+1}$ is sampled by successively applying Theorem 4 with $m=1,\ldots,k$. The resulting samples form the components of $\mathbf{x}_{n+1}$, thus,

$$\mathbf{x}_{n+1}^T = (r_1^*, \ldots, r_k^*) . \tag{39}$$

It may be noted that $Q_m(s)$ satisfies the normal requirement for a density function, viz.,

$$\int_a^b Q_m(s) \, ds = 1 . \tag{40}$$

5. COMPUTATIONAL CONSIDERATIONS

Theorems 3 and 4 together with the discussion in sections 3 and 4 form the basis for constructing a sampling algorithm for product regions. In this section some practical details of the implementation will be presented.

5.1 Solution of the linear systems

The procedure described in section 3 forms the simplest part of the algorithm to implement. The QR decomposition and subsequent updating from one vector to the next can be performed using standard packages such as those available under LINPACK [4] with suitable customisation to streamline for efficiency. The orthogonal matrix Q_i and triangular matrix R_i are overwritten with updated values from step to step so that storage requirements are modest. The structure of the procedure is such that considerable advantage could be taken of parallelisation although this has yet to be pursued.

It has been generally observed that the matrix computations are relatively stable probably due to the underlying orthogonal structure and the natural scaling of the matrix elements within (3).

5.2 Vector sampling

The procedure of section 5.1 makes available the coefficients w_i defined in (12) and the vector component sampling can commence using Theorem 4.

The implementation of Theorem 4 represents the more complex part of the implementation and it is where efficiency is of greatest importance. The procedure adopted, corresponding to (30), is to build the k x (d+1) matrix,

$$
B = \begin{bmatrix} Z_1(r_1^*) & \cdots & Z_{d+1}(r_1^*) \\ \vdots & & \vdots \\ Z_1(r_{m-1}^*) & \cdots & Z_{d+1}(r_{m-1}^*) \\ \vdots & & \vdots \end{bmatrix} \tag{41}
$$

as the successive vector components become known. Associated with this matrix and corresponding to (27), (29) and (31) is a k x n indexing matrix,

$$
X = \begin{bmatrix} v(1,1) & \cdots & v(1,n) \\ \vdots & & \vdots \\ v(k,1) & \cdots & v(k,n) \end{bmatrix} \tag{42}
$$

whose i^{th} row gives the orders of the orthonormal polynomials making up $\varphi_i(x)$.

Let us now suppose we are generating the sample component r_m. The matrix C, whose (i,j) component is defined by (36), is first generated. This is accomplished by scanning the matrix X and selecting all indices (i,j) such that,

$$
v(i,h) = v(j,h) , h = m+1, \ldots , k . \tag{43}
$$

For each selected index the quantity,

$$
w_i w_j Z_{v(i,h)}(r_h^*) Z_{v(j,h)}(r_h^*) \tag{44}
$$

is accumulated as the (i,j) element of C. We can now easily compute $q_m(s)$. Since $q_{m-1}(r_{m-1})$ is available from the last step of the previous iteration, $Q_m(s)$ is completely defined.

We must now consider the evaluation of the indefinite integral $P(r)$ defined by (38). It is clear from the definition of $q_m(s)$ that its maximum polynomial degree will be 2d and thus the integral of $Q_m(s)$ can be evaluated exactly using a Gaussian rule of degree $(d+1)$ with respect to the weight $p(s)$ in the interval $[a,b]$. In the actual implementation an automatic integrator can be used, with only slight loss in efficiency, to allow a wide range of weight functions to be conveniently processed.

It remains to consider the solution of (38). We note that the integral of $Q_m(s)$, being a cumulative distribution function, satisfies,

$$
0 \le \int_a^r Q_m(s) \, ds \le 1 , \quad a \le r \le b \tag{45}
$$

and increases monotonically over the interval $[a,b]$. There will be only one point at which it assumes the value σ_m corresponding to the required solution $r = r_*$. The derivative of $^m P(r)$ is, of course, simply $Q_m(r)$ and so Newton's method can be applied easily to quickly locate the root. Due to the monotone property of the integral a satisfactory initial approximation is,

$$r_0 = a + (b-a) \sigma_m \quad . \tag{46}$$

In the few cases when Newton's method goes astray a few bisections can be performed to isolate a more acceptable initial approximation. The monotonicity of (45) would also allow inverse interpolation to be applied.

6. SYMMETRISATION

It has been noted in section 4 that the number of basis functions associated with monomial degree d is,

$$n = \binom{k+d}{k} \tag{47}$$

which corresponds to the number of integrand evaluations per vector sample point. A selection of values of n for various values of d and k is given in Table 1 and indicates that the linear systems discussed in section 3 can be of quite large dimension. For a finite domain, the number of basis functions, and consequently labour, can be reduced when the one dimensional polynomials making up $\varphi_i(\mathbf{x})$ defined in (27) possess the common symmetry property,

$$Z_j(s) = (-1)^{j+1} Z_j(a+b-s) \tag{48}$$

with respect to reflection of their arguments about the mid-point of the interval of definition, $[a,b]$. That is, even degree (odd order) polynomials are symmetric and odd degree (even order) polynomials are antisymmetric.

To exploit this property we transform the integrand to,

$$F(\mathbf{x}) = \Sigma_s \, f(\mathbf{x})/2^k \tag{49}$$

where Σ_s denotes summation over all symmetric reflections of the components of $\mathbf{x}$ as indicated in (48). For example, in two dimensions and the interval $[0,1]$,

$$F(\mathbf{x}) = [f(x_1,x_2) + f(1-x_1,x_2) + f(x_1,1-x_2) + f(1-x_1,1-x_2)]/4 \quad . \tag{50}$$

It is clear that when (48) is satisfied then,

Table 1. Number of basis functions (equation 32) for degree d and
          dimension k

Degree (d) =	2	4	6	8	10
Dimension (k)					
2	6	15	28	45	66
3	10	35	84	165	286
4	15	70	210	495	1001
5	21	126	462	1287	3003
6	28	210	924	3003	8008
7	36	330	1716	6435	19448
8	45	495	3003	12870	43758
9	55	715	5005	24310	92378
10	66	1001	8008	43758	184756

Table 2. Number of symmetric basis functions (equation 52) for
          degree d and dimension k. Number of integrand
          evaluations per sample point is given in parentheses

Degree (d) =	2	4	6	8	10
Dimension (k)					
2	3	6	10	15	21
	(12)	(24)	(40)	(60)	(84)
3	4	10	20	35	56
	(32)	(80)	(160)	(280)	(448)
4	5	15	35	70	126
	(80)	(240)	(560)	(1120)	(2016)
5	6	21	56	126	252
	(192)	(672)	(1792)	(4032)	(8064)
6	7	28	84	210	462
	(448)	(1792)	(5376)	(13440)	(29568)
7	8	36	120	330	792
	(1024)	(4608)	(15360)	(42240)	(101376)
8	9	45	165	495	1287
	(2304)	(11520)	(42240)	(126720)	(329472)
9	10	55	220	715	2002
	(5120)	(28160)	(112640)	(366080)	(1025024)
10	11	66	286	1001	3003
	(11264)	(67584)	(292864)	(1025024)	(3075072)

$$\int_{\Omega_k} F(x) \, dx = \int_{\Omega_k} f(x) \, dx \quad . \tag{51}$$

Further, all $\varphi_i(x)$ defined in (27), having at least one antisymmetric polynomial component, disappear from $F(x)$ and hence can be removed from the basis representation. Compared to (47), the number of basis functions required diminishes to,

$$n' = \begin{pmatrix} k + [d/2] \\ k \end{pmatrix} \tag{52}$$

where $[d/2]$ denotes the integer part of $d/2$.

Symmetrisation is usually adopted in practice whenever possible although a penalty must be incurred. The number of integrand evaluations required per sample point is increased from n to $n' 2^k$ as given by (52) and implied by (49).

7. IMPLEMENTATION

The techniques discussed in previous sections have been implemented in general form as two procedures which we shall refer to as procedure A and procedure B. In essence, the former generates the sample vectors while the latter uses these vectors to evaluate an integral. It is advantageous to subdivide the computational work in this manner so that a "database" of sample vectors can be established (just as one does for Gaussian quadrature rules) for subsequent application to a wide class of multiple integrals. In the computations, all required uniformly distributed samples were supplied by a pseudo-random congruential-type generator.

7.1 Procedure A

The arguments to procedure A allow considerable flexibility in specifying the sampling requirements. In particular, in addition to specifying the dimension, monomial degree, etc., the user defines the product region (25), the basis representation (27), the appropriate orthogonality weight (26), the choice of starting vectors, and the symmetry (48). After initialisation of the various scalars and matrices for a particular choice of parameters each subsequent call of procedure A generates a new set of n sample vectors which can be stored in the "database" for future use.

It may be noted that given a set of sample vectors (or random nodes) $x_1, \ldots, x_n$ we can express the estimators a_i defined in (1) in the form of n cubature rules,

$$a_i = \sum_{j=1}^{n} H_{i,j} \, f(x_j) \quad i=1, \ldots, n \tag{53}$$

where $H_{i,j}$ is the ith row of the inverse of the matrix,

Table 3. Summary of parameters used in Tables 4,5 and 6

Dimension (k)	4	6	8
Degree (d)	6	4	4
No. of basis functions (n) (symmetric)	35	28	45
No. of starting iterates	10	10	10
No. of sample vectors	10	10	10
No. of integrand evaluations	5600	17920	115200

**Table 4. Evaluation of integrands 1 to 9 for 4 dimensions
(See Table 3 and the Appendix)**

Integrand no.	Ermakov-Zolotukhin	Uniform sample	Crude Monte-Carlo
1	1.8(-7)*	6.9(-5)	1.8(-3)
	0.0	1.1(-4)	4.5(-3)
2	2.6(-3)	6.8(-2)	3.9(-4)
	1.0(-2)	4.0(-2)	8.1(-3)
3	7.0(-7)	7.3(-5)	1.8(-4)
	0.0	0.0	3.8(-4)
4	3.1(-4)	6.6(-2)	1.6(-3)
	8.7(-3)	4.8(-2)	3.8(-3)
5	3.5(-5)	4.0(-4)	6.1(-3)
	9.0(-5)	4.0(-4)	6.5(-2)
6	6.8(-7)	6.3(-5)	1.6(-5)
	1.4(-4)	1.4(-4)	2.1(-3)
7	8.2(-3)	2.7(-2)	1.8(-3)
	1.1(-2)	3.2(-2)	4.4(-2)
8	1.6(-2)	2.9(-2)	7.2(-3)
	1.3(-2)	3.2(-2)	3.2(-2)
9	3.8(-4)	1.6(-3)	6.7(-3)
	1.6(-3)	7.3(-3)	1.3(-2)

Notation: Number in parentheses denotes the power of 10.
* The first number in each pair is the actual relative error.
Underneath is the estimated relative error.

$$
A = \begin{bmatrix} \varphi_1(x_1) & \cdots & \varphi_n(x_1) \\ \cdot & & \\ \cdot & & \\ \varphi_1(x_n) & \cdots & \varphi_n(x_n) \end{bmatrix} .
$$ (54)

In particular, we have, for the weighted integral of $f(x)$ itself,

$$
a_1 = \sum_{j=1}^{n} H_{1,j} \, f(x_j) .
$$ (55)

The integration weights $H_{1,j}$ can be stored in the database each along with the nodes $x_1, \ldots, x_n$ to make up each random cubature rule.

The initial vectors are normally chosen with randomly distributed components and the question immediately arises of how many starting iterates should be generated to comply with Theorem 2. We note that any set of vectors for which (3) is non-zero is in principle a valid sample from the required distribution and so no iteration should be necessary. In practice, that sample point may be unlikely to occur and the subsequent sequence generated may be untypical of the overall distribution pattern. The consistency of the results for a large number of experiments suggests that the number of starting iterates is not critical and generally about ten iterations appear sufficient to wipe out the "memory" of the originating random set.

7.2 Procedure B

Procedure B simply selects the appropriate random nodes and corresponding weights from the database and applies (55), or in general (53), to compute the Monte Carlo estimators. The usual statistical information is accumulated providing estimates of the variances and hence the estimated standard errors in the results.

7.3 Constant weight function for the hypercube

Most attention has been given to simplest case of generating sample vectors for the k dimensional hypercube in [0,1] with a constant weight function. The basis functions defined by (27) are then formed as products of suitably normalised Legendre polynomials.

Table 2 gives a summary of the sets of sample vectors which have been generated and applied to the test integrals given in the Appendix. These integrals, most of which are due to Genz [3], are parametrised by dimension and contain adjustable factors to control the difficulty of evaluation. A symmetric basis (section 6) has been taken in all cases to reduce the computational effort in generating samples. Tables 4, 5 and 6 give the calculated actual relative errors and the estimated relative errors (as determined from the standard statistical errors) for the particular set of random parameters indicated in the Appendix. For comparison, the crude Monte-Carlo

Table 5. Evaluation of integrands 1 to 9 for 6 dimensions
(See Table 3 and the Appendix)

Integrand no.	Ermakov-Zolotukhin	Uniform	Crude Monte-Carlo
1	1.8(-6)*	3.2(-4)	3.4(-4)
	1.2(-4)	3.4(-4)	2.6(-3)
2	1.3(-2)	3.1(-1)	1.7(-3)
	3.5(-2)	2.0(-1)	5.8(-3)
3	2.1(-6)	3.4(-4)	2.3(-5)
	1.0(-4)	3.1(-4)	9.0(-5)
4	5.3(-3)	1.1(-1)	1.4(-3)
	4.6(-3)	1.4(-1)	1.7(-3)
5	7.6(-5)	6.1(-4)	1.3(-3)
	1.3(-4)	5.7(-4)	5.7(-3)
6	0.0	3.1(-4)	7.9(-5)
	1.6(-4)	2.9(-4)	2.1(-4)
7	7.2(-3)	1.6(-2)	4.5(-3)
	6.2(-3)	1.4(-2)	1.3(-2)
8	1.3(-2)	1.4(-2)	2.5(-3)
	8.5(-3)	1.6(-2)	1.1(-2)
9	7.5(-3)	1.1(-2)	3.7(-3)
	5.2(-3)	1.4(-2)	4.3(-3)

Notation: Number in parentheses denotes the power of 10.
* The first number in each pair is the actual relative error.
  Underneath is the estimated relative error.

Table 6. Evaluation of integrands 1 to 9 for 8 dimensions
(See Table 3 and the Appendix)

Integrand no.	Ermakov-Zolotukhin	Uniform	Crude Monte-Carlo
1	7.6(-8)*	3.7(-5)	2.6(-4)
	0.0	9.3(-5)	1.1(-3)
2	7.8(-3)	6.5(-1)	2.9(-4)
	2.8(-2)	6.3(-1)	2.7(-3)
3	1.2(-7)	2.1(-5)	2.1(-4)
	0.0	1.4(-4)	4.3(-5)
4	5.3(-3)	9.3(-3)	1.2(-4)
	5.6(-3)	6.4(-2)	6.0(-4)
5	3.4(-7)	1.2(-5)	6.2(-4)
	8.3(-5)	8.3(-5)	4.9(-4)
6	3.2(-7)	2.0(-5)	1.6(-5)
	0.0	0.0	0.0
7	1.9(-5)	3.6(-3)	3.5(-3)
	2.0(-4)	1.6(-3)	3.2(-3)
8	1.7(-1)	1.7(-1)	1.6(-1)
	2.8(-4)	3.4(-3)	2.5(-3)
9	3.9(-4)	4.7(-3)	1.0(-3)
	4.9(-4)	3.8(-3)	1.0(-3)

Notation: Number in parentheses denotes the power of 10.
* The first number in each pair is the actual relative error.
Underneath is the estimated relative error.

result using the same number of integrand evaluations is given.
Results are also given for the case when the sample vectors have
their components distributed uniformly in [0,1] which should identify
the beneficial effect, if any, of choosing the more complex
Ermakov-Zolotukhin distribution. All calculations were carried
out in single precision floating point arithmetic on a VAX 8600
computer.
        Cases where the estimated error is given as zero indicate
variances which vanish to full machine accuracy. Although these tests
are by no means exhaustive, it is clear that the
Ermakov-Zolutukhin procedure produces superior accuracy in most cases
and the estimated errors are generally reliable within a factor of
about two. There are of course some notable exceptions.
Integrand 8 in high dimension proves difficult for all methods. The
crude Monte-Carlo method generally does well for integrands which
are not easily represented by low degree monomials such as for
integrands 2, 4 and 8. A feature of the Ermakov-Zolotukhin
technique is that the results tend to be reasonably consistent
with respect to different sets of sample vectors. On the contrary,
the results for the uniform samples have been found to vary
erratically, sometimes by several orders of magnitude, corresponding
to unfortunate choices of random vectors which cause (3) to be
nearly singular.
        The form of the partial cumulative distribution functions may be
of some interest and Figure 1 presents a typical
mid-calculation "snapshot" of all components. for the six dimensional
case. The curve labelled DIMENSION = m gives the distribution of
component r_m and plots the indefinite integral of Q_m(eq.(37)). This
corresponds to the integral of (33) with $r_1,\ldots,r_{m-1}$ already determined
(null for m=1) and integrated over all values of components
$r_{m+1},\ldots,r_k$. The symmetry of the curves is evident and as discussed
earlier is a consequence of (48).

7.4 Other weight functions

Application to other weighted finite domains is relatively
straightforward with the appropriate Gauss rule being applied to
evaluate the integral of the partial density, (38). For example,
with the weight,

$$p(x) = (1-x^2)^{-1/2} \qquad (56)$$

over [-1,1] one uses the Gauss-Chebyshev rules.
        On the other hand, dealing with infinite domains requires more
ingenuity in that some judicious cut-off limit must be placed on the
region defining the partial densities. The tail of the sampling
distribution tends to be a fairly flat plateau making some roots of
(38) insensitive to σ_m.
        Sample vectors for the weight e^{-x} over $[0,\infty)$ have been
successfully computed for low dimension and degree. Advantage

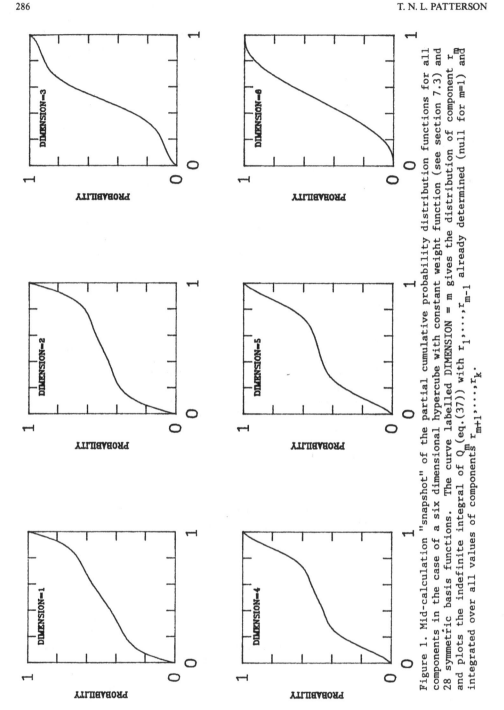

Figure 1. Mid-calculation "snapshot" of the partial cumulative probability distribution functions for all components in the case of a six dimensional hypercube with constant weight function (see section 7.3) and 28 symmetric basis functions. The curve labelled DIMENSION = m gives the distribution of component r_m and plots the indefinite integral of $Q_m(\text{eq.}(37))$ with $r_1, \ldots, r_{m-1}$ already determined (null for m=1) and integrated over all values of components $r_{m+1}, \ldots, r_k$.

cannot be taken of symmetry (section 6) and so the linear systems can become of large order (Table 1) for realistic cases.

8. CONCLUSIONS

At least for finite regions of moderate dimension it would appear that the computational refinements described in this paper of Bogues, Morrow and Patterson [2] iterative sampling technique can yield a reasonably practical procedure. It would now seem feasible to establish a database of sample vectors for various dimensions and monomial degree for a range of useful weight functions. This could simplify the evaluation of integrals containing standard types of singularities. Parallelisation has not been pursued in this paper but the structure of the procedures indicates that it could be exploited to good effect.

REFERENCES

1. Ermakov,S.M., and Zolotukhin,V.G., 'Polynomial approximations and the Monte Carlo method', Theor. Probability Appl., 5,428-431,1960.

2. Bogues,K., Morrow,C.R., and Patterson,T.N.L. 'An implementation of the method of Ermakov and Zolotukhin for multidimensional integration and interpolation', Numer. Math. 37,49-60,1981.

3. Genz,A.C. Private communication.

4. Dongarra,J., Bunch,J.R., Moler,C.B. and Stewart,G.W. 'LINPACK User's Guide', SIAM Publications, Philadelphia, 1978.

Appendix - Test integrals for dimension k

Some of these integrals contain parameters,

$$u_i, v_j \ , \ i=1,..,k$$

which are chosen randomly in [0,1]. The specific values chosen for the results quoted in Tables 4,5 and 6 were:

i	u_i	v_i
1	.02218956	.4927502
2	.6122712	.7660227
3	.9657478	.4207648
4	.2368536	.8059282
5	.2453519	.6531745
6	.2086521	.1134470

$$
\begin{array}{lll}
7 & .3965626 & .6688034 \\
8 & .1859371 & .5827188
\end{array}
$$

In some cases u_i are then normalised such that,

$$k^e \sum_{i=1}^{k} u_i^g = d \qquad (A1)$$

where the constants e, g and d are selected to control the degree of difficulty.

Integral 1

$$\prod_{i=1}^{k} u_i \exp(-u_i x_i)$$

Over $[0,1]^k$.   Exact $= \prod_{i=1}^{k} [1 - \exp(-u_i)]$

Integral 2

$$\exp(\sum_{i=1}^{k} |x_i|)$$

Over $[-1,1]^k$.   Exact $= 2^k (e - 1)^k$

Integral 3

$$\exp(a \prod_{i=1}^{n} x_i) \quad , a = 0.3$$

Over $[0,1]^k$.   Exact $= \sum_{i=1}^{\infty} a^i /(i + 1)^k / i!$

Integral 4

$$\sum_{i=1}^{k} |x_i - .5|$$

Over $[0,1]^k$.   Exact $= k/4$

Integral 5

$$\cos(2\pi r + \sum_{i=1}^{k} u_i x_i) \quad , r = 0.3$$

Over $[0,1]^k$.

$u_1, \ldots, u_k$ are normalised in (A1) such that,

$$e = 1, \; g = 1, \; d = 25$$

$$\text{Exact} = 2^k \; \cos(2\pi r + 0.5 {\textstyle\sum_{i=1}^{k}} u_i) \; {\textstyle\prod_{i=1}^{k}} \sin(u_i/2)/u_i$$

Integral 6

$$\prod_{i=1}^{k} 1/[u_i^2 + (x_i - v_i)^2]$$

Over $[0,1]^k$.

$u_1, \ldots, u_k$ are normalised in (A1) such that,

$$e = 7/2, \; g = -2, \; d = 170$$

Over $[0,1]^k$. $\text{Exact} = {\textstyle\prod_{i=1}^{k}}[\tan^{-1}\{(1 - v_i)/u_i\} + \tan^{-1}\{v_i/u_i\}]/u_i$

Integral 7

$$\exp({\textstyle\sum_{i=1}^{k}} u_i x_i)$$

Over $[0,1]^k$.

$u_1, \ldots, u_k$ are normalised in (A1) such that,

$$e = 1, \; g = 1, \; d = 60$$

$$\text{Exact} = \prod_{i=1}^{k}[\exp(u_i) - 1]/u_i$$

Integral 8

$$1/(1 + {\textstyle\sum_{i=1}^{k}} u_i x_i)^{r+k} \qquad , \; r = 0.3$$

Over $[0,1]^k$.

$u_1, \ldots, u_k$ are normalised in (A1) such that,

$$e = 2, \; g = 1, \; d = 80$$

$$\text{Exact} = (1/A) \, (1/{\textstyle\prod_{i=1}^{k}} u_i) \sum_{i_1=0}^{1} \cdots \sum_{i_k=0}^{1} (-1)^{i_1 + \ldots + i_k}/(1 + {\textstyle\sum_{j=1}^{k}} u_j i_j)^r$$

with, $A = r(r+1)...(r+k-1)$.

Integral 9

$$\exp[-\sum_{i=1}^{k} u_i^2 (x_i - v_i)^2]$$

Over $[0,1]^k$.

$u_1,...,u_k$ normalised in (A1) such that,

$$e = 3/2, \quad g = 2, \quad d = 140$$

$$\text{Exact} = (\pi^{1/2}/2)^k \prod_{i=1}^{k} \{ \text{erf}[u_i(1 - v_i)] + \text{erf}[u_i v_i] \} .$$

INTERACTIVE NUMERICAL QUADRATURE

A.H. Stroud
Department of Mathematics
Texas A&M University
College Station, Texas   77843

ABSTRACT. This describes a collection of algorithms for numerical
quadrature that can be used interactively from a computer terminal.
By prompting, the system assists the user in entering an integral and
the other arguments needed to use an algorithm. Alternatively, each
algorithm can be used non-interactively in the usual way as a subroutine appended to a larger program.

1. INTRODUCTION

1.1. Origin

    Some years ago I learned about a computer system for data
analysis called the Computer-Assisted Data Analysis Monitor, or CADA,
[3], [4]. Important features of this system are:
    1- It makes available to the relatively inexpert user a large
        number of methods for the statistical analysis of data.
    2- It is made up of various blocks that can be overlaid or
        chained together as needed for a particular computation.
    3- It is interactive in the sense that it leads the user step by
        step through an analysis by prompting.
That system led me to think about something analogous for numerical
integration. This talk is about a system of quadrature algorithms
that has come out of this.
    In its present form this system contains only about forty
algorithms. It should not be considered as a finished system. A
finished system would have every known quadrature algorithm. The
present system is only a small example of what might be done.

1.2. Desirable Features

    One should keep in mind a basic difference in the nature of the
problems of data analysis and numerical integration.
    In analyzing data the output from an analysis is usually one or
more numbers. Usually these numbers are not needed as part of a
larger calculation. On the other hand with numerical integration the

291

P. Keast and G. Fairweather (eds.), Numerical Integration, 291–306.
© 1987 by D. Reidel Publishing Company.

end result that is being sought is not usually a table of values of
integrals. Usually, values of integrals are needed as part of a
larger calculation. It follows that, whereas a data analysis
algorithm can be a stand-alone program, an algorithm for numerical
integration should be a subroutine that can be appended to a larger
program.

Even though this is true, one may still want to use an
integration algorithm as a stand-alone program. One may just want to
see for oneself, or to show someone else, how a particular algorithm
works on particular integrals. After all, even an expert on one
algorithm may be inexpert on another.

With the above remarks in mind we can list some features that we
would want to have in an interactive system for integration:

1-   A large number of different algorithms, including algorithms
     for one-dimensional and for multi-dimensional integrals.

2-   A systematic scheme for classifying the algorithms.

3-   Each algorithm should be a self-contained subroutine that can
     be appended to another program.

4-   An interactive feature whereby each algorithm can be used as
     a stand-alone program. When used in this way one should be
     led by prompting through the process of calling up a
     particular algorithm and entering an integral.

5-   When used interactively one should be able to enter a
     subroutine that computes the integrand without having to
     compile and link.

6-   A manual that is available on-line to the user.

1.3.   Choice of Language

The first decision to be made is what language to use. One wants
a scientific language so one will probably end up selecting one of:

                FORTRAN,   PASCAL,   C,   BASIC.

Of course each of these is available in different versions. One
should not rule out BASIC as being too simple-minded; the Computer-
Assisted Data Analysis Monitor is written in BASIC. If one wants a
compiled language FORTRAN is still a good choice. If one wants an
interpreted language the only real choice is BASIC. The advantages
and disadvantages of these two languages are summarized in Table 1.

We have chosen interpreted BASIC, specifically BASIC-11 for the
Digital Equipment Corporation PDP-11 computers running under operating
system RT-11. This BASIC has the following features that are
important for an interactive system:

1-   One can halt a program, enter a subroutine from the terminal,
     and continue running without compiling. This is important
     for a system that requires a function as part of the data.

2-   One can segment a large program by
     a)   the CHAIN statement; and
     b)   The OVERLAY statement.

TABLE 1. Some Differences of FORTRAN and BASIC

FORTRAN (Compiled)	BASIC (Interpreted)
Advantages:	
1- Widely used in scientific work.	1- Widely used on small computers.
2- Has indepenent subroutines.	2- Subroutines can be appended at run-time because compiling and linking are not needed.
3- Relatively fast running because it is compiled.	
Disadvantages:	
1- Subroutines must be compiled and linked.	1- Subroutines are not independent.
	2- Relatively slow running because it is interpreted.

The CHAIN statement replaces an entire program running in
memory with another program from an external file. The
OVERLAY statement replaces only a segment of the running
program - usually a subroutine - with another program
segment. The CHAIN statement is available in most versions
of BASIC; the OVERLAY is less widely available. The value of
the OVERLAY will be discussed further in Section 3.2.
BASIC is far from an ideal language. Features that an ideal language
for our system should have will be discussed in Section 4.

2. DESCRIPTION OF THE SYSTEM

2.1. Getting Started

.The system is started by running the main program QUAD10. This
lists the Preface of the manual. Then one can list other pages in the
manual or one can call up one of the algorithms.
     Chapter 0 of the manual describes the general features of the
system. Chapter 1 describes the individual one-variable algorithms.
Chapter 2 describes two variables; etc.
     For one variable we write the integrals as

     INTEGRAL[an interval] W(X; P)*F(X; P) DX.

     A particular problem may involve several intervals that depend on
an integer I = 1, 2, ... Also, the problem may depend on one or
more sets of parameters

     P = (P(1), P(2), ...)

that depend on

     I = 1, 2, ... and J = 1, 2, ...

     For a particular problem one must enter, along with other
arguments:
     1- A subroutine that generates the limits of integration, given
         I. If the interval is the entire line this is not needed.
     2- A subroutine that generates P(1), P(2), ... given I and
         J.
     3- A subroutine that computes F(X; P) given X and P.
This is done in response to prompting.
     Each algorithm is specified by a set of four integers

     N, N1, N2, N3

where
     N is the dimension of the integral;
     N1 denotes the type of interval or type of region of
         integration;

N2   denotes the type of weight function or the type
     of integrand;
N3   indicates the particular algorithm under classification
     N, N1, N2.
For N = 1 the integers N1 and N2 have the following
possible values and meanings:

N1   Interval

0   Finite length interval [C, D]
1   [C, INF)
2   (-INF, INF) INF = Infinity

N2   Integrand

1   Constant W(X)
2   Classical W(X)
3   Other W(X)
4   Integral Transforms
5   Principal Value Integrals
6   Rapidly Oscillating Integrands
7   Tabular Data

A list of the algorithms that are available under a classification N,
N1, N2 is given on the page determined as follows. For example if
N, N1, N2 = 1, 0, 2, the page is 1.02. In other words the digit
before the decimal point is the dimension N; the first digit to the
right of the decimal point is N1; the next digit is N2.

2.2. A One-Dimensional Example

Suppose, for example, that one wishes to evaluate an integral
such as

$$\text{INTEGRAL}[0, 1] \ (1-X)^B * X^B * \cos(A*X) \ DX$$

for several values of the constants A and B. What sort of thing
might one find in this system pertaining to this?
Here we can assume that

$$W(X; P) = (1-X)^{P(1)} * X^{P(2)}, \quad P(1) = P(2) = B$$

$$F(X; P) = \cos(P(3)*X), \quad P(3) = A.$$

Since this W(X; P) is the classical Jacobi weight function we are
led to look under classification

$$N, N1, N2 \ = \ 1, \ 0, \ 2.$$

On Page 1.02 we find five algorithms listed:

N3	W(X)	ALGORITHM NAME
1	(D-X)^P(1)*(X-C)^P(2)	ERF-MIDPOINT
2	(D-X)^P(1)*(X-C)^P(2)	IMT-LEGENDRE
3	((D-X)*(X-C))^(-0.5)	GAUSS-CHEBYSHEV (1-ST)
4	((D-X)*(X-C))^0.5	GAUSS-CHEBYSHEV (2-ND)
5	((D-X)/(X-C))^0.5	GAUSS-CHEBYSHEV (3-RD)

If one is interested in learning more about algorithm

IMT-LEGENDRE (1, 0, 2, 2)

one can find a description of it on pages

1.0202,   1.0202-1   and   1.0202-2.

In essence, one finds there the following information:
To evaluate

$$\int_0^1 w(x)f(x) \, dx$$

we start by making a non-linear change of variable

$$x = \psi(u) = \int_0^u \phi(t) \, dt$$

where

$$\phi(t) = K \exp\left(\frac{-1}{t(1-t)}\right)$$

and where the constant K is defined by the condition that

$$\psi(1) = 1.$$

In other words

$$K = \left[\int_0^1 \exp\left(-1/(t-t^2)\right) \, dt \right]^{-1}$$

This transforms the original integral with respect to x into

$$\int_0^1 w(\psi(u))f(\psi(u)) \, \phi(u) \, du.$$

This integral is approximated by a μ-point Gauss-Legendre formula

$$x_{k,\mu}^*, \ A_{k,\mu}^* \quad k = 1, \ldots, \mu.$$

This amounts to approximating the original integral by

$$\sum_{k=1}^{\mu} A_{k,\mu} w(x_{k,\mu}) f(x_{k,\mu})$$

where

$$x_{k,\mu} = \psi(x_{k,\mu}^*)$$

$$A_{k,\mu} = A_{k,\mu}^* \phi(x_{k,\mu}^*)$$

The $x_{k,\mu}$ are symmetric with respect to the midpoint of the interval and bunch up very much near the endpoints. For example, for $\mu = 24$ the $x_{k,24}$ that lie in $0 \leq x \leq 0.5$ are listed in Table 2, together with the corresponding $A_{k,24}$.

TABLE 2. The points and coefficients
in a 24-point IMT-LEGENDRE formula.

k	$x_{k,24}$	$A_{k,24}$
1	.1006579E-183	.1077759E-180
2	.3430724E-36	.3141382E-34
3	.3864224E-15	.9511917E-14
4	.3241625E-08	.3292674E-07
5	.4949276E-05	.2597820E-04
6	.2839854E-03	.8830889E-03
7	.3484851E-02	.7019251E-02
8	.1857401E-01	.2572822E-01
9	.6008540E-01	.5953592E-01
10	.1406003	.1020033
11	.2628862	.1408630
12	.4170083	.1639412

Since the the smallest positive floating point number that can be used in BASIC-11 is about $1.0E{-}38$ we cannot use all 24 of these points. Here we discard

$$x_{1,24}, \quad x_{2,24}, \quad x_{23,24}, \quad x_{24,24}$$

and use the above approximation as a 20-point formula.

In general let us assume we start with a μ-point Gauss-Legendre formula and discard the $x_{k,\mu}$ that are closer than about $1.0E{-}35$ to an endpoint of $[0,1]$. Assume that M_μ points remain. The values of μ and M_μ that are available in this system are listed in Table 3.

For

$$\text{INTEGRAL}[0,1] \ (1-X)^{\wedge}B*X^{\wedge}B*\cos(A*X) \ DX \qquad (1)$$

this algorithm gives the results listed in Table 4.

TABLE 3. Available Values of μ and M_μ

M_μ	μ	M_μ	μ	M_μ	μ	M_μ	μ
8	8	20	24	40	46	120	140
10	12	22	26	50	58	140	164
12	14	24	28	60	70	160	188
14	16	26	30	80	92	180	210
16	18	28	32	100	116	200	234
18	20	30	34				

TABLE 4. Approximate Values of Integral (1) using the IMT-LEGENDRE
Algorithm

M_μ	B = -.5 A = 5	15	B = .5 5	15
10	.122308	.157126	-.124718	-.0399343
14	.121762	.280313	-.125109	-.000158041
20	.121775	.290032	-.125112	.00490069
40	.121776	.290041	-.125112	.00490945
60	.121776	.290041	-.125112	.00490945
Exact	.121776	.290040	-.125112	.00490946

2.3. Two and Three Dimensions

At the present time there is available essentially only one
algorithm for two dimensions and only one for three dimensions. In
each case this is the algorithm based on generalized product formulas.
We review what we mean by a 2-dimensional generalized product
formula.
Consider an iterated integral

$$\int_C^D \int_{C(x)}^{D(x)} w(x,y)f(x,y) \ dydx.$$

If w(x,y) is a product

$$w(x,y) = w_1(x)w_2(x,y)$$

the above integral can be written as

$$\int_C^D w_1(x)G(x) \ dx \qquad (2)$$

where

$$G(x) = \int_{C(x)}^{D(x)} w_2(x,y)f(x,y)\ dy. \tag{3}$$

Assume that we choose an integration formula that is appropriate for integral (2); let us write this formula as

$$\int_C^D w_1(x)G(x)\ dx \approx \sum_{j=1}^{M_1} A_j G(x_j) \tag{4}$$

$$G(x_j) = \int_{C(x_j)}^{D(x_j)} w_2(x_j, y)f(x_j, y)\ dy.$$

Also assume that we can find another integration formula that, when suitably transformed, is appropriate for approximating $G(x_j)$ for each x_j, $j = 1, \ldots, M_1$. Let us write this formula as

$$G(x_j) \approx \sum_{k=1}^{M_2} B_{j,k} f(x_j, y_{j,k}) \tag{5}$$

$$j = 1, \ldots, M_1.$$

Then the original iterated integral is approximated by the double sum

$$\sum_{j=1}^{M_1} \sum_{k=1}^{M_2} A_j B_{j,k} f(x_j, y_{j,k}) .$$

This approximation is called a generalized product formula. (See Davis & Rabinowitz [2; Sec. 5.6.1].)

Formula (4) for the x variable will be called the first factor of the generalized product and formula (5) for the y variable will be called the second factor.

The generalized product is considered to be only one algorithm in each dimension. The factors are arguments to be specified. Currently one has a choice of about 20 different formulas for each factor.

If the weight function in the integral for the second factor contains parameters, then these parameters are allowed to depend on x. Consider, for example, the integral

$$\int_1^2 \int_2^3 (3 - y)^{x-3/2}(y - 2)^{x-3/2} \cos(xy)\ dydx.$$

We can take

$$w_1(x) = 1, \quad F(x,y) = \cos(xy)$$

$$w_2(x,y) = (3 - y)^\alpha(y - 2)^\beta, \quad \alpha = \beta = x-3/2.$$

Since, for each x, $w_2(x,y)$ is a Jacobi weight function on the
interval [2,3], we can use an IMT–Legendre formula for the second
factor. The first factor can be, say, a Gauss–Legendre formula.
The above discussion generalizes to higher dimensions. However,
these approximations are currently available only in two and three
dimensions.

3. SYSTEM SEGMENTATION

3.1. Memory Organization

Here we describe how the system is organized by line numbers of
the BASIC programs.
Line numbers 1000–3999:
        This is the main program QUAD10. This program remains in
memory and is never overlaid. Its main functions are
        i)   to list a specified page from the manual; and
        ii) to call an algorithm to be run.
When calling an algorithm this program prompts for the input
arguments and overlays the required subroutines. These overlays
are
        i)   the output routine, line numbers 4000–5999; and
        ii) one or more overlays for the algorithm, line numbers
             6000–8799.
Line numbers 4000–5999:
        This is the output routine. Different types of algorithms
require different output routines. However, algorithms of
similar type share the same output routine. The output routine
takes care of calling the algorithm and printing the results.
This routine is overlaid as necessary.
Line numbers 6000–8799:
        This is the subroutine for the algorithm that is being
used. This subroutine can be appended to any BASIC program for
use apart from this system. For a one-dimensional algorithm this
subroutine consists of one overlay. Future versions of this
system will also have multi-dimensional algorithms that also
consist of one overlay. However, a generalized product formula
consists of one overlay for each factor as follows:

        Line numbers 6000–6699:
            The overlay for the first factor.

        Line numbers 6700–7399:
            The overlay for the second factor.

      ⹀ Line numbers 7400–8099:
            If N > 2, the overlay for the third fator.

A brief flow diagram for

Figure 1

System Flow Chart, Part 1.

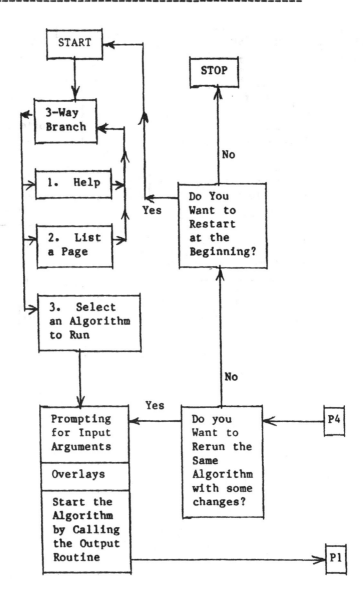

MAIN PROGRAM QUAD10

Figure 2

System Flow Chart, Part 2.

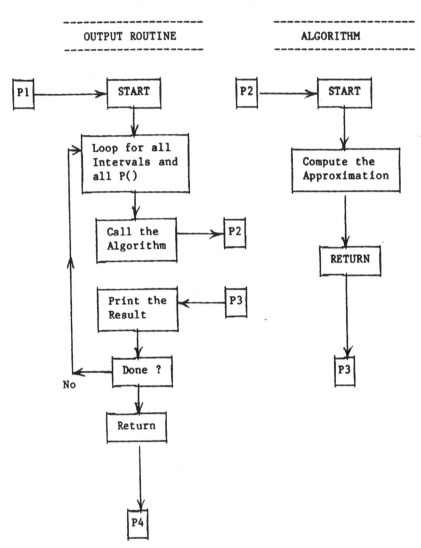

1-   main program QUAD10;
2-   an output routine; and
3-   an algorithm.
is given in Figures 1 and 2.

The program segments that reside in line numers 9100 to 9999
consist of subroutines that the user must enter as arguments for the
algorithm. These are as follows.

Line numbers 9100-9199:
> This is a subroutine that returns C and D, the endpoints
> of the interval of integration, given $I = 1, 2, \ldots$. If the
> algorithm is a generalized product then C and D are the
> endpoints of the interval on x.

Line numbers 9200-9299:
> For a generalized product algorithm this is a subroutine
> that returns $C(x)$ and $D(x)$, the endpoints of the interval on
> y, given x and I.

Line numbers 9300-9399:
> For a 3-dimensional generalized product this is a subroutine
> that returns the endpoints of the interval on z, given x, y
> and I.

Line numbers 9600-9799:
> A subroutine that returns the integrand F given its
> arguments.

Line numbers 9800-9999:
> A subroutine that, given I and J, returns

$$P(1), P(2), \ldots$$

the parameters in the integrand, if there are any.

The memory organization that we have just described is summarized
in Table 5.

TABLE 5.   Memory Organization Summary

Line Numbers	Program
1000-3999	Main program QUAD10.
4000-5999	Output routine.
6000-6699	First factor of generalized product.
6700-7399	Second factor of generalized product.
7400-8099	Third factor of generalized product.
9100-9199	Interval for first factor.
9200-9299	Interval for second factor.
9300-9399	Interval for third factor.
9600-9799	Integrand F.
9800-9999	Parameters P( ), if any.

3.2. Overlays

The importance of the OVERLAY statement can be seen from the above description of how memory is organized.
With only the CHAIN statement a one-dimensional algorithm, or a multi-dimensional algorithm that consists of only one subroutine, would be possible, but would be more cumbersome. In such a case one would need to have the output routine as part of the algorithm. This would be undesirable for two reasons:

1- It would be awkward to append the algorithm to a BASIC program for use apart from the system.

2- If a user found that, after running an algorithm he wanted to change some arguments and re-run it he would have to chain back to QUAD10 in order to input the arguments to be changed. One would then chain to the algorithm again. These chains would be very time consuming. With the OVERLAY no programs have to be reloaded.

Another restrictive feature of the CHAIN statement has to do with generalized product algorithms. Such an algorithm, as we now have it with the factors being arguments of the algorithm, seems impossible. It seems that such an algorithm would need to have its factors preassigned and fixed. Every combination of factors would be a separate algorithm. This would greatly limit the possibilities for generalized products.

4. AN IDEAL LANGUAGE

As mentioned above, there is no existing ideal language for this sytem. Here we mention some features that an ideal language would have.

An ideal language would be a cross between a compiled language such as FORTRAN, and an interpreted language such a BASIC. We will assume we are talking about a computer program that consists of a main routine and one or more subroutines. Each subroutine and the main routine will be called a program block.

An ideal language would be such that

1- Each block could be either compiled or interpreted; once debugged, an interpreted block could be compiled, if so desired.

2- Each subroutine would have independent variable names and FORTRAN-type arguments.

3- Each interpreted block could either
   a) be loaded into memory from an ASCII file in external storage; or
   b) be entered directly into memory from a terminal.

4- A running program could be stopped and modifications could be made to an interpreted block; the program could then be continued or restarted.

5- Each subroutine, either interpreted or compiled, could be appended to the main program before run-time or could be

overlaid at run-time with no linking.
The advantage that a language with these features would have for
our system is that the main parts namely

the main program QUAD10,
the output subroutines,
the subroutines for the algorithms,

could be compiled. The subroutines that are required as data for an
algorithm could be entered without compiling and linking as
interpreted subroutines, as is now the case.

An interpreted language has an advantage over a compiled language
that we have not yet mentioned, but which has a bearing on this
discussion. Coleman and Nasater [1] cite data on the man-hours
required to write and debug a large accounting program. The same
program was written by two groups of programmers; one group used a
compiled BASIC and the other an interpreted BASIC. It was expected
from the start that the compiled BASIC would take two to three times
as long to write as the interpreted. They were surprised to learn
that the compiled actually took seven to eight times as long!

It follows that a language with the above features would be
useful in writing many, if not all, large programs. A program block
that has been debugged could be compiled; a block that is not yet
debugged could be interpreted.

5. REMARKS

In the usual mode of operation of this system the arguments that
must be input to use an algorithm are entered in response to
prompting. This mode of operation is called the P-mode or the
prompting mode. If one uses this system a lot, the prompting may
become unnecessary and even a nuisance. Because of this, another mode
of operation is provided; it is called the Q-mode or the no-prompting
mode. In the Q-mode the main program QUAD10 is completely replaced by
another program QMOD10. One can use the CHAIN statement to transfer
from either of these modes to the other.

Whenever one writes a large program one becomes a little wiser.
Some things that I learned writing this system are:
1- How nice interpreted BASIC is.
2- How nice it is to have algorithms that do not require tables
    of constants.
3- Unfortunately, most of the better algorithms do require
    tables.
4- How nice the generalized product formulas are.

6. REFERENCES

[1] J. Coleman and D. Nasater, 'Interpretive Business BASIC with
     RMS-11K,' BASIC-SIG Newsletter, Digital Equipment Corp., August
     1983, pp. 76-103.

[2] P.J. Davis and P. Rabinowitz, Methods of Numerical Integraton,
     2nd edition, Academic Press, 1984.
[3] M.R. Novick, G.L. Isaacs and D.F. DeKeyrel, Computer-Assisted
     Data Analysis - 1977, Manual for the Computer-Assisted Data
     Analysis (CADA) Monitor, Iowa Testing Programs, Iowa City, Iowa,
     1977.
[4] M.R. Novick, G.L. Isaacs and D.F. DeKeyrel, 'Modularity,
     Readability, and Transportability of the Computer-Assisted Data
     Analysis (CADA) Monitor,' Proceedings of Computer Science and
     Statistics: Eleventh Annual Symposium on the Interface; A.R.
     Gallant & T.M. Gerig, editors; Institute of Statistics, North
     Carolina State University, 1978, pp. 176-180.

THE DESIGN OF A GENERIC QUADRATURE LIBRARY IN ADA

L.M.Delves
Centre for Mathematical Software Research
and Department of Statistics and Computational Mathematics

B.G.S Doman
Department of Statistics and Computational Mathematics

University of Liverpool, Liverpool, Merseyside L69 3BX
England

ABSTRACT We describe the design of a numerical quadrature library in
the programming language Ada. The design takes account of the ability
to write generic code in Ada, and we utilise this facility to provide
unusually powerful facilities with a relatively small number of
routines.

1. INTRODUCTION

There are a number of language facilities in Ada which are likely to be
of value in constructing general and easy-to-use scientific software.
One such facility is that of writing generic procedures, and, more
generally, generic packages. It is commonly accepted that generic
packages yield an appropriate mechanism for accessing the variable-
precision real facilities of Ada; but we have argued elsewhere [1] that
the generic mechanism is of much wider utility than this, and that its
wider applicability must not be ignored when libraries of mathematical
modules are being designed and constructed.

The argument rests on the observation that many standard numerical
algorithms are themselves generic in type. For example, a straightforward matrix-matrix multiply algorithm may be implemented to multiply
two real matrices (of any precision); two complex matrices; two block
matrices whose elements are themselves real matrices; or more generally, two matrices whose elements are elements of any set of objects
which can be added and multiplied to yield another element of the set.
The algorithm itself is independent of the particular set; and a generic implementation is possible which can be instantiated for any such
(computer representable) set, with the basic +, * operations being
imported into the generic package.

This example, though useful in its own right, is too simple to
show either the possible range of applicability of such a design
approach, or its possible difficulties. Two other algorithms were given
outline consideration in [1]:

P. Keast and G. Fairweather (eds.), Numerical Integration, 307–319.
© 1987 by D. Reidel Publishing Company.

1) Gauss Elimination for the solution of sets of linear equations.

2) Numerical Quadrature.

We are interested here in the latter. The discussion in [1] was restricted to routines in which the user specifies the number of points, and the integration rule, to be used ("non-automatic routines"). We here extend the discussion to cover also automatic routines, in which the user specifies the required accuracy but not the points to be used; and consider the design of a library quadrature chapter in which the Ada generic facility is used to provide a broad coverage of problems with as few routines as possible.

This work was carried out with EEC support as part of a pilot project (MAP750: Pilot Implementation of Numerical Libraries in Ada) designed to study the use of Ada in a traditional scientific library context; in particular, a serial computing environment is the main target. No attempt is therefore made here to utilise the real-time facilities of Ada, although the library being constructed will run in a realtime environment, and the error mechanism used (not discussed here) uses tasking to guarantee error flag integrity.

2. NON - AUTOMATIC ROUTINES

A numerical quadrature algorithm is one which makes the approximate replacement:

$$\int f(x)dx \;\Rightarrow\; \sum_{i=1}^{N} w_i f(x_i) \qquad\qquad (1)$$

We consider first the case when the value of N, and the rule to be used, are specified by the user; Gauss quadrature routines are commonly of this type. Most library routines for such problems assume explicitly that x is on the real axis; that $f(x)$ is real; and that so are the points and weights x_i, w_i. But other cases are quite common in practice. The function f may be complex; or vector valued; while x may also be complex (contour integrals) or vector valued (multiple integrals). The weights w_i are normally real but in principle need not be; and we can therefore distinguish three relevant mathematical sets in this problem:

1) The set $\{x\}$

2) The set $\{f(x)\}$

3) The set $\{w\}$

A generic quadrature package may attempt to be generic with respect to any or to all of the sets: how generic should we get? This must depend on how neat, and hence how easy to use, the resultant module is; and on whether efficiency is reduced in the process of going generic: we must balance the advantages against any disadvantages which we find in the process of generic module design.

The outline code given in [1], demonstrated that a module generic with respect to {f(x)}, is both possible and neat. We have pursued the details further, and have under development a completely generic Gauss Quadrature package in Ada; that is, one which is generic with respect to all three of the sets listed above. The resulting generic package contains a *single* quadrature routine, instantiations of which may be used to integrate both single- and multi-dimensional integrands, over a wide variety of regions; these include:

finite or infinite portions of the real axis.

contour and line integrals in the complex plane.

rectangular or curvilinear regions in q-dimensional space.

The integrands themselves may be of any (numeric) type; for example: Real (of any precision; complex; rational (given an Ada implementation of rational arithmetic); a vector of complex values (simultaneous integration of a set of functions); etc. Provision is also made for weights of arbitrary type.

The structure of the package is modelled after that in the Quadrature chapter D01 of the NAG Algol68 library; this chapter makes available a wide variety of Gauss rules, in a form in which they are available both to quadrature routines and for other purposes. The Ada implementation retains the main features of the Algol68 implementation, including the concept of library mapping procedures to widen the scope of the rules; it cannot follow the implementation in detail, since in Algol68 use was made of the facility to store procedures within structures; the lack of this facility significantly complicates the Ada code, with library (generic) packages replacing the structures in question. At the same time, we have as noted above addressed the generic nature of the underlying algorithm; the result is a set of facilities of significantly wider applicability than those in the Algol68 library.

3. ADA IMPLEMENTATION OF GENERIC QUADRATURE ROUTINES

It is of some interest to summarise the main features of the generic Gauss package. We start by introducing Private types for each of the three main types identified as part of the problem definition:

```
TYPE DOMAIN_TYPE IS PRIVATE;-- the set {x}
TYPE RANGE_TYPE  IS PRIVATE;    -- the set of values returned by
                                -- f(x)
TYPE WEIGHT_TYPE IS PRIVATE; -- the set {w}
```

In addition, we note that in practice, the points $\{x_i\}$ in (1) may not be from the set $\{x\}$; for example, a contour integral is normally performed using *real* points from a Gauss-Legendre (say) rule. Moreover, as in the Algol68 library, we distinguish between *basic* and *mapped* rules. Trivial examples of the difference are given by the set of Gauss-Legendre points and weights. These *basic* rules provide approximate integrals on the region [-1,1]. But the user will normally want to integrate over some region [a,b]; we refer to the Gauss-Legendre rule, adjusted to do this, as a *mapped* rule. Although the difference seems trivial in this case, it is important in treating the general case to retain this distinction. We therefore make explicit provision for storing and using a set of library or user-provided *mappings* from a basic region to a more general region; and provide for these maps to contain one or more *parameters*.

We are therefore led to the introduction of a further two basic types:

```
        TYPE POINT_TYPE IS PRIVATE;-- the set {x_i}
        TYPE PARAM_TYPE IS PRIVATE;-- parameters for the mappings
```

We can now define the structures needed to store a set of basic rules: we need to store a set of points and weights for a given N-point rule, and to arrange to store sets of these for various values of N. We introduce the declarations

```
TYPE NUM_SETS IS RANGE < >;
TYPE NUM_POINTS IS RANGE < >;
TYPE POINTS_VECTOR  IS ARRAY(NUM_POINTS) OF POINT_TYPE;
TYPE WEIGHTS_VECTOR IS ARRAY(NUM_POINTS) OF WEIGHT_TYPE;
TYPE PARAMS_VECTOR IS ARRAY(RANGE< >) OF PARAM_TYPE;

TYPE REF_POINTS IS ACCESS POINTS_VECTOR;
TYPE REF_WEIGHTS IS ACCESS WEIGHTS_VECTOR;

TYPE QUADRULE IS
RECORD
        terms : NATURAL;
        points : REF_POINTS;
        weights : REF_WEIGHTS;
END RECORD;

TYPE QUADRULE_VECTOR IS ARRAY(NUM_SETS) OF QUADRULE;

TYPE REF_QUADRULE_VECTOR IS ACCESS QUADRULE_VECTOR;
```

so that a single N-point rule will be stored in a QUADRULE, with sets of them in a QUADRULE_VECTOR. Note the use of pointers (ACCESS types) for efficiency; different mapped rules will use a given set of basic points and weights, and by using pointers we avoid copying what may be quite bulky sets of data.

We now introduce the mapped rules. Suppose that we have a mapping z: x = $x(z)$ which maps the region C to a region D for which we have a basic rule. Then we have

$$I = \int_C f(x)dx = \int_D F(z)J(z)dz \qquad (2)$$

where $F(z) = f(x(z))$ and $J(z)$ is the Jacobian of the map. Now, if an integration rule for D has points z_i, and weights w_i, we may approximate (2) as

$$I = \sum_i f(x(z_i))J(z_i)w_i$$

and this identifies

$$w'_i = w_i J(z_i)$$

$$x'_i = x(z_i)$$

as the weights and points of a *mapped rule* for the region C. The rule is defined by the choice of basic rule, together with the mapping function $x(z)$ and Jacobian $J(z)$ (for which an additional type declaration is needed). In Ada, these will be generic parameters to a Map package; we make individual values available via a procedure:

```
PROCEDURE mapping (t : POINT_TYPE;
                   params: PARAMS_TYPE;
                   x_map: OUT DOMAIN_TYPE;
                   x_jac: OUT JACOB_TYPE) IS < >;
```

A library of maps is intended to be provided, via such Mapping functions; these are used as parameters to the generic quadrature routine (package), along with the function to be integrated, and the necessary basic arithmetic operators needed to carry out the quadrature sum. The basic package contains the quadrature routine QUAD, and a routine GET_RULE which gives access to an individual mapped rule, for uses other than numerical integration; a skeletal version is:


```
WITH quadhead;--contains the basic type declarations outlined above

GENERIC

     TYPE DOMAIN_TYPE IS PRIVATE;
     TYPE RANGE_TYPE IS PRIVATE;
     TYPE POINT_TYPE IS PRIVATE;
     TYPE WEIGHT_TYPE IS PRIVATE;
     TYPE PARAMS_TYPE IS PRIVATE;
     PACKAGE localhead IS NEW quadhead(DOMAIN_TYPE, RANGE_TYPE,
                          POINT_TYPE, WEIGHT_TYPE, PARAMS_TYP
--the above is actually not legal Ada, but illustrates briefly
--what is wanted; the final version uses nested generics.
```

```
USE localhead;
WITH FUNCTION fun(x : DOMAIN_TYPE) RETURN RANGE_TYPE;
WITH FUNCTION map(T : POINT_TYPE; params: PARAM_VECTOR)
                                  RETURN MAP_TYPE IS < >;
WITH FUNCTION "*"(a : WEIGHT_TYPE; b : RANGE_TYPE)
                  RETURN RANGE_TYPE IS < >;
WITH FUNCTION "*"(a, b : WEIGHT_TYPE) RETURN WEIGHT_TYPE IS < >;

PACKAGE quadlib IS

    FUNCTION get_rule(n: NATURAL; rules: REF_QUADRULE_VECTOR)
                      RETURN QUADRULE;
    FUNCTION quad(n : NATURAL; rules: REF_QUADRULE_VECTOR;
                  params: PARAMS_VECTOR)
                  RETURN RANGE_TYPE;

END quadlib;
```

Note that the quadrature routine quad has as parameter a basic
integration rule (parameter rules); the mapping function which it
requires is provided via the generic parameter map. Note also the use
of overloaded versions of the operator "*". Ada handles such
overloading quite happily, and is even able, from the context, to
distinguish between two versions which have the same argument types but
return a different result type (human readers however are less good at
this).
    The code given above is illustrative only; the actual implementa-
tion uses a two-level generic structure to allow the production (by
instantiation of the outer level package) of a single generic quadrature routine for a given class of regions. See the code listings for
details.
    As an example of the use of these facilities, we consider the
"trivial" case: we want to integrate the function $4/(1+x^2)$ over an
interval $(0,1)$. The Gauss Legendre rule is appropriate; we assume that
a set of basic Gauss Legendre points and weights are stored in the
package LEGENDRE, and that this package also contains the function
legendre.map mapping the basic interval $(-1,1)$ onto the "user" interval
(a,b); a and b are the parameters to the map. Given that function
MY_FUN defines the integrand, and that One_Dim_Quadlib is an
instantiation of the outer level package for real, one dimensional
functions, the essential code needed to perform the integration with N
points is:

```
WITH One_Dim_Quadlib,Legendre; USE One_Dim_Quadlib;
-- Legendre is a package containing points, weights and a
-- standard mapping for one dimensional real intervals
PACKAGE MY_QUAD IS NEW Generic_Quad_Package(
                              FUN        = > MY_FUN;
                              MAP        = > Legendre.map);

    interval: PARAMS_VECTOR(1..2):=(0.0,1.0);
```

```
USE MY_QUAD;
ANS := QUAD(N,Legendre.rules,interval));
```

4. AUTOMATIC ROUTINES

Gauss quadrature rules are useful in a number of areas, of which numerical quadrature is only one. However, they do not satisfy every numerical quadrature need, and we look to provide a wider range of
facilities, including routines which return a result accurate (it is
hoped) to within a user-provided tolerance ("automatic routines").
There are a wide variety of these in the NAG FORTRAN library; again, we
hope to provide a more powerful set of facilities, with fewer routines,
by providing generic facilities. However, it is not easy to provide a
generic description of current automatic routines, at the same level as
is possible for non-automatic routines. The major point at which these
routines tend to differ from the Gauss routines discussed above, is in
the subdivision of the region, which most automatic routines undertake
when the quadrature proves difficult. A dimension-independent description of this process is required if routines of the same generality as
thse Gauss routines are to be provided; even within a one-dimensional
framework, the routines in the FORTRAN library differ widely in the way
they subdivide. At this stage in the project, therefore, we accept the
range of geometries which the available Fortran routines handle, limit-
ing generic generalisation of the geometry to the precision with which
it is described. Within this limitation, we have been developing
generic versions of a number of automatic routines, each of which will
have the range of the integrand as a generic parameter. Thus, as for
the Gauss routine, the user will be able to instantiate versions for
REAL (any precision); COMPLEX; VECTOR of REAL; etc, functions. The
routines which we have included in the initial chapter are based upon
the algorithms coded in the following NAG Algol68 or FORTRAN routines:

a) D01GAB :A routine requiring only the function on a pre-given point
set, based on an algorithm by Gill and Miller; an error estimate is
returned.

b) D01AGB: An automatic routine based on an underlying Clenshaw
   Curtis approach.

c) D01GBB: a multi-dimensional Monte Carlo routine, for hyper-rectangular regions; non-rectangular regions can be treated by pretending the
integrand is zero outside the given region.

d) D01ARB: a one-dimensional Romberg routine.

e) D01FDF: integration over an n- sphere by the method of Sag -
Szekeres; the routine can handle singularities on the surface, or at
the centre of, the sphere.

f) D01PAF: Integration over an n-simplex.

g) D01APF: a one-dimensional automatic integrator for functions with end-point algebraic singularities.

All of these are well-tested and current library routines; they have been chosen to give a reasonable coverage of problems while staying within the manpower available. A large number of other quadrature routines are available in the NAG library, including a set based on Quadpack; we expect to add further routines as effort permits. For those who know the NAG FORTRAN quadrature chapter, we note that the Gauss quadrature package outlined above provides also (more than) the functionality of the NAG FORTRAN routines:

D01AMF; D01BAF; D01BBF; D01FBF.

It would have been a relatively simple matter to directly translate the Fortran code into Ada. However, in order to make use of the generic facility, some re-designing is necessary. For example: the Fortran routines expect a real-valued integrand, and deal with a real error estimate. In the generic code, the type of the integrand is a generic parameter; but (we have assumed) the error estimate remains real. Hence, it is necessary to sort out within the code which variables remain real, and which become generic; and to provide conversions between these where necessary. While doing this, we have taken the opportunity to completely restructure the code, making use of Ada facilities not available in Fortran - such as recursion- to provide shorter and more readily comprehensible implementations of the under-lying algorithms. At the same time, we have made minor improvements in functionality: the user can specify the accuracy required in a more flexible and uniform manner than in the Fortran codes, and default values are provided for the accuracy and also for a number of other parameters. A restart facility as in the Fortran Monte Carlo routine, has also been provided for other routines.

5. DEVELOPMENT OF GENERIC AUTOMATIC ROUTINES

The automatic routines listed above cover a variety of problem areas; in designing generic versions, we have tried to retain uniformity of style, and, where possible, also of detail, in the user interfaces. The factors to be considered include the following:

5.1 Generic Parameters

In some aspects life is easier than for the Gauss routines. First, of course, we are not aiming to make the routines fully generic with respect to the set {x}, although the precision of x remains as a generic parameter. Second, in these routines the user does not need to see, or to know about, the choice of weights w_i or of integration points x_i; these are hardwired in for the given region. The user must however supply:

The type of the function values $f(x)$; we refer to this type as SCAL here.

The precision of x (or of its components, for multi-dimensional routines)

Operators to carry out the basic multiply and add operations needed by any numerical quadrature rule.

The function to be integrated. In Ada this has regrettably to be supplied as an instantiation parameter, and not, as in Fortran, as a parameter of the quadrature routine, since procedure parameters to procedures and functions are not allowed in Ada.

In addition, since automatic routines try to control the error achieved, they need an estimate of size for objects of type SCAL. In the general case (eg, vector vauled integrands) we treat the error estimate as representing the norm of the error; we therefore need an operator which will return a (real) norm for objects of type SCAL. We have used the operator ABS as a way of importing a norm for SCAL types; a suitable ABS must be available for each instantiation.

Finally, some algorithms use standard functions such as log, sqrt, for REAL objects; since the precision of REAL is left as a generic parameter these functions need to be imported too, although Ada allows default provision of such functions if suitable versions are directly visible at the time of instantiation.

5.2 Procedure Parameters

We have made particular efforts to keep the calling sequences of the routines as close as possible. Input parameters required include:

i) The region of integration. For the regions covered, a vector or matrix of REAL components can carry this information, but currently we use upper and lower bounds, which for multidimensional routines are vectors. More complicated domains can be described by a sequence of functions as in the Fortran routine D01FDF; and we plan to introduce general routines for non-product regions via the use of mappings as for the Gauss routines.

ii) The required accuracy: currently, a REAL value which represents an upper bound on the norm of the SCAL-valued error, and in addition an indication of the type of error criterion to be used. We provide default values for those users who want to see how much accuracy can be achieved for a given number of function evaluations. The accuracy criterion may currently be *relative* or *absolute*; we use an enumerated type:

TYPE ERRTYPE IS (relative, absolute);

with a default of absolute. Provision of a mixed criterion is under
consideration.

iii) The allowed minimum and maximum number of function evaluations:
default values are again provided.

iv) A parameter to set, or not set, a restart facility. Such a facility
may save having to redo the first part of a calculation, if an initial
trial ended with the maximum number of function evaluations. In such a
case, a restart needs information which depends strongly on the algorithm; we can however hide the detail needed from the user, with the
routine generating its own storage and hooking on to a user provided
parameter to retain it between calls.

5.3 Termination Criteria

We also try to make these uniform across routines. A routine will run
until the minimum mumber of function evaluations has been made, and
then continue until one of the following occurs:

i) The requested relative or absolute accuracy has been achieved.

ii) The default relative or absolute accuracy has been achieved. This
can happen if an inappropriate type of error estimate has been demand-
ed: for example, a relative acccuracy criterion on a function with
integral close to zero but large positive and negative values.

iii) The error estimate starts increasing - this can occur due to a
buildup of rounding errors.

iv) The maximum number of function evaluations has been achieved. The
algorithms distinguish two cases:

a) The integral is converging, but was harder than the user guessed:
The restart facility is intended to help this case. We do not flag an
error; the number of function evaluations, and the current accuracy
estimate, are returned normally.

b) The integrand has a singularity, which may or may not be integrable.

We assume it is helpful to warn the user of a possibly non-integrable
singularity. It is also helpful to the user if we can distinguish
integrable from non-integrable singularities. In a one dimensional
integral a simple but apparently new test (see Appendix A) can pick out
a singularity of the form $|x-a|^{-b}$ for $b \geq 1$, for values of a anywhere in
the domain other than at an integration point. The test will also flag
a singularity of the form $|x-a|^{-b}$ for $0 < b < 1$, or $\log|x-a|$, if a is
sufficiently close to a function evaluation point. We have utilised the
test in the one dimensional routines: the finding of an apparent singularity is treated as a soft error, so that the user can use the
restart facility to force an integral estimate if he believes the

singularity to be integrable. For users who worry about the possibility of false alarms: the test *can* signal a singularity if a finite but narrow peak occurs in the integrand, of width less than the minimum point spacing used; the alarm should then be used as an indication that too few points were insisted upon, and the calculation repeated with a larger minimum point count parameter.

c) A function overflow occurs.
This may stem from either an integrable, or a non-integrable, singularity. Again, we treat the event as a soft error so that the user who believes his integrand is integrable can then force continuation, in which case the contribution from the singularity is ignored (a standard, if crude, method of treating such integrable singularities).

d) The input data is plain faulty: for example, the dimensionality information associated with the region and with the function, is inconsistent. In such cases we raise a hard error and give in.

5.4 *Output Information*

Output from the routines will be via an object of standard type RESULT - a record most of whose components are identical in meaning for all routines. The fields of the record include:

i) The best available estimate of the integral

ii) An estimate of the error.

iii) The error estimate type: absolute or relative

iv) The number of function evaluations used

v) An error code

vi) A routine-dependent field: a sub-record containing restart information.

6. CURRENT STATUS OF THE CHAPTER

The current status of this work is as follows:

6.1 *Gauss Package*
This generic package is now complete, together with an associated library of points and weights and of mappings. Instantiations for real, complex, and vector valued functions, and for real, complex (contour) and two-dimensional regions, have been made, demonstrating the usefulness of the overall generic approach.

6.2 *Gill Miller and Clenshaw Curtis routines*
Code for a semi-generic version of these is available; by semi-generic, we mean that the code is formally written in terms of type SCAL for the range of the integrand, but does not use the generic facility and has

only been tested for SUBTYPE SCAL is FLOAT: the compiler under which
they were written, does not support generics. It will be necessary to
go back in due course and complete these routines.

6.3 Romberg routine
This is complete, in its generic form, and a FLOAT instantiation has
been well tested. Currently only a skeleton error mechanism is in; and
other instantiations need testing too.

Compared with the original FORTRAN/Algol68 routine, considerable work
has gone into this routine to include: generics; a restart mechanism;
the facility to detect and flag singular integrals. It has also been
totally re-structured.

6.4 Monte Carlo routine
This has also been totally re-structured, compared with the original
quite unstructured code. The routine is not simple, due to its
stratification strategy, and in particular the original subdivision
algorithm has no totally direct generic counterpart. We have therefore
modified this part of the algorithm. The revised, generic routine is
now complete, including restart facilities; at the time of writing it
is still being de-bugged.

Other Routines
Multi-dimensional Sag-Szekeres and Simplex routines, and a one-dimensional automatic routine for functions with endpoint singularities,
exist in prototype form.

Acknowledgements

The Gauss quadrature package was designed and implemented in collabora-
tion with C.J.Pursglove and G Howard; part of the work on automatic
routines was carried out by S Gould. We are grateful to these, and to
members of the Ada Europe Numerics Working Group, for useful
discussions. The work was partly funded by research grant MAP750 from
the EEC.

REFERENCE

L.M.Delves and C.Pursglove, The use of Generic Facilities in Ada
Scientific Subroutine Libraries, Working Paper A-ENWG2.1, Ada Europe
Numerics Working Group, 1984; to be published.

Appendix: Test for Singularity

We consider the situation in which a singular integrand is supplied by
the user; then the quadrature rule is likely to sample $f(x)$ at
positions at which $|f(x)|$ is large. Moreover, near such positions, $f(x)$
will vary rapidly.
Let $f_m, x_m, f_n,$ and x_n be the values and positions of the largest
and next largest evaluations of $|f(x)|$. We shall assume that x_m and x_n

are adjacent quadrature points, as is likely for a singularity of the form $|x-a|^{-\alpha}$. To simplify the algebra, we also suppose without loss of generality that $x_m = 0$ and $x_n > 0$; and we assume that the singularity, if it exists, lies between 0 and x_n. This last assumption may not be satisfied automatically; however, we may check it by evaluating $|f(-x_n)|$

If this is greater than f_m, we replace f_m by $|f(-x_n)|$ and f_n by f_m.

Suppose then that $x_m=0$ and $x_n>0$. Then since $f_m > f_n$, $0<\alpha<x_n/2$.

If we are close to a singularity of the form $|f(x)| = |x-a|^{-\alpha}$, and $0 < a < x_n/3$, then

$$(x_n - a)/a > 2;\ \text{and}$$

$$f_m/f_n > 2^{\alpha}.$$

Similarly, if $x_n/3 < a < x_n/2$, then

$$a/(x_n/2 - a) > 2;\ \text{and}$$

$$|f(x_n/2)|/f_m > 2^{\alpha}$$

The singularity detection algorithm thus consists firstly of finding points x_m and x_n which bracket any singularity, and evaluating f_m and f_n as above. If $f_m > 2*f_n$, a singularity is flagged. If not, then $|f((x_m + x_n)/2)|$ is evaluated; if this is greater than $2*f_m$, a singularity is flagged.

This test will normally detect a functional behaviour of the form $|x-a|^{-\alpha}$, $\alpha > 0$. It will pick out such a singularity for a anywhere in the range of integration (including near to the endpoints) for $\alpha \geq 1$, but will also be triggered by a logarithmic singularity of the form $\log|x-a|$, or a power law singularity with $0 < \alpha < 1$ if a is sufficiently close to a function evaluation point.

THE INTEGRATION OF THE MULTIVARIATE NORMAL DENSITY FUNCTION FOR THE TRIANGULAR METHOD

John A. Kapenga [1]
Computer Science Department
Western Michigan University
Kalamazoo, MI 49008, U.S.A.

Elise de Doncker [1]
Computer Science Department
Western Michigan University
Kalamazoo, MI. 49008, U.S.A.

Kenneth Mullen
Department of Mathematics and Statistics
University of Guelph
Guelph, Ontario, N1G 2W1, Canada

Daniel M. Ennis
Philip Morris Research Center
Commerce Road
Richmond, Virginia, 23261

ABSTRACT. The model involved in the triangular method is presented, which leads to the need for evaluating a multidimensional integral of the multidimensional normal density function over an irregular region. Work done on the numerical evaluation of this integral is discussed.

1. INTRODUCTION

The triangular method is widely used in the analysis of sensory data. In each trial three stimuli are selected, two, S_1 and S_2, randomly chosen from one stimulus set X, and one, S, from another, Y. Given the three stimuli a subject is instructed to select the one that is perceptually different. The trial is repeated to collect data for analysis.

Theory for the triangular method was first developed for situations in which one assumes a univariate sensory response ([5,16]). The assumption of a multivariate sensory response leads to far more reliable results, for both estimation and hypothesis testing ([3,4,11]). The use of a multivariate model, with n sensory dimensions, requires the evaluation of a multidimensional integral of dimension $2n$. Furthermore, because of the number of parameters involved, a library program for the evaluation of the integral is desirable.

Under basic normality assumptions, the critical values for the multivariate model can be expressed in terms of a multidimensional integral, of dimension $2n$, of the multivariate normal density function. The $2n$-fold integral has boundaries of positive and negative infinity on the outer n limits. The inner n-fold integral is over the difference of two n-spheres, which depend on the outer n variables of integration.

---

[1]Sponsored in part by a grant from STW (the Netherlands) under the Megabit program.

P. Keast and G. Fairweather (eds.), Numerical Integration, 321–328.

In evaluating this integral a translation, a Householder transformation and a change to spherical coordinates, followed by a transformation on the r coordinate is done to the inner n-fold integral (The translation and Householder transformations both depend on the values of the outer n variables of integration). This takes the integral into one with constant limits of integration. For several reasons the resulting integral is still unsuited for numerical evaluation. To help overcome these problems the outer n-fold integral is changed to spherical coordinates and transformations to smooth out the integrand are done.

2. THE ORIGIN OF THE INTEGRAL

The following assumptions, made in Ennis and Mullen [4], are a generalization of the univariate assumptions of Frijters [5].

1. The stimuli S_1, S_2 and S give rise to corresponding sensory values $\bar{s}_1$, $\bar{s}_2$ and $\bar{s}$, which are vectors of size n. The momentary sensory values are assumed to be random variables, independently distributed, with $\bar{s}_1$ and $\bar{s}_2$ having density function $f(x)$ and $\bar{s}$ having density function $f(y)$.

2. The probability density function $f(x)$ and $f(y)$ are multivariate normal, with means $\bar{\mu}_x = (\mu_{x1}, \mu_{x2}, \cdots, \mu_{xn})^\tau$ and $\bar{\mu}_y = (\mu_{y1}, \mu_{y2}, \cdots, \mu_{yn})^\tau$ and variance-covariance matrices V_x and V_y respectively. If $\bar{\mu}_x$ and $\bar{\mu}_y$ are in standard units the distance between them is the Euclidian norm.

3. In a particular trial, a correct response will be obtained if

$$\|\bar{s}_1 - \bar{s}_2\| \le \|\bar{s}_1 - \bar{s}\| \quad \text{and} \quad \|\bar{s}_1 - \bar{s}_2\| \le \|\bar{s}_2 - \bar{s}\|$$

for triangles composed of S_1, S_2 and S.

4. There are no response–preferences due to spatial or temporal positions of the three alternate stimuli.

Let the vectors $\bar{u}$, $\bar{v}$ and $\bar{z}$ be defined by

$$\bar{u} = \bar{s}_1 - \bar{s}_2, \tag{1}$$
$$\bar{v} = \bar{s}_1 - \bar{s} \quad \text{and}$$
$$\bar{z}^\tau = (u_1, u_2, \cdots, u_n, v_1, v_2, \cdots, v_n)^\tau.$$

From standard multivariate results ([7]), $\bar{z}$ has a multivariate normal distribution with mean

$$\bar{\mu}^\tau = (0, 0, \cdots, 0, \mu_{x1} - \mu_{y1}, \mu_{x2} - \mu_{y2}, \cdots, \mu_{xn} - \mu_{yn})^\tau. \tag{2}$$

and variance-covariance matrix

$$V = \begin{bmatrix} V_1 & V_2 \\ V_2 & V_3 \end{bmatrix}. \tag{3}$$

where $V_1 = 2V_x, V_2 = V_x$ and $V_3 = V_x + V_y$.

If P_c is the probability of a correct decision then $P_c = 2P_1$, where

$$P_1 = \Pr[\, \|\bar{u}\| < \|\bar{v}\| \quad \text{for} \quad \|\bar{v}\| < \|\bar{u} - \bar{v}\| \,]. \tag{4}$$

This probability may be expressed as the integral

$$P_1 = \int_{R^n} \int_D f(\bar{u}, \bar{v})\, d\bar{u}\, d\bar{v}, \tag{5}$$

where f is the function

$$f(\bar{u}, \bar{v}) = \frac{\exp[-1/2(\bar{z} - \bar{\mu})^\tau V^{-1}(\bar{z} - \bar{\mu})]}{(2\pi)^n |V|^{1/2}}, \tag{6}$$

$$d\bar{u} = du_1\, du_2 \cdots du_n,$$

$$d\bar{v} = dv_1\, dv_2 \cdots dv_n$$

and

$$\mathcal{D} = \mathcal{S}_n(\bar{0}, \|\bar{v}\|) - \mathcal{S}_n(\bar{v}, \|\bar{v}\|) \tag{7}$$

is the set difference of the two hyperspheres in $\bar{u}$,

$$\|\bar{u}\| < \|\bar{v}\| \quad \text{and} \quad \|\bar{u} - \bar{v}\| < \|\bar{v}\|. \tag{8}$$

The value P_c can also be written somewhat more simply as a $3n$-fold integral of the joint probability density function of $\bar{u}$, $\bar{v}$ and $\bar{w}$ where $\bar{w} = \bar{s}_2 - \bar{s}$. The outer $2n$-fold integral for $d\bar{v}$ and $d\bar{w}$ is over R^{2n} and the inner n-fold integral for $d\bar{u}$ is over the sphere $\|\bar{u}\| < \min(\|\bar{v}\|, \|\bar{w}\|)$. However, with regard to the numerical evaluation of the integral, the difficulty of the higher dimensionality does not offset the simplification.

3. INITIAL CONSIDERATIONS

The initial investigations of this problem used Monte–Carlo methods to approximate P_c directly. These roughly correspond to Monte–Carlo integration. This approach worked somewhat better than several crude Monte–Carlo integration methods tested.

The integrand in (5) was also given a zero extension to spherical and rectangular regions. These extended functions were integrated by several integrators. The performance was slightly better than that of the Monte–Carlo integration procedures.

In general these methods required several hundred thousand function evaluations to get close to 2 digits of accuracy. They have the advantage of being very easy to implement and serve as convenient checks on more complex procedures during development. It should also be noted that the evaluation of the integrand is far faster than the evaluation of the integrand after the transformations introduced in the next section. For this reason the raw number of function evaluations is not a totally fair value on which to base a comparison of these methods.

The use of several variance reduction methods was considered and some testing done (e. g. [13]). Though not a totally dead end, these were not pursued.

4. REDUCTION TO CONSTANT LIMITS

To apply one of the standard multiple dimension integration methods (e. g. [6,12,14,15]) the first step is to transform the integral (5) into an integral with constant finite limits of integration. The following series of transformations accomplishes this for the inner n-fold integral. In all cases the full Jacobian of $\bar{z}$, rather then just the Jacobian of $\bar{u}$, meets the conditions for a valid transformation with respect to the indicated limits.

All the following transformations are on $\bar{u}$ and are functions of $\bar{v}$. The Jacobians will be denoted as $|J|$. For each transformation a prime (') will be added to indicate the new variable.

1. The vector $\bar{v}$ in $\bar{u}$-space is translated to $\bar{0}$ by

$$\bar{u} = \bar{u}' - \bar{v} \quad \text{with} \quad |J'| = 1. \tag{9}$$

2. The vector $-\bar{v}$ in $\bar{u}$-space is taken by an orthogonal transformation onto the $-u_n$ axis by

$$\bar{u}' = T\bar{u}'' \quad \text{with} \quad |J''| = 1. \tag{10}$$

Two general types of transformations T were considered. Givens rotations may be built up to zero out $u_1', u_2', \cdots, u_{n-1}'$ in that order or, a single Householder transformation may be applied ([8]). Formally there is no difference between the two approaches with regard to the goal stated above. However, there are $n - 2$ degrees of freedom with respect to rotations alone in choosing T, and these choices are important from a numerical standpoint, which will be discussed in the next section.

3. The inner n-fold integral is transformed into spherical coordinates by

$$
\begin{aligned}
u_n'' &= r\cos\theta_1, & (11) \\
u_{n-1}'' &= r\sin\theta_1\cos\theta_2, \\
u_{n-2}'' &= r\sin\theta_1\sin\theta_2, \\
&\cdots \\
u_2'' &= r\sin\theta_1\sin\theta_2 \cdots \sin\theta_{n-2}\cos\theta_{n-1}, \\
u_1'' &= r\sin\theta_1\sin\theta_2 \cdots \sin\theta_{n-2}\sin\theta_{n-1}, & (12)
\end{aligned}
$$

where

$$|J'''| = (-1)^{n(n-1)/2} r^{n-1} \sin^{n-2}\theta_1 \sin^{n-3}\theta_2 \cdots \sin\theta_{n-2}. \tag{13}$$

4. Finally, all the limits of integration except those on r are constants. the last transformation remedies this. Let

$$r = \|\bar{v}\|(1 + (2\cos\theta_1 - 1)r') \quad \text{with} \quad |J''''| = \|\bar{v}\|(2\cos\theta_1). \tag{14}$$

n a	1	2	3	4	5	6	7	8
2	527	527	527	527	527	527	527	527
3	561	561	1947	2079	2541	3729	3795	3267
4	513	1425	9519	14877	12597	26391	21489	1083
5	651	4557	56079	99789	72633	99789	99789	99789
6	745	13857	99681	99681	99681	99681	99681	99681
7	2651	723	99533	99533	99533	99533	99533	99533
8	12431	99047	99047	99047	99047	99047	99047	99047

TABLE 1. MINPTS for the integral $I(n, a)$ with $EPS = .01$.

At this point let $\bar{t}^\tau = (r', \theta_1, \theta_2, \cdots, \theta_{n-1})^\tau$ and $d\bar{t} = dr'\, d\theta_1\, d\theta_2 \cdots d\theta_{n-1}$. The integral (5) is now in the form

$$P_1 = \int_{R^n} \int_0^{2\pi} \int_0^{\pi} \cdots \int_0^{\pi} \int_0^{\pi/3} \int_0^1 f(g(\bar{t}, \bar{v}), \bar{v}) \frac{|J'''|}{r^{n-1}} \times \qquad (15)$$
$$\times (1 + (2\cos\theta_1 - 1)r')^{n-1} |J''''| \, d\bar{t}\, d\bar{v},$$

where g is the accumulated transformation $\bar{u} = g(\bar{t}, \bar{v})$. The limit $\pi/3$ on the second inner integral follows from (7) and was the critical observation in the selection of this sequence of transformations.

5. NUMERICAL CONSIDERATIONS

Consideration of a singular value decomposition, SVD, of V in (6) indicates that if the outer limits of integration in (5) are set at (-8,8) then at least 7 digits of accuracy should result (for $\bar{\mu} = 0$). Tests indicate that a range of (-5,5) is too small, affecting the second digit of the result in some cases.

Initially, to find out how intrinsically difficult the integral (5) is to calculate, a simpler form can be considered, so that the series of transformations done in the last section is not a concern. The simplest candidate is

$$I(n, a) = \int_{-8}^8 \int_{-8}^8 \cdots \int_{-8}^8 e^{(-1/2 \sum_{i=1}^n x_i^2)}\, dx_1\, dx_2 \cdots dx_n. \qquad (16)$$

Again by applying an SVD to V in (6), the integral (5) can be seen to be a somewhat more difficult problem then calculating $I(2n, 8)$. Values for n desired are 3-10. Several integrators were compared on this problem with similar results.

TABLE 1 is a summary of the results from ADAPT [6,12] showing the number of function evaluations used. The requested relative error was .01 , the minimum number of function evaluations was set to 100,000 . Single and double precision versions of ADAPT gave similar results.

The entries in TABLE 1 which exceed 99,000 function evaluations all correspond to tests which returned an error indicator, $IFAIL \neq 0$. As expected, the actual relative errors were smallest for trials in the upper left and largest for trials in the lower right portions of the table. In most cases of the evaluation of $I(n, a)$ and other related integrals during this work, the estimated relative error returned by ADAPT was found dependable.

It can be seen that the problem as stated causes concern. The evaluation of the integral (15), with $\bar{\mu} = 0$ and uncorrelated x and y, behaved much like that of (16). Of course a change to spherical coordinates in (16) effectively reduces the problem to a one dimensional integral, which is easily handled. The same transformation cannot reduce either of the integrals (5) or (15) into lower dimensional integrals. However, much of the same effect can be had.

By carefully selecting the transformation T in the previous section the inner n-fold integral shows a great deal of spherical symmetry with respect to the outer variables, $\bar{v}$. Thus a change to spherical coordinates in the outer n-fold integral can be expected to have a beneficial effect. Call these transformed outer variables

$$\bar{t}_0^T = (r_0, \psi_1, \psi_2, \cdots, \psi_{n-1})^T \quad \text{with Jacobian} \quad |J_0|. \tag{17}$$

Since r_0 is positive we have that $\|\bar{v}\|$, which appears in the product of the Jacobians in (15), is replaced by r_0. This gives a factor of r_0^{2n-1} in the integrand after the transformation.

The selection of the integration range for the outer n-fold integral is important. The fact that the ratio of the n-cube volume to n-sphere volume increases quickly with n means that selecting the r_0 range as $(0,8)$ gives the transformed integral an immediate advantage over (15) with the outer n limits set at (-8,8).

The choice of the limits of integration on the ψ_is is also worth mentioning. The inner n-fold integral can exhibit irregular behavior as a function of $\bar{v}$ for $\bar{v}$ near the negative v_n axis. If the outer limits of integration are chosen so that the negative v_n axis is transformed to the outer limits of the integration range this behavior is smoothed out.

An adaptive integrator will find the inner n-fold integral is 0 for $r_0 = 0$ and gets exponentially small as r_0 increases. The behavior is very regular with respect to the ψ coordinates, so that subdivisions are done mainly in the r_0 direction.

One additional transformation which has proved useful is a translation of the outer integral before changing to spherical coordinates. This places $\bar{\mu}/2$ at the origin in an attempt to keep the limits of integration for r_0 small and improve the symmetry.

A careful selection of Givens transformations has resulted in a successful T transformation. The more efficient Householder transformation produces similar results for $\bar{\mu} = 0$, but does not function well as $\bar{\mu}$ increases.

At the present time our program evaluates P_1 with about 4,000 function evaluations when $n = 3$, a 6-fold integral, and 34,000 evaluations when $n = 4$. ADAPT is used with a relative error request of .01. The results returned usually have a relative error between .002 and .0002. The performance degrades gracefully as $\bar{\mu}$ varies from zero.

6. FURTHER WORK AND SOME TOOLS

The product of the Jacobians in (15) contains polynomial factors which must be smoothed out before any additional improvement can be expected. Simple integrands with the same behavior as the product of the Jacobians, used in the same manner as $I(n, a)$, have shown this point. This problem can be treated in a straightforward manner for the next few n s.

The behavior of the Householder transformations is under study. There are also very general adjustments which can be made when $\bar{\mu}$ becomes large since then P_c rapidly approaches zero.

Currently a QR-decomposition of V is computed which is used in evaluating the quadratic form in the integrand's exponential as well as providing $|V|$ and a condition check ([2]). The need for going to an SVD when the V matrix takes on an unusual structure has not been fully investigated. There are clear advantages in efficiency and stability; however, at this point it is not clear how important these will be.

The question of the use of an adaptive integrator in a project such as this can be raised([10]). Our current feeling is that the transformations being used are the same sort which are likely to make a non-adaptive method work and if enough smoothing is done a fixed rule might be advantageous at some later point.

As this work was being done the question of what was really happening often arose. Thus,·thought was given to the question "What tools would help indicate where the integrator was having trouble and what transformations might help?". For $1D$ integration such a tool was provided by Kahaner and Wyman [9] in their GLAQ program. In higher dimensions the need for such aids is even more evident. Yet, the nature of what such an aid should be is not easy to define.

Generally, if the points in R^n and the function values an adaptive program (not just an integrator) asks for are saved, then they should be able to be analyzed by a post-processor, hopefully using interactive graphics. Generally, the density of the points in a region is a indication of the difficulty of the region and any transformation which gives these points a more uniform distribution should be considered as likely to improve the performance of the program; as should transformations which move the variations so that they tend to be parallel to the directions the adaptive program divides in.

A system for viewing sectioned projections of the points saved from a run of an adaptive procedure, with color indicating function magnitude, has been implemented on a microcomputer. We expect a full system with built in transformations and measures of variation could be implemented as an effective tool for improving the performance of certain multidimensional programs. We plan to actively pursue this development.

7. REFERENCES

1. DAVIS, P.J. and RABINOWITZ, P.: *Methods of Numerical Integration* , Academic Press, (1984).

2. DONGARRA, J.J., BUNCH, J.R., MOLER, C.B. and STEWART, G.W.: *LINPACK Users Guide*, SIAM, (1979).

3. ENNIS, D.M. and MULLEN, K.: 'The effect of dimensionality on results from the triangular method', *Chemical Senses*, **10**, 4 (1985), 605-608.

4. ENNIS, D.M. and MULLEN, K.: 'A multivariate model for discrimination methods', *J. Math. Psych.*, **30**, 4 (1987), 206-219.

5. FRIJTERS, J.E.R.: 'Variation of the triangular method and the relationship of its unidimensional probabilistic models to three alternative forced choice-signal detection theory models', *Br. J. Math. Stat. Phych.*, **32** (1979), 229-241.

6. GENZ, A.C. and MALIK, A.A.: 'Remarks on Algorithm 006: an adaptive algorithm for numerical integration over an N-dimensional rectangular region', *J. Comp. and Appl. Math.*, **6**, 4 (1980), 295-301.

7. GRAYBILL, F.A.: *Theory and Application of the Linear Model*, North Scituate, Mass. Duxbury Press, 1976.

8. GOLUB, G.H. and VAN LOAN, C.F.: *Matrix Computations*, Baltimore, Maryland, The Johns Hopkins Univ. Press, 1983.

9. KAHANER, D. and WYMAN, W.: 'Mathematical software in Basic - - evaluation of definite integrals', *IEEE MICRO*, Oct., (1983).

10. LYNESS, J.N.: 'When not to use an automatic quadrature rule', *SIAM Review*, **25** (1983), 63-87.

11. MULLEN, K. and ENNIS, D.M.: 'Mathematical formulation of multivariate psychological models for the triangular and duo-trio methods', *Psychometrika*, in press.

12. NAG: *NAG FORTRAN Library Manual Mark 11*, Numerical Algorithms Group, chapter D01, (1984).

13. PATTERSON, T.N.L.: 'On the construction of a practical Ermakov-Zolotukhin variance reducing Monte-Carlo multiple integrator and interpolator', this conference.

14. SAG, T.W. and SZEKERES, G.: 'Numerical evaluation of high-dimensional integrals', *Math. Comp.*, **18**, 86 (1964), 245-253.

15. STROUD A.H.: *Approximate Calculation of Multiple Integrals*, Englewood Cliffs, New Jersey, Prentice Hall, 1971.

16. URA, S.: 'Pair, triangle and duo-trio tests', *Rep. Stat. Appl. Res., JUSE*, **7** (1960), 107-119.

NUMERICAL INTEGRATION IN SCALAR WAVE SCATTERING, WITH APPLICATION TO ACOUSTIC SCATTERING BY FISH

Kenneth G. Foote
Institute of Marine Research
P.O. Box 1870
5011 Bergen
Norway

ABSTRACT

Only a relatively few problems in scalar wave scattering allow exact or near-exact solution. These generally involve shapes with a high degree of symmetry or wavelengths which are quite long or quite short compared to the characteristic scatterer dimensions. The bulk of interesting problems involves integrals which must be numerically addressed.

The essential problem is to solve the scalar wave equation in the exterior of a finite surface in a homogeneous and infinite medium. Equivalently, it is desired to solve the Helmholtz equation,

$$\left(\nabla^2 + k^2 \right) \psi = 0 \ ,$$

for the scalar field ψ, given its definition on a finite surface S. The wavenumber k is generally complex.

For present purposes it is sufficient to assume that ψ vanishes on S. Three solutions to this so-called homogeneous Dirichlet problem are described: those by Kirchhoff approximation, T-matrix method, and variational method. The common integral type is oscillatory. Evaluation is typically accomplished by Riemann summation.

Solution by the Kirchhoff approximation is illustrated for the problem of acoustic scattering by swimbladdered fish. The nature of the swimbladder as a nearly ideal pressure-release surface justifies modeling of the fish by the swimbladder surface, with the usual Dirichlet condition. Computational results obtained by Riemann summation are compared with physical measurements on fish specimens for which the swimbladder surface has been triangulated. Planned uses of the same data to examine the other solution methods are also described.

P. Keast and G. Fairweather (eds.), Numerical Integration, 329.

TESTING MULTIPLE INTEGRATORS

T.N.L. Patterson
Department of Applied Mathematics & Theoretical Physics
The Queen's University of Belfast
Belfast, BT7 1NN
N .Ireland

ABSTRACT: Some simple techniques are described which have been used to construct multidimensional test integrals for the hypercube. The constructions are examined in terms of their ability to control the properties of the integrand, the availability of exact values of the integral, the efficiency of computation and the multivariate behaviour.

1. INTRODUCTION

To assess a multiple integrator effectively it is essential to have a well designed set of test integrals. Among the minimum requirements would be:

(a) the integrator should be exercised with respect to some clearly defined and controllable properties of the integrand (such as degree of smoothness, strength of singularities, order of differentiability, etc.),

(b) the exact values of the integrals should be known or easily computable,

(c) the calculation of the integrands should not require an inordinate amount of computational labour,

(d) the integrand should have true multivariate properties.

The literature abounds with test integrals whose assessment objectives are not entirely clear but appear frequently to be motivated by requirement (b). This may perhaps be associated with the difficulties of reliably calculating even simple integrals let alone pathological ones.

This paper describes some simple procedures which have been used to construct malleable test integrands in a reasonably logical manner. Attention has been restricted to the product region hypercube although some of the ideas may be generalised to other regions.

P. Keast and G. Fairweather (eds.), Numerical Integration, 331–335.
© 1987 by D. Reidel Publishing Company.

2. SPECIAL FORMS OF TEST INTEGRALS

2.1 Generation from differentiation

This forms   the mathematically simplest   but not the   most easily
implemented   approach.   Let $f(x_1,\ldots,x_k)$ be   any   k-dimensional
function differentiable at least once in each variable and let,

$$g(x_1,\ldots, x_k) = \frac{\partial^n f(x_1,\ldots,x_k)}{\partial x_1 \cdots \cdots \partial x_k} .$$ (1)

Then the proposed test integral is,

$$\int_0^1 dx_1 \ldots \int_0^1 dx_k \, g(x_1,\ldots,x_k) =$$ (2)

$$(-1)^k \sum_{i_1=0}^1 \ldots \sum_{i_k=0}^1 (-1)^{i_1+\ldots+i_k} f(i_1,\ldots,i_k) .$$

Requirement (a)   can   be   met   but   not   altogether   directly.
Requirement (b) is easily satisfied. Requirement (c) is unlikely to be
achieved   since   repeated   differentiation   would   almost   certainly
yield   very unwieldy expressions.   Requirement (d) is satisfied.

On the whole   this is not a test integral   which one would relish
dealing with although an algebraic manipulation package might be used
for automatic generation of code.

2.2 One dimensional combinations

Let,

$$f_j(x) , \quad j = 0,\ldots,n$$ (3)

be a set of one dimensional functions whose integrals are known over
[0,1].   Standard sets of one dimensional test integrands could fill
this role although some might be too difficult. The proposed test
integrand which combines m products is,

$$g(x_1,\ldots,x_k) = \sum_{i=1}^m w_i \prod_{j=1}^k f_{s(i,j)}(x_j)$$ (4)

where, w_i are the combination weights and,

$$0 \le s(i,j) \le n .$$ (5)

This fills requirements (a)-(c) without too much difficulty but the
separability of its variables violates requirement (d). Thus, its

behaviour with respect to any one variable is not influenced by
the values of other variables. However, by judiciously choosing
the basic functions it would be possible to engineer various
pathological situations in a controlled way either in specific parts
of the region or distributed generally.

2.3 Permutationally symmetric integrals

The function $g(x_1,\ldots,x_k)$ is said to permutationally symmetric if its
value remains unchanged when its variables are permuted. (e.g.
$\sin[x_1+x_2]$). This is a true multidimensional function
(requirement (d)) although of special structural form.
Requirements (a) and (c) can be satisfied although (b) may cause some
difficulty.

Since we may write,

$$\int_0^1 dx_1 \ldots \int_0^1 dx_k \; g(x_1,\ldots,x_k) = n! \int_0^1 dx_1 \int_0^{x_1} dx_2 \ldots \int_0^{x_{k-1}} dx_k \; g(x_1,\ldots,x_k) \tag{6}$$

a powerful simplex integration rule could be used for ("brute
force") numerical evaluation. A product m-point Gauss rule may also
be effective in the form,

$$\int_0^1 dx_1 \ldots \int_0^1 dx_k \; g(x_1,\ldots,x_k)$$

$$= \sum_{i_1=1}^{m} w_{i_1} \sum_{i_2=i_1}^{m} w_{i_2} \ldots \sum_{i_k=i_{k-1}}^{m} w_{i_k} \; N(a_{i_1},\ldots,a_{i_k}) \; g(a_{i_1},\ldots,a_{i_k}) \tag{7}$$

where w_i and a_i are the appropriate weights and nodes for the
interval $[0,1]$ and $N(\ldots)$ is the number of distinct permutations
of the nodes.

The number of integrand evaluations for (7) is,

$$\binom{k+m-1}{n} \tag{8}$$

which is substantially smaller than the m^n evaluations for the full
product Gauss application.

The main criticism with this procedure is that by definition the
function is the same in each of the k! simplices $[0,x_i]$, $i=1,\ldots,k$. This
difficulty can be removed as discussed in section 3.

2.4 Multinomial orthogonal expansion

Combined with the transformations described later in section 3 this
forms an attractive procedure with the convenient property that all

test   integrals   automatically   have   the   same   value.
Singularities   can   be   introduced   by   using   appropriate   weight
functions.   All   requirements   for   section   1   are   met   although   to
achieve (c) may involve careful coding.

Let $\varphi_0(x),\ldots,\varphi_m(x)$ be   a   set   of   k-dimensional   algebraic
monomials, orthonormal with respect to the weight $w(x)$ in $[0,1]$. That
is,

$$\int_0^1 dx_1 \ldots \int_0^1 dx_k \, w(x)\varphi_i(x)\varphi_j(x) = \delta_{i,j} \tag{9}$$

where the vector variable $x = (x_1,\ldots,x_k)^T$ and $\delta_{i,j}$ is the Kronecker δ.

The   test   integral   is   then   taken   to   be   the   multivariate
polynomial,

$$T(x) = \varphi_0(x) \, w(x) \sum_{j=0}^m a_j \, \varphi_j(x) \tag{10}$$

the   value   of   whose   integral   is a_0.   The   more   rapidly a_j decreases
with increasing j the "smoother" is the function being represented.
For example the coefficients might take the form,

$$a_j = e^{-jr} \tag{11}$$

where r dictates the rate of fall-off of a_j.

Applying   the   transformations   of   section   3   destroys   the   underlying
polynomial   form   and   allows   various   pathological   situations   to   be
introduced.

In   practical   terms,   with $w(x) = 1$,   (10)   can   be   taken   to   be   a
polynomial of multivariate degree d. Then,

$$m = \binom{k + d}{d} \tag{12}$$

and,

$$\varphi_i(x) = S_{i_1}(x_1)\ldots S_{i_k}(x_k) \tag{13}$$

where,

$$0 \le i_1 + \ldots + i_k \le d \tag{14}$$

and,

$$S_i(x) = (2i+1)^{1/2} P_i(2x-1) \tag{15}$$

with $P_i(t)$ the ith degree Legendre polynomial.

Without loss of generality a_0 may be chosen as unity giving all integrals the same value.

3. TRANSFORMATIONS

An important aspect of one dimensional integration involves finding transformations which make the integration easier. However this can be exploited in reverse to make the integration tougher to a varying degree. In the multidimensional case there is much more flexibility since a different transformation can be applied to every variable. Of course the transformation does not change the value of the integral so requirement (b) is retained. Formally then let,

$$G_i(t) = \int_0^t Z_i(s)ds \ , \ i = 1,\ldots,k \tag{16}$$

with,

$$\int_0^1 Z_i(s)ds = 1 \tag{17}$$

and the test integral transforms to,

$$\int_0^1 dx_1 \ \ldots \ \int_0^1 dx_k \ g(x_1,\ldots,x_k) =$$

$$\int_0^1 dt_1 \ldots \int_0^1 dt_k \ g(G_1(t_1),\ldots,G_k(t_k)) \ Z_1(t_1)\ldots Z_k(t_k) \tag{18}$$

This procedure destroys the undesirable permutational symmetry of the integral defined in section 2.3 and converts the integral of section 2.4 into a more general function.

4. EXAMPLE SCHEME

In conclusion we give a simple example based on the technique of section 2.4. Let test integral (10) be used with a_j defined by (11). The integrand is thus parametrised by k, r and d (which determines m through (12)).

Taking the simple transformation form,

$$Z_i(s) = (1+q) \ (1-s)^q \ , \ 0 < q < 1 \tag{19}$$

gives,

$$G_i(t) = 1 - (1-t)^{q+1} \tag{20}$$

for all i. Thus q becomes an additional behaviour parameter. The controlling effect of the various parameters would be ,

    d -   multivariate complexity
    r -   smoothness (the larger r, the smoother the integrand),
    q -   boundary steepness and derivative singularities.

A PACKAGE FOR TESTING MULTIPLE
INTEGRATION SUBROUTINES

Alan Genz
Computer Science Department
Washington State University
Pullman, WA 99164-1210
U. S. A.

1. INTRODUCTION

During the years 1960-1980 a significant number of methods were proposed for the approximate calculation of multiple integrals (see the books by Stroud [11] and Davis and Rabinowitz [4]). Usually the publication of the description of one of these methods was accompanied by results from a few selected test problems. Users of these methods discovered that although good results could occasionally be obtained in a short time with a particular method, it was often the case that it took hours of computer time to produce inaccurate results. The original description of the methods usually provided only limited information about the practical range of application of the method, and it was not clear if there could ever be any practical means of testing the different methods.

During the same period, standards were developed for the testing of one dimensional integration software and highly robust and reliable FORTRAN subroutines were produced. One of the key aspects of the careful testing of one dimensional integration software was the use of the performance profile method. This was developed in a series of papers by Lyness [7] and Lyness and Kaganove [8,9]. An important feature of the performance profile method is the use of families of test integrands. The test families replace the single integrals that had been previously used with battery tests. A series of sample integrands are drawn from these families, and the results obtained when an integration subroutine is applied to the sample integrands are used to produce performance statistics for a specific family. The statistics can be plotted in various ways to produce different kinds of performance profiles, and the profiles for different integration routines can be used to compare the software.

A paper by the present author [5] contained a description of an extension of the performance profile method to the testing of multidimensional integration subroutines. In the rest of this paper, we briefly describe the software package that was produced for that paper, summarize some of the results of that paper and report on some recent uses of the package.

2. THE TEST PACKAGE

The test package was designed for testing subroutines that have a FORTRAN calling sequence in the form

CALL INTGRL(N, REGION, FUNCTN, EPS, ACCUR, VALUE, MAXVLS, ...),

P. Keast and G. Fairweather (eds.), Numerical Integration, 337–340.
© 1987 by D. Reidel Publishing Company.

for an N dimensional integral with integrand FUNCTN over a region defined by REGION (which might be a set of parameters or a subroutine). For subroutines of this type the user supplies a requested accuracy EPS and a maximum allowed number of integrand evaluations MAXVLS, and the subroutine should return an integral value VALUE with estimated accuracy ACCUR. The test package that was developed assumed that the region was a hyper-rectangular region and the errors were relative ones. Minor changes to the package would allow it to be used for other regions and/or integration subroutines that worked with absolute errors.

A necessary part of a software testing package that is based on the performance profile method is a collection of test problem families. In order to help distinguish between strengths and weaknesses of the tested software each family should have a specific attribute that distinguishes the family from the other families. The families used in the package were as follows:

Integrand Family	Attribute
$f_1(\mathbf{x}) = cos(2\pi u_1 + \sum\limits_{i=1}^{n} a_i x_i)$	*Oscillatory*
$f_2(\mathbf{x}) = \prod\limits_{i=1}^{n}(a_i^{-2} + (x_i-u_i)^2)^{-1}$	*Product Peak*
$f_3(\mathbf{x}) = (1 + \sum\limits_{i=1}^{n} a_i x_i)^{-(n+1)}$	*Corner Peak*
$f_4(\mathbf{x}) = exp(-\sum\limits_{i=1}^{n} a_i^2(x_i-u_i)^2)$	*Gaussian*
$f_5(\mathbf{x}) = exp(-\sum\limits_{i=1}^{n} a_i |x_i-u_i|)$	C^0 *Function*
$f_6(\mathbf{x}) = \begin{cases} 0 & \text{if } x_1 > u_1 \text{ or } x_2 > u_2 \\ exp(\sum\limits_{i=1}^{n} a_i x_i) & \text{otherwise} \end{cases}$	*Discontinuous*

The integration region for these test families is the unit n-cube. The u parameters are parameters that should not affect the difficulty of the integration problem as long as $0 \le u_i \le 1$, so these parameters can be varied randomly in order to generate sample test integrals. The a parameters are assumed to be positive numbers and their overall size should affect the degree of difficulty of integrals in a particular problem family. Increasing $||\mathbf{a}||$ will in general make the integrands more difficult. In order to fix the difficulty level for a series of tests, the package uses a scaling which is given by

$$n^{e_j}\sum\limits_{i=1}^{n} a_i = h_j.$$

Here j indexes the integrand family, and the numbers n, e_j and h_j are fixed for a series of tests. For each test in the series an n-vector $\mathbf{a}'$ is computed with random components chosen from [0,1]. It is then scaled by a constant c so that $\mathbf{a} = c\,\mathbf{a}'$ satisfies the scaling equation. The integrand families were chosen so that they could be easily evaluated analytically. This is important for testing multiple integration algorithms, which can often take hours of computer time to produce results with unreliable error estimates.

The test package was written in Standard FORTRAN and includes a main subroutine MULTST which calls various supporting subroutines, including the subroutine that is being tested. MULTST takes as input the name of the subroutine to be tested, a set of j values to select test families, h and e vectors associated with the families, a set of values of n, a sample size S, a limit on the integrand calls allowed for each call of the tested subroutine and a requested relative accuracy for the tested subroutine. For each j and value of n, MULTST makes S calls of the tested subroutine. For each of these calls the u and a parameters are randomly selected and the correct value for the sample integral is computed. While the S tests are being carried out MULTST accumulates results of the actual and estimated errors for each test, and the actual number of integrand calls required by the tested subroutine for each test. When all of the tests have been completed for each test family and each value of n, MULTST prints a table of results for that series of tests. For each family and value of n the table gives confidence intervals for the median of the estimated errors, actual errors and wrong digits. The number of wrong digits is defined as the difference between the number of estimated correct decimal digits and the actual correct decimal digits when this difference is positive for a particular test. If the difference is negative then zero is used. The test result table also includes a reliability ratio and a median for the actual number of integrand calls for each test family and value of n. The reliability ratio is the the number of times the tested routine had zero wrong digits divided by S.

3. USE OF THE TESTING PACKAGE

For the original report the package was used for a series of tests with three automatic integration methods, a globally adaptive method [6,12], a number theoretic method based on Korobov rules (see Stroud [11]) and a simple Monte-Carlo method. All six test families were used with n = 2, 3, 4, 6, 8. Other input parameters were S = 20, requested accuracy EPS = 10^{-14}, h = (110, 600, 600, 100, 150, 100) and e = (1.5, 2, 2, 1, 2, 2). The limit on the number of integrand calls was set at 5000, 10000, 20000 and 40000, so that four tables of results were produced for each integration method. In general, the results showed the Monte-Carlo method to be the most reliable but the least accurate of the methods. The adaptive method was reasonably reliable for the first four integrand families and more efficient than the other two methods for n = 2, 3, 4. For n = 6 the number theoretic method reliably produced results that were as good as the adaptive method and for n = 8 the number theoretic method usually produced the best results. While the results were somewhat limited, it seemed clear that a globally adaptive method like the one tested could be reasonably reliable and very much more efficient than the other two methods for moderate size n, as long as the integrands were smooth. For larger values of n the number theoretic method could be much better than a simple Monte-Carlo method.

Since the original tests were carried out the package has been used by the present author to test modifications to the error estimating procedure in the globally adaptive method. Bernsten [1,3] and Bernsten and Espelid [2] have used the package to test different combinations of basic rules for the globally adaptive method for n = 2, 3. They have shown that the use of certain good rule pairs can substantially increase the reliability of this method. A modified version of the package has also been used by the NAG [10] Library integration committee to compare different subroutines for integration over the n-sphere.

4. REFERENCES

[1]   J. Bernsten, A Test of the NAG Software for Automatic Numerical Integration over the Cube, Reports in Informatics No. 15, University of Bergen, 1985.

[2]   J. Bernsten and T. Espelid, On the Construction of Minimum Point Symmetric Three Dimensional Embedded Integration Rules for Adaptive Integration Schemes, Reports in Informatics No. 16, University of Bergen, 1985.

[3]   J. Bernsten, Cautious Adaptive Numerical Integration Over the Cube, Reports in Informatics No. 17, University of Bergen, 1985.

[4]   P. J. Davis and P. Rabinowitz, *Methods of Numerical Integration*, Academic Press, New York, 1984.

[5]   A. C. Genz, Testing Multiple Integration Software, in *Tools, Methods and Languages for Scientific and Engineering Computation*, B.Ford, J-C. Rault and F. Thomasset (Eds.), North Holland, New York, 1984, pp. 208-217.

[6]   A. C. Genz and A. A. Malik, An Adaptive Algorithm for Numerical Integration over an N-Dimensional Rectangular Region, J. Comp. Appl. Math., 6(1980), pp. 295-302.

[7]   J. N. Lyness, Performance Profiles and Software Evaluation, Argonne National Laboratory Applied Mathematics Division Technical Memorandum 343 (1979).

[8]   J. N. Lyness and J. J. Kaganove, Comments on the Nature of Automatic Quadrature Routines, ACM Trans. Math. Software, 2 (1976), pp. 65-81.

[9]   J. N. Lyness and J. J. Kaganove, A Technique for Comparing Automatic Quadrature Routines, Comp. J., 20 (1977), pp. 170-177.

[10]   Numerical Algorithms Group Limited, Mayfield House, 256 Banbury Road, Oxford OX2 7DE, United Kingdom.

[11]   A. H. Stroud, *Approximate Calculation of Multiple Integrals*, Prentice-Hall, New Jersey, 1971.

[12]   P. van Dooren and L. de Ridder An Adaptive Algorithm for Numerical Integration over an N-Dimensional Rectangular Region, J. Comp. Appl. Math. 2(1976), pp. 207-217.

A TEST OF SOME WELL KNOWN ONE DIMENSIONAL GENERAL PURPOSE AUTOMATIC QUADRATURE ROUTINES

Jarle Berntsen
Department of Informatics
Universitetet I Bergen
Allegt. 55
N-5000 Bergen, Norway

ABSTRACT

The purpose of the report is to provide numerical evidence that may help potential users to choose between well known one dimensional general purpose automatic quadrature routines (see [1]). The tested routines are chosen from well known libraries and include the QUADPACK routines QAG and QAGS [2], the NAG routines D01AHF, D01AJF and D01AKF [3], and the IMSL routine DCADRE [4]. The testing techniques used are the performance profile method described by Lyness and Kaganove, the performance profile method used in QUADPACK and a classical battery test.

Ten test families of integrands including singular, discontinuous, peak and oscillatory problems are selected for the performance profile tests. We regard the Lyness and Kaganove type performance profile test to be the main ingredient. Here some random parameter is varied in order to provide statistical evidence on how the routines perform with respect to reliability and efficiency. In the other performance profile test we vary some difficulty parameter, and we thereby get information on how the routines perform when the difficulty of a problem is varied. From the tests a considerable amount of numerical evidence on these routines performances is produced. We also include tables of error messages. However, the experience we get by running the performance profile tests is restricted to our 10 test families, and we find the battery test where we have chosen 20 integrands, to be a useful supplement.

The main conclusion is that the QUADPACK routines QAG and QAGS (or the NAG variants D01AJF and D01AKF) are reliable, robust and reasonably efficient routines. QAGS should provide the user with a warning that the routine may be unreliable on internal singularities, and the users should be advised to split the integration range at trouble spots. If this is done, QAGS is a reliable and efficient routine on singular problems. On oscillatory problems we recommend QAG combined with the Gauss-Kronrod 61 point rule and on peak problems QAG with the Gauss-Kronrod 15 or 21 point rule. The

P. Keast and G. Fairweather (eds.), Numerical Integration, 341–342.

NAG routine D01AHF and the IMSL routine DCADRE are in general less reliable and less robust than the QUADPACK routines.

REFERENCES

1. J. Berntsen, A test of some well known one dimensional general purpose automatic quadrature routines, Report in Informatics, No. 11, (1984), Department of Informatics, University of Bergen, Norway.

2. R. Piessens, E. de Doncker-Kapenga, C.W. Uberhuber and D.K. Kahaner, QUADPACK, A Subroutine Package for Automatic Integration, Springer Series in Comp. Math. 1., Springer-Verlag, New York, 1983.

3. NAG FORTRAN Library Manual, Mark 10, The Numerical Algorithms Group, Oxford, England, 1983.

4. IMSL Library, Edition 9.2, International Mathematical and Statistical Libraries, Inc., Houston, TX, 1984.

DEVELOPMENT OF USEFUL QUADRATURE SOFTWARE, WITH PARTICULAR EMPHASIS ON MICROCOMPUTERS

David K. Kahaner
National Bureau of Standards
Technology Building, Room A161
Gaithersburg, Maryland 20899
U.S.A.

1. INTRODUCTION

The purpose of this paper is to address several issues which are associated with the development of software for quadrature. The emphasis here is on the delivery of service to users rather than on the development of new mathematics, methods or algorithms. Of these latter items, the need (intellectual and practical) has been amply demonstrated at this meeting. But one of my tasks is to help users solve problems and their needs are often very different from those of us who create the programs that they use. Another task, less pleasant, is to organize, catalog, and generally manage the software which scientists ultimately access. Both tasks merge in practice, questions and ideas emerge from each and should be fed back into the community of "producers". Three main ideas will stand out: (1) the need for standardization (2) the need for better documentation and (3) the need for programs which capitalize on new computing technology.

In thinking about this workshop, I was reminded of another workshop I was privileged to run in 1975, almost 11 years ago at Los Alamos. Many of you were there, but some of our new stars were not. We didn't have a proceedings but Fred Fritsch, Ron Jones and James Lyness did chair a panel discussion on Testing and Certifying Quadrature Software. This was taped and published in SIGNUM [1]. Some of us made remarks that retrospectively were amusing - - like wondering if quadrature routines shouldn't always give maximum possible precision (as elementary functions do) rather than having a tolerance parameter; or that the object of certifying a quadrature subroutine "would be to prove a theorem and to have every algorithm that is certified to be accompanied by a theorem that states the accuracy that can be gained and for which functions". But most of the discussion has held up very well when reviewed today, questions of who are our users, what problems we are trying to solve and how we should best "market" our wares. I will draw from this when necessary but I urge you all to obtain a copy of this discussion.

Some things have changed since that workshop. That was limited to a discussion of finite interval integrals because "there is still not enough in the way of software existing to talk about ... multidimensional integration". Today we have several good multidimensional routines in the major commercial libraries (and these are the sources that most users refer to first) as well as many other programs privately circulating.

In 1975, it was remarked that "the world isn't ready for a universal quadrature

343

P. Keast and G. Fairweather (eds.), Numerical Integration, 343–369.
© 1987 by D. Reidel Publishing Company.

package,...., systematized collection" although that was disputed as evidence "that our standards are too high". Today, we have QUADPACK, which, while far from perfect, is heavily used and much appreciated.

But several troublesome problems which bothered us then, still seem unresolved today. We still have no consistent standards for naming, documenting, or testing, no attempt to seriously monitor the problems that people try to solve, and no visible progress utilizing nonnumeric, or symbolic, methods. I want to discuss some of these things and make a few concrete suggestions.

2. CLASSIFICATION AND NAMING

When one of us writes a quadrature program for more-or-less general use, the usual procedure in developing the routine seems to be to pick a name very early on and stick with it. For obvious reasons, it has to be called something right away. Usually at the end (or nearly the end) of the development we rewrite the user documentation in the form of comments in our Fortran. A test program, if there is any, is the one we have been using while debugging. Since we have expended great efforts to produce software which is solid, there is no doubt that if we are queried five years from now for a good 3-D quadrature program for polytopes, the name of our program will immediately spring to mind. We will be able to tell the requestor where to locate the program and we might even remember something about the call sequence. It is unlikely that we will remember anything about any competing routines.

This is fine for the user who knows about your expertise, but what about everybody else. They will naturally inquire at the Computer Center where perhaps they will be referred to the NAG [2], IMSL [3], PORT [4] or some other manual. Each of these commercial software vendors has its routines carefully organized according to some functional form and can, with very little difficulty, find your routine if it has been included in one of these libraries. If our user is unsuccessful, he might be directed to the Collected Algorithms of the ACM, the algorithms published in Computer Physics Communications, or the Journal of Computational and Applied Mathematics. Here the situation is more difficult since the organization within these collections is mostly chronological rather than functional. We ignore the important question of how to actually get a copy of one of these programs even after it has been located. This is all very tedious.

Finally, consider the situation from the perspective of the Computer Center. Its staff has to maintain correct copies of everything and know the organization which is used by each collection in order to provide effective consultation. Further, the staff must know the applicable machine types and precision for each routine. If the routines are not in a proprietary library, but can be copied to another computer, the staff must know the procedure and exactly what must be altered in order for the routine to run correctly. Further, if there are external procedures which are required, as there frequently are (for example a linear equation solver) it must be possible to deal with these in an expeditious way. These are formidable problems which grow in difficulty each year.

Remarkably, there are several concrete things we can do to simplify these problems. One possibility would be to establish a consistent naming scheme, but this seems unrealistic, especially in view of the fact that vendors such as NAG [2] and C. Abaci [5] already have their own such conventions. It also presupposes the ability to agree upon what is an important and somewhat personal stylistic issue. As someone who was involved, for months, in similar discussions on ODE software with people having many divergent viewpoints, I believe that naming is doomed.

Two alternatives are keywords and classification. Keywords are the most natural because they correspond to looking things up in a dictionary. Unfortunately, they seem fairly difficult to implement. All ACM algorithms require keywords but these often appear to be poorly chosen, without much thought. In the long run, however, we hope to improve the tools for retrieving and manipulating these, and every author should make a serious effort to select appropriate keywords and include them in his documentation.

Classification is the poor-man's key words. Just as the Dewey Decimal System is used to classify library books, we can use a similar system to classify our software. As part of a major software organization project at NBS, we developed a detailed classification scheme (GAMS) for mathematical software. The existing ACM classes lumped almost all mathematical software in a very few classes and the older Bolstad scheme [6] was not rich enough to deal with the profusion of new routines which have appeared since its creation. The GAMS classes were developed after consultation with many numerical analysts. We tried to accommodate all the relevant suggestions and feel that we succeeded to the extent that we were able to classify all the routines from the collections that were made available to us. The scheme is not meant to be permanent, but we hope that it lasts until 1996, or at least until the next workshop.

A complete listing of the GAMS classes appears in [7]. In Appendix II, we list the top level classes and, in Appendix III, the subclasses associated with quadrature. Each class is a sequence of letter-digit combinations; classes go from simple arithmetic operations (A) toward more complex algorithms - partial differential equations follow ordinary differential equations, multidimensional quadrature follows one dimensional quadrature. Within a major classification such as H, subclasses provide finer detail. One major goal in our design was that the classes actually partition the routines which were available; thus if too many routines appeared in a class we considered reorganizing it into subclasses. As a practical matter, we think that any class with more than about 10 routines is confusing for users. We also tried not to create too many classes which were obviously empty. Our experience suggests that software developers frequently wanted classes to reflect routines they felt ought to be created, but unless these (routines) were likely to appear within the next few years we usually didn't create a new class.

The purpose of the classes is to provide a convenient way for software librarians to manage and collect programs, and we hope that everyone who writes mathematical software will endeavor to include one or more relevant classes in the documentation of their programs. This is generally easier for the author; the librarian, who may not be a specialist in quadrature, ought to check the given classifications to be sure that they are consistent with those in similar routines. We encourage authors to communicate with us if they feel that some classes are poorly developed or new classes need to be provided.

Several years ago, I discussed a preliminary version of the GAMS classes at a math library workshop (at Oak Ridge) and discovered that some of the commercial library vendors misinterpreted my remarks as suggesting that routine names be organized around the classes. While this can certainly be done [5], it was never our intention to propose any such thing; including one or more class designations in the program documentation would be sufficient.

Concerning routine names, it has been remarked that "they shouldn't be too easy to remember else it is difficult to replace them" (NAG's names satisfy that requirement) but we feel this is a question of individual taste. On the other hand, all routines which are in the same "package" should share a consistent naming convention. To avoid potential problems, it is useful to try to use names which are unlikely to conflict; thus use QPSLV instead of SOLVE, etc. This applies not only to those which are user-callable, but even more emphatically to subsidiary routines. With respect to versions in single and double precision, there appears to be no reasonable justification not to use different names; this vastly simplifies handling of the software and does not seem to represent an inconvenience to users. Within the body of the routine, authors should use Fortran intrinsics in their generic form. Thus always use ABS, SQRT and MAX, rather than DABS, DSQRT, or DMAX1. This puts the burden of type conversion on the compiler; converting from AMAX1 to DMAX1 is easy for a user to miss.

3. DOCUMENTATION

Every Fortran subroutine should have machine readable documentation. By now, all of us have accepted this necessity. Authors seem to have a natural inclination to organize this in very individualistic ways. But is is possible to maintain that individualism and still adhere to a simple standardized format. One of the best thought out is the SLATEC Prologue format [8]. Each module contains a Prologue in the form of Fortran comments delimited by

C***BEGIN PROLOGUE ...
        .
        .
        .
C***END PROLOGUE ...

For those routines which are not normally called directly by a user, or for which the documentation is naturally subsumed by that of another routine there may only be one additional line between these:

C***REFER TO ...

For user callable routines the Prologue format appears in Appendix I. A dozen fields are included, each beginning with C***. Within most of these, the author is free to use any convenient structure to describe the software. A few, such as

C***DATE WRITTEN

have a specific syntax. Thus the Prologue is not restrictive while assuring that all relevant information is included (it is amazing how often the developer's name and affiliation are otherwise missing) and allowing easy machine processing of the Prologues. Some restrictions are still recommended; experience suggests that any description of call sequence parameters be in exactly the same order as the actual call sequence and that arguments be identified not only as to type and value, but also as to whether or not they are altered on exit from the called routine. Of course, a software vendor may elect to alter or even remove the Prologue to conform to a format it has established, but even this is easier to do when the documentation is structured.

The Prologue contains fields for classifications as well as keywords. Within the keyword field, a convention has been established for those cases where the routine is part of a larger collection, e.g. QUADPACK, of the form PACKAGE=QUADPACK, again allowing automated processing.

The Prologue contains both Description and Long Description fields, the latter to contain information of secondary importance. We think that the Description field should also contain a sample problem. The sample should be a complete, but hopefully short, Fortran program. It should include data (if necessary) as well as output. Users like these programs, as can be attested by their success in the NAG library where they often form a template for user programs. In this context, we suggest that parameter names should be exactly the same as those in the description of arguments and that authors not take shortcuts, for example,

$$A = 0.0$$
$$B = 1.0$$
$$CALL\ QUAD(A,B,...)$$

rather than

$$CALL\ QUAD(0.0,1.0,...)$$

It is obvious, but is worth repeating, that the Prologue must provide enough information to run the routine and interpret the results. It need not be a complete description of the mathematical or algorithmic details. References to technical reports and papers can be used as pointers to provide this additional background information but cannot be viewed as essential to call the routine. Documentation requiring or even emphasizing some technical report is counterproductive. Average users never have handy access to these reports and experts rarely use software which is not their own except for the purpose of testing it against a newer procedure. In fact, during the process of classifying quadrature routines, we discovered by asking experts that most of them were strangely unaware of the detailed contents of standard software collections.

Documentation should also be written in terminology which is understandable. There is not much point in describing a subroutine as "one for computing the integral of a function f from the class HP(DD), p > 1, over the interval (a,b)", and that "D = a parameter of the class HP(DD), 0 < D < = pi/2". In [1] Huddleston noted that "96% of the computing jobs at Sandia National Laboratory take less than 5 minutes and we often get to consult on the remaining 4%". That is, most problems are easy (this is even more true on microcomputers); we ought to respect that, and develop our software and documentation to emphasize reliability and clarity. For example, consider the following error return messages which appear in a well-publicized subroutine,

IER=2 means "The maximum number of function evaluations has been reached".
IER=3 means "It is presumed that the tolerance cannot be achieved within the specified number of integrand evaluations and that the result returned is the best which can be obtained within that limit".

It strikes me that even experts will have a very difficult time distinguishing these two cases; of course, real users are even less sophisticated with respect to the algorithms that we are employing.

While we are on the subject of documentation I wonder WHY WE SEEM TO INSIST THAT ALL OF OUR MACHINE READABLE DOCUMENTATION BE IN UPPER CASE? THIS MAKES IT MUCH MORE DIFFICULT TO READ. ON THE OTHER HAND, NOBODY CARES MUCH ABOUT THE CODE ITSELF, AS IT IS VERY RARELY READ BY USERS. At one time, it was difficult to use lower case on some computers; it is still true that CDC Cybers want to convert files to all upper case. But this is done automatically, whereas using upper and lower case effectively must be done by hand. We can go a long way towards making our programs friendlier if we will take this simple step.

4. PORTABILITY

By now it is accepted that our subroutines will be written in ANSI standard Fortran 77. Over the past few years, we have seen gratifyingly few routines which fail to meet this requirement. But other, more subtle issues are still frequently violated.

In [9] a model of computing is described which naturally leads to the definition of a number of "machine constants". These are available in response to an appropriate call to I1MACH, R1MACH, and D1MACH. These functions are available on every computer system that I know and should be the only way to access machine dependent information. The most frequently used machine constant, epmach, is defined as

$$\min_{x>0} \{1.0 + x > 1.0\}$$

which is either R1MACH(3) or D1MACH(3) depending on whether x is declared as REAL or DOUBLE PRECISION. Some quadrature subroutines require epmach as an input argument, but we doubt if users (other than numerical analysts) have any idea what this really means. It is often claimed that epmach can be approximated to within a factor of 2 by the following program:

```
      REAL X, XP1
      X = 1.0
    1 X = X/2.
      XP1 = X+1
      IF(XP1.GT.1.0) GOTO 1
      WRITE(*,*) 'EPMACH= ',X*2.0
      STOP
      END
```

Some program developers have used this to avoid referencing R1MACH. Unfortunately, this program returns different numbers depending upon the specific Fortran compiler used and compiler options which are activated. For example, the epmach value for two popular IBM Personal Computer PC compilers appear below:

IBM Professional Fortran Compiler (version <= 1.19)
& Ryan McFarland Fortran Compiler (version 2.0)

(optimizer on [This is the default])	epmach = 1.08420E-19
(optimizer off)	epmach = 1.19209E-07
Lahey Computer Systems F771 (version 2.0)	epmach = 1.08420E-19

The fantastic differences between these values is not a bug but reflects a more complex computational process than our program segment can distinguish. The PC does floating point computations on a second cpu, called a math coprocessor, or math chip. These are done in 80 bit (64 bit mantissa) form regardless of the precision of the input values. Normally, however, real (single precision) variables require four bytes (32 bits) to store. Thus any references to memory result in 32 bit numbers being transferred; arithmetic is in 80 bit mode. When the compiler tries to generate optimized code, it is clever enough to perform the entire computation (excluding the WRITE) on the math chip. Without optimization, there is more data transfer between the math chip and registers on the main cpu; the latter also stores reals as 32 bit numbers. Thus, from the perspective of the math coprocessor, the small value of epmach is exactly correct, but viewed with respect to memory and the main cpu it is not. If the smaller value of epmach is used in an iteration which runs until "convergence" it will never terminate normally. We should mention in passing that generating an incorrect value of epmach by a code segment like the one above was the cause of a spectacular failure in one of our singular value decomposition routines.

Another, less common, machine constant used in quadrature is oflow, defined as the largest non-overflowing floating point number. In this case too, its value is different on/off the math chip, but luckily it cannot be computed internally without causing some type of abort. The machine constant oflow is best determined as R1MACH(2). On the other hand, using oflow can be very dangerous, because many compilers, such as FTN5 on the Control Data Cyber 205, are very fussy about manipulating numbers near the end of their range; for this reason virtually all the QUADPACK tests abort on the Cyber 205. However, a careful study of these routines makes it clear that oflow is only used to initialize some (error) quantity to a "large" value. It seems better to add the special logic which is required in order to deal with the situation at the beginning of these computations and eliminate the need for oflow entirely. Some quadrature routines also use uflow, the smallest, positive, non-underflowing floating point number. On microcomputers, uflow may be as large as 1.E-36, which is far from zero for many scientific computations. But this is rarely a problem because most compilers will default to give zero if a computation is about to underflow. On the other hand, the F771 compiler for the IBM PC requires a special directive, otherwise it will abort. Again, R1MACH(1) will provide uflow.

Occasionally, we still see special numbers which are hardwired into many subroutines by people who should know better. Thus use

pi = atan(1.0)*4.0

instead of

pi = 3.14159265...

and use

$$e = \exp(1.0)$$

instead of

$$e = 2.71828...,$$

etc. Finally, we mention that many quadrature subroutines have built-in constants in the form of weights and nodes. The most portable way to include these is to the maximum number of digits to which they are known, including a D exponent if it is required. For example,

DATA W(1)/2.34659485940113764857746D-02/

which will compile correctly whether W(1) is declared REAL or DOUBLE (albeit, perhaps with a compilation warning message that some digits will be truncated). Unfortunately, in some subroutines, these values are not expressed to enough digits, and neither documentation nor error diagnostic gives any hint of possible problems. Given the rapid spread of microcomputers with 32 bit real numbers, and the desire of many scientists to have accurate results, it is common to see routines which were written in single precision converted automatically to double. In that case, the nodes and weights will be padded on the right with zeros, and the routine will happily give incorrect answers.

Developers of quadrature (or other mathematical) software would like to have a completely self-contained package which they can send to anyone who requests a copy. Often the quadrature algorithms require some subsidiary software such as one which solves a system of linear equations, sorts, etc. These must be included before the main subroutine can run. Among these routines are the machine constant functions I1MACH, etc., mentioned above. We think that unless there is an exceptionally good reason not to, every effort should be expended for the author to use standard software, rather than a "new" program which has no new functionality. Thus we cannot think of any justification for not using SGEFA and SGESL from LINPACK [10] for solving a full, non-symmetric set of linear equations, SDOT and SNRM2 from BLAS [11] for computing dot products and L2 norms, etc. The advantage of this to the library manager is obvious--fewer and more reliable routines requiring less testing.

Before I drop this problem of portability, I want to urge software authors to avoid the use of COMMON, either blank or named, unless it is absolutely required. In addition to the usual problem of possible conflict with the user's COMMON, there is the very real possibility that routines utilizing COMMON will not run correctly on some of the newer multiprocessing computers. Unfortunately, however, almost all problems which scientists want to solve in the real world involve functions which are defined in terms of parameters. Presently, we have no provision for passing these parameters to the integrator except by the use of COMMON. The integrand argument list in quadrature routines has not changed much in twenty years, but, in the era of multidimensional integrals, we must rethink this. A great deal can be learned by looking at software for ordinary differential equations or optimization. For both of these there is an analogous problem of a function or subroutine which the user has to provide. Examining several of these routines suggests that for quadrature, the procedure for computing the integrand, F, ought to include as parameters, not only the location of the point at which the integrand is to be evaluated but also two arrays, RARGS and IARGS, real and integer respectively. These can be used to pass parameters from the user's calling program to F. In addition, we feel that an output parameter, ISTOP, is also needed which the user can set in F to affect the course of the quadrature. The arrays RARGS and IARGS remove the need for COMMON, and ISTOP gives the user the ability to halt the computation if it appears to be going out of control. Not allowing this capability was a serious omission even for 1-D routines, but is essential in the multidimensional case.

A final modification to the integrand parameter list has been suggested to me by Ian Gladwell at this conference, and, with his consent, I am repeating it here. Gladwell notes that, for most quadrature routines, F is called within a loop. This has been necessitated by our tradition of requiring F to be a Fortran function which evaluates the value of the integrand at one input point P. His tests indicate that substantial time savings can be gained by requiring F to return an array of values. To accommodate this, the interface to the integrand should be of the form

SUBROUTINE F(NP,P,FVALS,RARGS,IARGS,ISTOP)

where the input NP is the number of points at which the integrand is to be evaluated, P is a list of these points, and FVALS is the output array of integrand values.

5. TESTING

The test programs that we write are rarely useful to anyone but the original author. We call these "full" tests to indicate that they have been written to thread through as many of the program options as possible. The integrands we use are difficult, and the error tolerances we request are as near as possible to epmach. Output usually consists of pages (sometimes many) of numbers, error codes, variables and their values. The first version of the tests for QUADPACK were of this form. In testing LINPACK [10], the output was nearly 100 pages. It is a mistake to think that others are as interested in our results as we are. In [1] the problems of testing were discussed at some length, and the hope expressed for some general testing programs. I am happy to acknowledge that real progress has been made beginning with work of Lyness, Keast and Fairweather, and now current efforts of Berntsen, Genz, and Patterson which are reported at this workshop. However, the testing problem as seen by the recipient of a piece of quadrature software is very different. The user, or program librarian, has no particular expertise in quadrature. He will not have the capability of appreciating the results of such full tests and no recourse, except to contact the author, if there is a suspicion that a failure has occurred. What these people require is a much simpler type of test. This "quick" test only has one purpose, to verify that all parts of the package have been installed correctly. The output of this test should not be numerical; rather it should be some statement like

Test of QAGE passed

We almost never receive tests like these from authors. They are very easy to write and make portable; parameters like eps can be taken to be .001, or as SQRT(R1MACH(4)). Results can be compared automatically against known values. Users who care to run full tests should obviously be encouraged. But we have seen very few users who, given the alternative of trying to sort out voluminous test output or accepting a routine, will not choose the former. Since most of the time when a full test fails in a new environment it is the fault of the test, not the underlying integrator, this seems a rational strategy.

Of course, the author of a quadrature program will want to test this routine as carefully as possible. That these full tests are very important is obvious. What is not so obvious is that the tests must not only concentrate on the performance of the routine but also on details of the user interface. By that I mean that these tests should anticipate, as much as possible, common human foibles. Thus we should test not only for unreasonable accuracy requests, but blunders such as wrong regions, recursive calls, and alteration of input parameters. If a two dimensional routine requires four vertices to specify a rectangle, how will it perform when the last point is mistyped? How will a routine react if the integrand evaluation point, P, is altered? For one dimensional programs this is usually not a problem, as the call to F usually involves a temporary, scalar, variable. But in more than one dimension P is passed as an array pointer and a statement such as $P=P+1$ inside F is likely to bomb further along in the routine. Recursion can be protected against by the use of a scalar flag which is turned on when the routine is entered and off just before exit. My experience is that users are clever but rarely do what you expect.

6. SPECIAL CHARACTERISTICS OF NEW MICROCOMPUTERS

With a very few exceptions, all the quadrature software I am aware of has been written for scientists who are Fortran programmers. (Pointers to some of the few other software products are given in Appendix IV.) I have even used this explicitly in some of my remarks above. This model was true for many years when our major computing was done on mainframes. Today, the computing mix is much more complex, involving supercomputers, multiprocessors, vector computers, midi, mini and micro computers and even hand-held computers. We think that it is still true that quadrature, in particular, is usually done in the context of some other mathematical or computational operation, and in such cases the most effective mode is via a subroutine call. We note the new interest from the scientific community in the use of languages other than Fortran; this is evident from the growing popularity of "higher" languages such as Pascal, C and Ada. It is gratifying to see at least one paper at this workshop [15] in which the capabilities of one of these new languages are being explored. We also note that other researchers have begun to address the question of the use of advanced architectures such as parallel, vector or hypercube processors.

Desktop computers seem to be conspicuously absent. Perhaps because they have evolved from a very different environment. Often these were purchased originally for use in a laboratory where they were used to direct an experiment. Later, scientists discovered that they could process some of the data which had been gathered without going to a mainframe. As more and more of the data processing was done locally, there was the inclination to perform subsidiary calculating there too. Thus we see a progression from a special purpose lab computer to general purpose computer, with a natural increase in the tasks which this device was asked to perform. This has occurred at the same time that desktop computers have become more and more capable computing engines. A desktop IBM AT is never less than about 10% of a VAX 11/780, often much more. These computers already have more memory than many large mainframes, and within the very near term it will be commonplace for them to have more power and memory than a current VAX.

Many experts in numerical software are supported by the large computing organization at their laboratory. This leads to an independence from the day to day difficulties associated with specific project duties, but also results in a lack of concern for computing resources which are often regarded as free. Further, all but the most progressive of these organizations have a vested interest in large computers, and their management has little experience with the new micro technology, except as word processors, or for project management. Coming at a time when new supercomputers are also being introduced at a rapid rate, desktops are often thought of as an annoyance by the "troops". It is no wonder that there is much less expertise in the use of these new computers in the mathematics and computing sections of the national labs, than in the physics, chemistry, and engineering sections, and in less entrenched academic circles. This is both a criticism of these organizations and an opportunity for others. We must realize that desktops are not just big computers made small, but represent a real revolution in the way people solve problems. This applies not only to traditional numerical processing but also to every aspect of the computing environment. Text editors which would be considered very sophisticated by mainframe standards barely rate a C- when compared to their micro counterparts. Multiple screens, color, graphics, hierarchical files and UNIX-like operating systems are common and expected. Desk calculators, spelling checkers, spreadsheets and lightning-fast Pascal compilers with integrated editor, are being generated with the speed that only market forces can produce.

As far as I can tell, this has only had a minor ripple effect within our community. It has been left to the engineers to produce software which takes advantage of this new environment. For example, Asyst [12] lets a scientist display his data on the screen, use the cursor to block out those points of interest (eliminating data before the experiment has been turned on, and after it has returned to steady state) and then compute the integral between two arbitrary limits. The Scientific Desk [5] includes a Basic program which employs adaptive quadrature and then uses the intervals generated in order to plot the integrand. Clever users have developed graphical interfaces to quadrature software which give them the opportunity to insert information about difficulties in their problem. On a micro, it is easy to interrupt a running computation in the following manner:

> While integral is not finished
>   Try to improve integral estimate
>   If user touches keyboard
>   Give user opportunity to look at current results
> End while

and provides the kind of interaction that can make the difference between solving a problem or not.

Further, there is another trend which has not yet reached the mainframe community in full force, the pressure to avoid programming. In [1] Atkinson remarked that, as the number of computer users grows, the average level of competence will decrease. Perhaps this has been the fuel, but scientists no longer expect to have to write Fortran to solve their problems. We still have a long way to go before the flexibility of "real" programming languages can be matched, but systems such as MATLAB [13], Asyst, and the problem solvers in The Scientific Desk are convincing scientists that it is not necessary to write programs. Arthur Stroud's work [16], presented at this workshop, as well as the Ada project of Delves and Doman [15] mentioned above, are very good steps in this direction, but we have much further to go. For simple quadratures, there is no earthly reason why a user should have to do more than type in his integrand and describe the region in the most direct terms. On the other hand, for those users who would like to make use of the full capabilities of the routines in QUADPACK, a complete, tested and very clearly documented version is available as QUADLIB from [14].

7. SUMMARY

This workshop has enabled us to come together and see what topics are "hot". In this way we look to the future. I have tried to emphasize that it is important to think of the ultimate user of our efforts and make our software palatable for them. I have also tried to suggest that we have neglected some areas which are of great importance in applications. This was vividly brought home to me just last week. A colleague, a mathematics professor at a Midwestern university with a large engineering department, sent me a note on electronic mail asking for "programs for single variable integration of empirical data". He wasn't able to find any, and a look through the routines listed in the GAMS classes in Appendix IV will turn up only one. This is a problem which occurs in virtually every laboratory, almost daily. So let us remember that we've got to pay our dues here and now before we can expect anyone to subsidize us to look toward tomorrow.

REFERENCES

1. Los Alamos Workshop on Quadrature Algorithms, Panel Discussion on Certification and Testing of Quadrature Codes, F. Fritsch chairman, ACM SIGNUM Newsletter, Vol. 11 # 1, May 1976.

2. Numerical Algorithms Group, 1131 Warren Avenue, Downers Grove, Illinois 60515.

3. IMSL Inc., 2500 Park West Tower One, 2500 City West Blvd, Houston, Texas, 77042-3020.

4. PORT, Bell Laboratories, 600 Mountain Ave., Murray Hill, New Jersey 07974.

5. C. Abaci Inc., The Scientific Desk, 208 St. Mary's Street, Raleigh, North Carolina 27650.

6. Bolstad, J., A proposed classification scheme for computer program libraries, ACM SIGNUM Newsletter, Vol. 10 ## 2-3, Nov 1975.

7. Boisvert, R., Howe, S., Kahaner, D., GAMS: A framework for management of scientific software, ACM Trans. Math. Software, Vol. 11, 1985, pp. 313-355.

8. Fong, K., Jefferson, T., Suyehiro, T., SLATEC Common Mathematical Library Source File Format, Lawrence Livermore National Laboratory, Livermore California, 94550, Technical Report, UCRL-53313, July 1982.

9. Fox, P., Hall, A., Schryer, N., Algorithm 528: A framework for a portable library, ACM Trans. Math. Software, Vol. 4, 1978, pp. 177-188.

10. Dongarra, J., Moler, C.B., Bunch, J.R., Stewart, G.W., LINPACK User's Guide, SIAM, Philadelphia, Pennsylvania, 1979.

11. Lawson, C., Hanson, R., Kincaid, D., Krogh, F., Basic linear algebra subprograms for Fortran usage, ACM Trans. Math. Software, Vol. 5, 1979, pp. 308-325.

12. Asyst, Macmillan Software Company, 866 Third Ave, New York, New York 10022.

13. PC-MATLAB, The Math Works, Inc., 124 Foxwood Rd., Portola Valley, California 94025.

14. QUADLIB, PC Scientific Inc., 4710 Debra Lane, Shoreview, New Mexico 55112.

15. Delves, L.M., Doman, B.G.S., The design of a generic quadrature library in ADA, these proceedings.

16. Stroud, A.H., Interactive numerical quadrature, these proceedings.

APPENDIX I: SLATEC COMMON MATHEMATICAL LIBRARY SOURCE FILE FORMAT

What follows is an abbreviated description of the fields which should appear in a typical user callable Fortran Subroutine. All fields are required, except for Long Description and Common Blocks, but entries can be blank.

```
      SUBROUTINE QAG(F,A,B,EPSABS,EPSREL,KEY,RESULT,ABSERR,
     *  NEVAL,IER,LIMIT,LENW,LAST,IWORK,WORK)
C***BEGIN PROLOGUE  QAG
C***DATE WRITTEN  800101   (YYMMDD)
C***REVISION DATE 830518   (YYMMDD)
C***CATEGORY NO.  H2A1A1   (More than one category is allowed)
C***KEYWORDS  AUTOMATIC INTEGRATOR,GAUSS-KRONROD,
C      GENERAL-PURPOSE,GLOBALLY ADAPTIVE,
C      QUADRATURE, PACKAGE=QUADPACK
C***AUTHOR
C
C      [Names and addresses]
C
C***PURPOSE
C
C      [Brief description of what this routine does]
C
C***DESCRIPTION
C
C      [Call sequence parameters are given here]
C
C***LONG DESCRIPTION
C
C      [Optional section describing the algorithm]
C
C***REFERENCES
C
C      [Paper or book can be referenced here]
C
C***ROUTINES CALLED  QAGE,XERROR
C
```

```
C     [All external names except Fortran's]
C
C***COMMON BLOCKS
C
C     [Required if and only if Common is used]
C
C***END PROLOGUE QAG
```

[Declarations, externals, etc. appear here]

```
C***FIRST EXECUTABLE STATEMENT QAG
```

APPENDIX II: TOP LEVEL GAMS CLASSES

A. ARITHMETIC, ERROR ANALYSIS
B. NUMBER THEORY
C. ELEMENTARY AND SPECIAL FUNCTIONS
D. LINEAR ALGEBRA
E. INTERPOLATION
F. SOLUTION OF NONLINEAR EQUATIONS
G. OPTIMIZATION
H. DIFFERENTIATION, INTEGRATION
I. DIFFERENTIAL AND INTEGRAL EQUATIONS
J. INTEGRAL TRANSFORMS
K. APPROXIMATION
L. STATISTICS, PROBABILITY
M. SIMULATION, STOCHASTIC MODELING
N. DATA HANDLING
O. SYMBOLIC COMPUTATION
P. COMPUTATIONAL GEOMETRY
Q. GRAPHICS
R. SERVICE ROUTINES
S. SOFTWARE DEVELOPMENT TOOLS

APPENDIX III: THE H PORTION OF THE GAMS CLASSIFICATION SCHEME

H. DIFFERENTIATION, INTEGRATION

H1.   Numerical Differentiation
H2.   Quadrature (numerical evaluation of definite integrals)
H2a.   One-dimensional integrals
H2a1.   Finite interval (general integrand)
H2a1a.   Integrand available via user-defined procedure
H2a1a1.   Automatic (user need only specify required accuracy)
H2a1a2.   Nonautomatic
H2a1b.   Integrand available only on grid
H2a1b1.   Automatic (user need only specify required accuracy)
H2a1b2.   Nonautomatic
```
```

H2a2.	Finite interval (specific or special type integrand including weight functions, oscillating and singular integrands, principal value integrals, splines, etc.)
H2a2a.	Integrand available via user-defined procedure
H2a2a1.	Automatic (user need only specify required accuracy)
H2a2a2.	Nonautomatic
H2a2b.	Integrand available only on grid
H2a2b1.	Automatic (user need only specify required accuracy)
H2a2b2.	Nonautomatic
H2a3.	Semi-infinite interval (including e**(-x) weight function)
H2a3a.	Integrand available via user-defined procedure
H2a3a1.	Automatic (user need only specify required accuracy)
H2a3a2.	Nonautomatic
H2a4.	Infinite interval (including e**(-x**2)) weight function)
H2a4a.	Integrand available via user-defined procedure
H2a4a1.	Automatic (user need only specify required accuracy)
H2a4a2.	Nonautomatic
H2b.	Multidimensional integrals
H2b1.	One or more hyper-rectangular regions
H2b1a.	Integrand available via user-defined procedure
H2b1a1.	Automatic (user need only specify required accuracy)
H2b1a2.	Nonautomatic
H2b1b.	Integrand available only on grid
H2b1b1.	Automatic (user need only specify required accuracy)
H2b1b2.	Nonautomatic
H2b2.	Nonrectangular region, general region
H2b2a.	Integrand available via user-defined procedure
H2b2a1.	Automatic (user need only specify required accuracy)
H2b2a2.	Nonautomatic
H2b2b.	Integrand available only on grid
H2b2b1.	Automatic (user need only specify required accuracy)
H2b2b2.	Nonautomatic
H2c.	Service routines (compute weight and nodes for quadrature formulas)

APPENDIX IV: CONTENTS OF SOME POPULAR SOFTWARE COLLECTIONS
                BY GAMS CLASSES

Collection

CMLIB	Local NBS Core Math Library (SLATEC Math Library) QUADPACK routines are available for the IBM PC as QUADLIB from [14]
IMSL	See [3] An IBM PC version is available
NAG	See [2] An IBM PC version is available
PORT	See [4]
SCI DESK	See [5] Only available for microcomputers

CLASSES	MODULE	LIBRARY	DESCRIPTION
(* indicates Easy to Use in that class)			
*H2a1a1/	H2A1	SCI DESK	Automatic adaptive integrator, will handle many non-smooth integrands, using Gauss Kronrod formulas, random initial subdivision.
H2a1a1/	H2A1U	SCI DESK	Automatic adaptive integrator, will handle many non-smooth integrands, using Gauss Kronrod formulas, user set initial subdivision, restarting feature, flexible.
H2a1a1/	QNG	CMLIB	Automatic non-adaptive integrator for smooth functions, using Gauss Kronrod Patterson formulas.
*H2a1a1/	QAG	CMLIB	Automatic adaptive integrator, will handle many non-smooth integrands using Gauss Kronrod formulas.
*H2a1a1/	QAGS	CMLIB	Automatic adaptive integrator, will handle most non-smooth integrands including those with endpoint singularities, uses extrapolation.
H2a1a1/	QAGE	CMLIB	Automatic adaptive integrator, can handle most non-smooth functions also provides more information than QAG.
H2a1a1/	QAGSE	CMLIB	Automatic adaptive integrator, can handle integrands with endpoint singularities provides more information than QAGS.

CLASSES	MODULE	LIBRARY	DESCRIPTION
(* indicates Easy to Use in that class)			
H2a1a1/	DQNG	CMLIB	Automatic non-adaptive integrator for smooth functions, using Gauss Kronrod Patterson formulas.
*H2a1a1/	DQAG	CMLIB	Automatic adaptive integrator, will handle many non-smooth integrands using Gauss Kronrod formulas.
*H2a1a1/	DQAGS	CMLIB	Automatic adaptive integrator, will handle most non-smooth integrands including those with endpoint singularities, uses extrapolation.
H2a1a1/	DQAGE	CMLIB	Automatic adaptive integrator, can handle most non-smooth functions also provides more information than DQAG.
H2a1a1/	DQAGSE	CMLIB	Automatic adaptive integrator, can handle integrands with endpoint singularities provides more information than DQAGS.
*H2a1a1/	Q1DA	CMLIB	Automatic evaluation of a user-defined function of one variable. Special features include randomization and singularity weakening.
H2a1a1/	Q1DAX	CMLIB	Flexible subroutine for the automatic evaluation of definite integrals of a user-defined function of one variable. Special features include randomization, singularity weakening, restarting, specification of an initial mesh (optional), and output of smallest and largest integrand values.
H2a1a1/	Q1DB	CMLIB	Automatic evaluation of a user-defined function of one variable. Integrand must be a Fortran Function but user may select name. Special features include randomization and singularity weakening. Intermediate in usage difficulty between Q1DA and Q1DAX.
*H2a1a1	GLAQ	SCI DESK	Automatic adaptive integration of a user-defined function. Interactive input, no programming, can plot integrand on demand.

CLASSES	MODULE	LIBRARY	DESCRIPTION
(* indicates Easy to Use in that class)			
H2a1a1/	DCADRE	IMSL	Numerical integration of a function using cautious adaptive Romberg extrapolation.
H2a1a1/	D01AHF	NAG	Quadrature for one-dimensional integrals, adaptive integration of a function over a finite interval suitable for well-behaved integrands.
H2a1a1/	D01AJF	NAG	Quadrature for one-dimensional integrals, adaptive integration of a function over a finite interval allowing for badly-behaved integrands.
H2a1a1/	D01ARF	NAG	Computes definite integral over a finite range to a specified relative or absolute accuracy, using
H2a1a1/	D01BDF	NAG	Quadrature for one-dimensional integrals, non-adaptive integration over a finite interval. Patterson's method.
H2a1a1/	ODEQ	PORT	Finds the integral of a set of functions over the same interval by using the differential equation solver ODES1. For smooth functions.
*H2a1a1/	QUAD	PORT	Finds the integral of a general user defined EXTERNAL function by an adaptive technique to given absolute accuracy.
H2a1a1/	DODEQ	PORT	Finds the integral of a set of functions over the same interval by using the differential equation solver ODES1. For smooth functions.
H2a1a1/	RQUAD	PORT	Finds the integral of a general user defined EXTERNAL function by an adaptive technique. Combined absolute and relative error control.
*H2a1a1/	DQUAD	PORT	Finds the integral of a general user defined EXTERNAL function by an adaptive technique to given absolute accuracy.
H2a1a1/	DRQUAD	PORT	Finds the integral of a general user defined EXTERNAL function by an adaptive technique. Combined absolute and relative error control.

CLASSES	MODULE	LIBRARY	DESCRIPTION
(* indicates Easy to Use in that class)			
H2a1a2/	QK15	CMLIB	Evaluates integral of given function on an interval with a 15 point Gauss Kronrod formula and returns error estimate.
H2a1a2/	QK21	CMLIB	Evaluates integral of given function on an interval with a 21 point Gauss Kronrod formula and returns error estimate.
H2a1a2/	QK31	CMLIB	Evaluates integral of given function on an interval with a 31 point Gauss Kronrod formula and returns error estimate.
H2a1a2/	QK41	CMLIB	Evaluates integral of given function on an interval with a 41 point Gauss Kronrod formula and returns error estimate.
H2a1a2/	QK51	CMLIB	Evaluates integral of given function on an interval with a 51 point Gauss Kronrod formula and returns error estimate.
H2a1a2/	QK61	CMLIB	Evaluates integral of given function on an interval with a 61 point Gauss Kronrod formula and returns error estimate.
H2a1a2/	DQK15	CMLIB	Evaluates integral of given function on an interval with a 15 point Gauss Kronrod formula and returns error estimate.
H2a1a2/	DQK21	CMLIB	Evaluates integral of given function on an interval with a 21 point Gauss Kronrod formula and returns error estimate.
H2a1a2/	DQK31	CMLIB	Evaluates integral of given function on an interval with a 31 point Gauss Kronrod formula and returns error estimate.
H2a1a2/	DQK41	CMLIB	Evaluates integral of given function on an interval with a 41 point Gauss Kronrod formula and returns error estimate.
H2a1a2/	DQK51	CMLIB	Evaluates integral of given function on an interval with a 51 point Gauss Kronrod formula and returns error estimate.

CLASSES   MODULE   LIBRARY   DESCRIPTION

(* indicates Easy to Use in that class)

H2a1a2/	DQK61	CMLIB	Evaluates integral of given function on an interval with a 61 point Gauss Kronrod formula and returns error
H2a1a2/H2a3a2/H2a4a2	D01BAF	NAG	Quadrature for one-dimensional integrals, Gaussian rule-evaluation. estimate.
H2a1b2/	H2A1T	SCI DESK	Quadrature for one-dimensional integrals, integration of a function defined accurately by data values only.
H2a1b2/	D01GAF	NAG	Quadrature for one-dimensional integrals, integration of a function defined by data values only.
H2a1b2/	CSPQU	PORT	Finds the integral of a function defined by pairs (x,y) of input points. The x's can be unequally spaced. Uses spline interpolation.
H2a1b2/	DCSPQU	PORT	Finds the integral of a function defined by pairs (x,y) of input points. The x's can be unequally - spaced. Uses spline interpolation.
H2a2a1/E3/K6/	BFQAD	CMLIB	Integrates function times derivative of B-spline from X1 to X2. The B-spline is in "B" representation.
*H2a2a1/	QAGP	CMLIB	Automatic adaptive integrator, allows user to specify location of singularities or difficulties of integrand, uses extrapolation.
*H2a2a1/	QAWO	CMLIB	Automatic adaptive integrator for integrands with oscillatory sine or cosine factor.
*H2a2a1/	QAWS	CMLIB	Automatic integrator for functions with explicit algebraic and/or logarithmic endpoint singularities.
*H2a2a1/*J4/	QAWC	CMLIB	Cauchy principal value integrator, using adaptive Clenshaw Curtis method (real Hilbert transform).
H2a2a1/	QAGPE	CMLIB	Automatic adaptive integrator for function with user specified endpoint singularities, provides more information that QAGP.
H2a2a1/	QAWOE	CMLIB	Automatic integrator for integrands with explicit oscillatory sine or cosine factor, provides more information than QAWO.

CLASSES	MODULE	LIBRARY	DESCRIPTION
(* indicates Easy to Use in that class)			
H2a2a1/	QAWSE	CMLIB	Automatic integrator for integrands with explicit algebraic and/or logarithmic endpoint singularities, more information than QAWS.
H2a2a1/J4/	QAWCE	CMLIB	Cauchy principal value integrator, provides more information than QAWC (real Hilbert transform).
*H2a2a1/	DQAGP	CMLIB	Automatic adaptive integrator, allows user to specify location of singularities or difficulties of integrand, uses extrapolation.
*H2a2a1/	DQAWO	CMLIB	Automatic adaptive integrator for integrands with oscillatory sine or cosine factor.
*H2a2a1/	DQAWS	CMLIB	Automatic integrator for functions with explicit algebraic and/or logarithmic endpoint singularities.
*H2a2a1/*J4/	DQAWC	CMLIB	Cauchy principal value integrator, using adaptive Clenshaw Curtis method (real Hilbert transform).
H2a2a1/	DQAGPE	CMLIB	Automatic adaptive integrator for function with user specified endpoint singularities, provides more information that DQAGP.
H2a2a1/	DQAWOE	CMLIB	Automatic integrator for integrands with explicit oscillatory sine or cosine factor, provides more information than DQAWO.
H2a2a1/	DQAWSE	CMLIB	Automatic integrator for integrands with explicit algebraic and/or logarithmic endpoint singularities, more information than DQAWS.
H2a2a1/J4/	DQAWCE	CMLIB	Cauchy principal value integrator, provides more information than DQAWC (real Hilbert transform).
H2a2a1/C3a2/	DQMOMO	CMLIB	Computes integral of k-th degree Tchebycheff polynomial times selection of functions with various singularities.
H2a2a1/C3a2/	QMOMO	CMLIB	Computes integral of k-th degree Tchebycheff polynomial times selection of functions with various singularities.
H2a2a1/E3/K6/	DBFQAD	CMLIB	Integrates function times derivative of B-spline from X1 to X2. The B-spline is in "B" representation.

CLASSES	MODULE	LIBRARY	DESCRIPTION

(* indicates Easy to Use in that class)

CLASSES	MODULE	LIBRARY	DESCRIPTION
H2a2a1/E3/K6/	PFQAD	CMLIB	Computes integral on (X1,X2) of product of function and the ID-th derivative of B-spline which is in piecewise polynomial representation.
H2a2a1/E3/K6/	DPFQAD	CMLIB	Integrates function times derivative of B-spline from X1 to X2. The B-spline is in piecewise polynomial representation.
H2a2a1/	D01AKF	NAG	Quadrature for one-dimensional integrals, adaptive integration of a function over a finite interval, method suitable for oscillating functions.
H2a2a1/	D01ALF	NAG	Quadrature for one-dimensional integrals, adaptive integration of a function over a finite interval, allowing for singularities at user-specified points.
H2a2a1/	D01APF	NAG	Adaptive integration of a function of one variable over a finite interval with weight function with algebraico-logarithmic endpoint singularities.
H2a2a1/	D01ANF	NAG	Quadrature for one-dimensional integrals, adaptive integration of a function over a finite interval, weight function cos(wx) or sin(wx).
H2a2a1/J4/	D01AQF	NAG	Quadrature for one-dimensional integrals, adaptive integration of a function over a finite interval, weight function 1/(x-c) (Hilbert transform).
H2a2a1/	BQUAD	PORT	Adaptively integrates functions which have discontinuities in their derivatives. User can specify these points.
H2a2a1/	DBQUAD	PORT	Adaptively integrates functions which have discontinuities in their derivatives. User can specify these points.
H2a2a2/	QK15W	CMLIB	Evaluates integral of given function times arbitrary weight function on interval with 15 point Gauss Kronrod formula and gives error estimate.

CLASSES	MODULE	LIBRARY	DESCRIPTION
(* indicates Easy to Use in that class)			
H2a2a2/	QC25F	CMLIB	Clenshaw-Curtis integration rule for function with cos or sin factor, also uses Gauss Kronrod formula.
H2a2a2/	QC25S	CMLIB	Estimates integral of function with algebraico-logarithmic singularities with 25 point Clenshaw-Curtis formula and gives error estimate.
H2a2a2/J4/	QC25C	CMLIB	Uses 25 point Clenshaw-Curtis formula to estimate integral of F*W where W=1/(X-C).
H2a2a2/	DQK15W	CMLIB	Evaluates integral of given function times arbitrary weight function on interval with 15 point Gauss Kronrod formula and gives error estimate.
H2a2a2/	DQC25F	CMLIB	Clenshaw-Curtis integration rule for function with cos or sin factor, also uses Gauss Kronrod formula.
H2a2a2/	DQC25S	CMLIB	Estimates integral of function with algebraico-logarithmic singularities with 25 point Clenshaw-Curtis formula and gives error estimate.
H2a2a2/J4/	DQC25C	CMLIB	Uses 25 point Clenshaw-Curtis formula to estimate integral of F*W where W=1/(X-C).
H2a2b1/	DINT	SCI DESK	Computes integral of function defined by accurate data, interactive operation, plots smooth curve through the integrand on demand.
H2a2b1/E3/K6/	BSQAD	CMLIB	Computes the integral of a B-spline from X1 to X2. The B-spline must be in "B" representation.
H2a2b1/E3/K6/	DBSQAD	CMLIB	Computes the integral of a B-spline from X1 to X2. The B-spline must be in "B" representation.
H2a2b1/E3/K6/	DPPQAD	CMLIB	Computes the integral of a B-spline from X1 to X2. The B-spline must be in piecewise polynomial form.
H2a2b1/E3/K6/	PPQAD	CMLIB	Computes the integral of a B-spline from X1 to X2. The B-spline must be in piecewise polynomial representation.
H2a2b1/E3/K6/	DCSQDU	IMSL	Cubic spline quadrature.
H2a2b1/E3/K6/	E02BDF	NAG	Evaluation of fitted functions, cubic spline as E02BAF, definite integral.

CLASSES	MODULE	LIBRARY	DESCRIPTION
(* indicates Easy to Use in that class)			
H2a2b1/C3a2/K6/	E02AJF	NAG	Integral of fitted polynomial in Chebyshev series form.
H2a2b1/E3/K6/	SPLNI	PORT	Integrates a function described previously by an expansion in terms of B-splines. Several integrations can be performed in one call.
H2a2b1/E3/K6/	BSPLI	PORT	Obtains the integrals of basis splines, from the left-most mesh point to a specified set of points.
H2a2b1/E3/K6/	DBSPLI	PORT	Obtains the integrals of basis splines, from the left-most mesh point to a specified set of points.
H2a2b1/E3/K6/	DSPLNI	PORT	Integrates a function described previously by an expansion in terms of B-splines. Several integrations can be performed in one call.
*H2a3a1/ *H2a4a1/	QAGI	CMLIB	Automatic adaptive integrator for semi-infinite or infinite intervals. Uses nonlinear transformation and extrapolation.
*H2a3a1/	QAWF	CMLIB	Automatic integrator for Fourier integrals on (a, ∞) with factors SIN(OMEGA*X), COS(OMEGA*X) by integrating between zeros.
H2a3a1/H2a4a1/	QAGIE	CMLIB	Automatic integrator for semi-infinite or infinite intervals and general integrands, provides more information than QAGI.
H2a3a1/	QAWFE	CMLIB	Automatic integrator for Fourier integrals, with SIN(OMEGA*X) factor on (a, ∞), provides more information than QAWF.
*H2a3a1/ *H2a4a1/	DQAGI	CMLIB	Automatic adaptive integrator for semi-infinite or infinite intervals. Uses nonlinear transformation and extrapolation.
*H2a3a1/	DQAWF	CMLIB	Automatic integrator for Fourier integrals on (a, ∞) with factors SIN(OMEGA*X), COS(OMEGA*X) by integrating between zeros.
H2a3a1/H2a4a1/	DQAGIE	CMLIB	Automatic integrator for semi-infinite or infinite intervals and general integrands, provides more information than DQAGI.

CLASSES	MODULE	LIBRARY	DESCRIPTION
(* indicates Easy to Use in that class)			
H2a3a1/	DQAWFE	CMLIB	Automatic integrator for Fourier integrals, with SIN(OMEGA*X) factor on (a, ∞), provides more information than DQAWF.
H2a3a1/H2a4a1/	D01AMF	NAG	Quadrature for one-dimensional integrals, adaptive integration of a function over an infinite or semi-infinite interval.
H2a3a2/H2a4a2/	QK15I	CMLIB	Evaluates integral of given function on semi-infinite or infinite interval with a transformed 15 point Gauss Kronrod formula and gives error estimate.
H2a3a2/H2a4a2/	DQK15I	CMLIB	Evaluates integral of given function on semi-infinite or infinite interval with a transformed 15 point Gauss Kronrod formula and gives error estimate.
H2b1a1/H2b2a1/	H2B2A	SCI DESK	Automatic (adaptive) evaluation of a user specified function f(x,y) on one or more triangles to a prescribed relative or absolute accuracy.
H2b1a1/	ADAPT	CMLIB	Computes the definite integral of a user specified function over a hyper-rectangular region in dimension 2 through 20. User specifies tolerance. A restarting feature is useful for continuing a computation without wasting previous function values.
H2b1a1/H2b2a1/	TWODQ	CMLIB	Automatic (adaptive) evaluation of a user specified function f(x,y) on one or more triangles to a prescribed relative or absolute accuracy. Two different quadrature formulas are available within TWODQ. This enables a user to integrate functions with boundary singularities.
H2b1a1/	DMLIN	IMSL	Numerical integration of a function of several variables over a hyper-rectangle (Gaussian method).
H2b1a1/	D01FAF	NAG	Quadrature for multi-dimensional integrals over a hyper-rectangle, Monte Carlo method.

CLASSES	MODULE	LIBRARY	DESCRIPTION
(* indicates Easy to Use in that class)			
H2b1a1/	D01FCF	NAG	Quadrature for multi-dimensional integrals over a hyper-rectangle, adaptive method.
H2b1a1/	D01GBF	NAG	Calculates an approximation to the integral of a function over a hyper-rectangular region, using a Monte-Carlo method. An approximate relative estimate is also returned. Suitable for low accuracy work.
H2b1a2/	D01FBF	NAG	Quadrature for multidimensional integrals over a hyper-rectangle, Gaussian rule-evaluation.
H2b1a2/	D01FDF	NAG	Calculates an approximation to a definite integral in up to 30 dimensions, using the method of Sag and Szekeres. The region of integration is an n-sphere, or by built-in transformation via the unit n-cube, any product region.
H2b1a2/	D01GCF	NAG	Calculates an approximation to a definite integral in up to 20 dimensions, using the Korobov-Conroy number theoretic method. Returns a simple error estimate by repeating the computation with a different (randomized) set of points.
H2b1b2/	DBCQDU	IMSL	Bicubic spline quadrature.
H2b2a1/H2b1a1/	H2B2A	SCI DESK	Automatic (adaptive) evaluation of a user specified function f(x,y) on one or more triangles to a prescribed relative or absolute accuracy. Two different quadrature formulas are available within TWODQ. This enables a user to integrate functions with boundary singularities.
H2b2a1/	DBLIN	IMSL	Numerical integration of a function of two variables.
H2b2a1/	D01DAF	NAG	Quadrature for two-dimensional integrals over a finite region.
H2b2b1/	D01JAF	NAG	Attempts to evaluate an integral over an n-dimensional sphere (n=2,3,4), to a user specified absolute or relative accuracy, by means of a modified Sag-Szekeres method.

CLASSES	MODULE	LIBRARY	DESCRIPTION
(* indicates Easy to Use in that class)			
			Can handle singularities on the surface or at the center of the sphere. Returns an error estimate.
H2b2b2/	D01PAF	NAG	Returns a sequence of approximations to the integral of a function over a multi-dimensional simplex, together with an error estimate for the last approximation.
H2c/	D01BBF	NAG	Weights and abscissae for Gaussian quadrature rules, restricted choice of rule, using pre-computed weights and abscissae.
H2c/	D01BCF	NAG	Weights and abscissae for Gaussian quadrature rules, more general choice of rule calculating the weights and abscissae.
H2c/	DGQM11	PORT	Finds the abscissae and weights for Gauss Legendre quadrature on the interval (-1,1).
H2c/	GAUSQ	PORT	Finds the abscissae and weights for Gauss quadrature on the interval (a,b) for a general weight function with known moments.
H2c/	DGAUSQ	PORT	Finds the abscissae and weights for Gauss Legendre quadrature on the interval (a,b).
H2c/	GQM11	PORT	Finds the abscissae and weights for Gauss Legendre quadrature on the interval (-1,1).
H2c/	GQ0IN	PORT	Finds the abscissae and weights for Gauss Laguerre quadrature on the interval (0, +∞).
H2c/	DGQ0IN	PORT	Finds the abscissae and weights for Gauss Laguerre quadrature on the interval (0, +∞).

APPENDIX I: FINAL PROGRAM

Monday August 11   One-dimensional Integration (Software and Developments)

Morning Session - Chairman : P. Keast (Canada)

9:00 - 9:15   Opening Remarks:
              Dr. K. A. Dunn, Chairman
              Department of Mathematics, Statistics and Computing Science
              Dalhousie University

9:15 - 10:15 J.N. Lyness (USA)
              Some quadrature rules for finite trigonometric and related
              integrals.

10:45 - 11:15 L. Gatteschi (Italy)
               Bounds and approximations for the zeros of classical orthogonal
               polynomials. Their use in generating quadrature rules.

11:15 - 11:45 H. J. Schmid (Germany)
               On positive quadrature rules.

11:45 - 12:15 C. T. H. Baker (UK)
               Fast generation of quadrature rules with some special properties.

Afternoon Session - Chairman : R. Piessens (Belgium).

2:00 - 3:00   P. Rabinowitz (USA)
              On sequences of imbedded integration rules.

3:00 - 3:30   G. F. Corliss (USA)
              Performance of self-validating adaptive quadrature.

4:00 - 4:30   S. Haber (USA)
              Indefinite integration formulas based on the sinc expansion.

4:30 - 5:00   Discussion

Tuesday August 12   One-dimensional Integration (Applications)

Morning Session - Chairman : A. C. Genz (USA)

9:00 - 10:00   E. de Doncker-Kapenga (Netherlands)
               Parallelization of adaptive integration methods.

10:30 - 11:00   W. Squire (USA)
Comparison of Gauss-Legendre and mid-point quadrature
with application to the Voigt function.

11:00 - 11:30   G. Krenz (USA)
Using weight functions in self-validating quadrature.

11:30 - 12:30   R. Piessens (Belgium)
Modified Clenshaw-Curtis integration and applications to
numerical computation of integral transforms.

Afternoon Session - Chairman : D. Kahaner (USA)

2:00 - 3:00   P. Rabinowitz(USA)
The convergence of noninterpolatory product integration rules.

3:30 - 4:00   J. Berntsen (Norway)
A test of some well known one dimensional general purpose
automatic quadrature routines.

4:00 - 4:30   L. C. Hsu (USA)
Approximate computation of strongly oscillatory integrals
with compound precision.

4:30 - 5:00   Discussion

**Wednesday August 13   Multi-dimensional Integration (Software and
Developments)**

Morning Session - Chairman : P. Rabinowitz (USA)

9:00 - 10:00   E. de Doncker-Kapenga (Netherlands)
Asymptotic expansions and their applications in
numerical integration.

10:30 - 11:00   R. Cools (Belgium)
Construction of sequences of embedded cubature
formulae for circular symmetric planar regions.

11:00 - 11:30   A. Haegemans (Belgium)
Construction of three-dimensional cubature formulae
with points on regular and semi-regular polytopes.

11:30 - 12:00   T. N. L. Patterson (UK)
Testing multiple integrators.

12:00 - 12:30   A. C. Genz (USA)
A package for testing multiple integration subroutines.

Afternoon Session - Chairman : T. N. L. Patterson (UK)

2:00 - 3:00 H. M. Moeller (Germany)
             On the construction of cubature formulae with few
             nodes using Groebner bases.

3:30 - 4:00 B. G. S. Doman (UK)
             The design of a generic quadrature library in ADA.

4:00 - 4:30 F. Mantel (USA)
             Non-fortuitous, non-product, non-fully symmetric cubatures.

4:30 - 5:00 Discussion

Thursday August 14   Multi-dimensional Integration (Applications)

Morning Session - Chairman : A. Haegemans (Belgium)

9:00 - 10:00   A. C. Genz (USA)
               The numerical evaluation of multiple integrals on
               parallel computers.

10:30 - 11:00 T. O. Espelid (Norway)
               On the construction of higher degree three dimensional
               embedded integration rules.

11:00 - 11:30 T. Sorevik (Norway)
               Fully symmetric integration rules for the unit four-cube.

11:30 - 12:00 J. P. Lambert (USA)
               Quasi-random sequences for optimization and numerical
               integration.

12:00 - 12:30 J. A. Kapenga (USA)
               The integration of the multivariate normal density
               function for the triangular method.

Afternoon Session - Chairman : C. T. H. Baker (UK)

2:00 - 3:00 A.H. Stroud (USA)
             Interactive numerical quadrature.

3:30 - 4:30 T. N. L. Patterson (UK)
             On the construction of a practical Ermakov-Zolotukhin
             variance reducing Monte-Carlo multiple integrator
             and interpolator.

4:30 - 5:00 Discussion

Friday August 15   General Applications.

Morning Session - Chairman : P. Keast (Canada)

9:00 - 10:00   D. Kahaner (USA)
Development of useful quadrature software, with particular
emphasis on microcomputers.

10:30 - 11:00   I. Gladwell (UK)
Vectorisation of one-dimensional quadrature codes.

11:00 - 11:30   M. P. Carpentier (Portugal)
Computation of the index of an analytic function.

11:30 - 12:30   N.I. Ioakimidis (Greece)
Quadrature methods for the determination of zeros of
transcendental functions - a review.

Afternoon Session - Chairman : G. Fairweather (USA)

2:00 - 2:30   K. G. Foote (Norway)
Numerical integration in scalar wave scattering,
with application to acoustic scattering by fish.

2:30 - 3:00   W. Squire (USA)
Quadrature rules with end-point corrections - comments
on a paper by Garloff, Solak and Szydelko.

3:30 - 4:30   Discussion and Closing Remarks.

APPENDIX II: DISCUSSION

Time was set aside each afternoon for general discussion and questions. The discussion was at times centered on the talks of the day, and at other times was quite wide ranging. These sessions were not recorded verbatim, but notes were taken to provide at least a flavour of the debates. These notes follow, in edited form. The accuracy of statements attributed to individuals is not guaranteed, and the editors apologise for any unintended misrepresentation.

MONDAY AUGUST 11

Kahaner (to Lyness): What experience exists with the kind of problems which require calculation of Fourier coefficients?

Lyness: Most people actually want the Fourier Transform, and make the mistake of chopping the range to get a Fourier coefficient calculation. This is a process that requires great care in the choice of the finite interval.

Piessens (to Gatteschi): When Gaussian Quadrature rules are generated, there are two main methods used :
1. Special methods such as those described by Gatteschi.
2. Methods based on recurrence relations, and eigenvalue calculation.
What is the best approach to use in general?

Gatteschi: Probably best in practice is the eigenvalue approach. The first method only applies to some special weight functions.

Piessens (Comment): In addition, the classically based methods do not give results which are stable in the whole interval. The computations are less stable at the ends.

Haber (to Corliss): Has any thought been given to disseminating the software discussed in your talk?

Corliss: The work was done under contract to IBM. The software may be distributed as source code with the purchase of ACRITH.

Kahaner (to Corliss): What role does ACRITH play in the self-validating quadrature package? (That is, is the technique intimately tied to ACRITH or can it stand by itself?)

Corliss: ACRITH comes from IBM. It is a vital component of the package. Many of the techniques used in the package of self-validating quadrature routines are based on material in the review paper by U.W. Kulisch and W.L. Miranker in SIAM Review 28(1986), 1-40. Their techniques are used extensively in the package, for example in the computation of the scalar product with a single round-off error.

Kahaner (to Corliss): If the scalar product described in ACRITH is faster, and more accurate, why is it not generally available?

Corliss: I believe that it is implemented experimentally in Floating Point Systems, and also in some Pascal and C systems. But it uses fairly large amounts of storage. It is not implemented in Fortran. The ACRITH system represents the first time that interval arithmetic has been made available from a vendor.

Kahaner (A general remark): Could there be an interface with the BLAS? (The feeling was that this would be valuable, and that there should be.)

Smith (Remark): There is a critique written by W. Kahan and E. LeBlanc of the ACRITH package. It is a Department of Mathematics report, University of California at Berkeley, March 1985, entitled 'Anomalies in the IBM ACRITH package'.

Patterson (to Corliss): You have to compute the 2n-th derivative in the process you described. How is that done?

Corliss: It is done symbolically on each subinterval. It cannot be done using interpolation because of the restriction that the bounds be guaranteed.

Haber (to Corliss): But you seem to need maxima and minima. How can you compute these?

Corliss: We do not need extrema exactly, only bounds. The tightest bounds possible are not required, so long as they are guaranteed. If the integrand has a singularity in the range, an error is returned.

Kahaner (to Corliss): Why is there no portable Fortran implementation of the basic underlying components.

Corliss: The procedures would be very slow in Fortran. Probably they should be implemented in micro-code or assembly language. (This makes it unlikely they would be portable).

TUESDAY AUGUST 12

Keast (to Kapenga & de Doncker): From the results given in the paper, it appears that not a great deal of parallellisation is achieved, since only a few processors are utilised.

Kapenga: Only 8 processors were available. Not all were utilised, but it should be possible to do so.

Berntsen: Our group has achieved a speed-up in adaptive integration algorithms of up to 30 times, using 32 loosely coupled processors.

Kahaner (to Berntsen): Could you comment a little on the functions used for those experiments.

Berntsen: The same functions as were discussed in the paper on comparisons of adaptive methods. These included functions with discontinuities, with discontinuous derivatives and oscillatory functions.

Patterson: I have a comment on what to look at in the performance of integration methods on parallel machines. With one processor machines one usually looks at the number of function evaluations and ignores the overhead. This means we consider the function evaluation as the most significant thing. On parallel machines the opposite is true. Overhead becomes important. We should therefore be looking at complicated functions.

Kapenga: In fact, easy (uncomplicated) functions are hardest for these machines. It seems that the more difficult the function, the greater the speed-up.

Baker: As a novice in the world of parallel computers, may I ask on what sort of machines I can expect the kind of speed-up mentioned here, using standard Fortran?

Kapenga: You need help from macros. It is not good enough, for example, to simply use the optimizer on the Cyber 205. The greatest speed-ups are obtained using macros on MIMD machines such as HEP, Cray, Alliant.

Baker: This is of course an entirely different philosophy from the one developed by NAG. The macros talked about here are not likely to be portable.

Kapenga: The macros used in Argonne are portable, to MIMD machines. Work is being done to make them portable to new generations of machines.

Haber (to Berntsen): Families of functions with discontinuities give most trouble to Gauss-type codes. Why did your tests show that the Gauss rules did better than CADRE?

Berntsen: We have no idea.

Lyness: CADRE is an extrapolation code that tries to get high degree by extrapolating. If it does not realise there is a discontinuity it will do as badly as ordinary Gauss.

Lyness (A comment on Berntsen's paper): The performance profiles shown in the paper would all be shifted to the left if one re-ran the experiments with epsilon replaced by epsilon/100. The requested tolerance is a kind of 'volume control' which can be turned. The effect of epsilon must in some sense be factored out.

Gladwell: This in done in the Toronto DETEST package for testing ODE routines.

Lyness: There is also a suggestion in the paper by Lyness and Kaganove, TOMS 2(1976), 65-81.

Haber (Comment): Some years ago the following suggestion was made at NBS. For one month, all quadrature on the machines would be done free. Each routine used would be rated on the fraction of failures it suffered. The question was : would the same scores be obtained if the experiment was done again a few months later? Most people said that the scores would be different. The feeling among the proposers of the experiment was that the kind of integrals which would be attempted would be more realistic ones, a more 'natural' set of integrands than are usually chosen in testing programs.

Lyness: For required accuracies of 0.001 to 0.0001 the conclusions would not change.

Kahaner (to Krenz): Why does the user of the self-validating package have to type in the integrand interactively? Is it not possible to use a Fortran function?

Krenz: There are two reasons why not. One is that I arrived too late in the project to change things. The second is that ISPF panels had to be used to ensure a consistent interface: ISPF is an IBM interactive interface to ACRITH. That is, this is an ACRITH restriction.

WEDNESDAY AUGUST 13

Lyness (to Genz): One minor comment on testing in general. The cost of the actual runs is a once only cost. The output from the experiments should be kept, on disk or tape, and the analysis done later.

Rabinowitz (to Genz): The type of function not included in the tests is the set of nice functions with no real difficulties. If only bad functions are looked at we get a distorted view.

Genz: Note that the functions had 'difficulty parameters' included. Some of the functions were in fact easy for certain choices of these parameters.

Haber (Comment): The reason for emphasising difficult functions in 1 dimension is that the customer will always use some version of Simpson for easy functions. There is no parallel to Simpson's method in multiple dimensions, i.e. no simple easily used rule.

Kahaner (to Doman): Who is envisaged as the future user of ADA, outside of those who are required to do so in certain laboratories?

Doman: Perhaps most users of Fortran, eventually.

Genz (to Doman): Why was Romberg the routine chosen from NAG to be converted to ADA? Why not an adaptive routine?

Doman: We did in fact do an adaptive routine, but not the best in the library. We plan to do the others eventually.

Lyness (to Doman): The procedure you described could stop if the estimated error increases. Can you restart using all the information available?

Doman: Yes.

de Doncker (to Doman): What happens when one specifies an accuracy? Is the precision of the calculation changed accordingly?

Doman: In principle you may specify the precision.

Patterson (to Moeller): Is there any hope of applying the theory described in your paper in 3 dimensions?

Moeller: So far things have been done very much by hand. But if computer algebra systems are developed far enough, yes, 3 dimensions will be possible. There has been much dissatisfaction with the performance of MACSYMA and REDUCE on zero finding problems for polynomials of several variables. New polynomial handling routines are under development in Maple.

Rabinowitz (Comment): Finding the roots of sets of moment equations is a non-trivial problem. You can set up the equations, and ensure that they are at least linearly consistent, but unless you use ad hoc techniques you cannot solve them. In addition, they tend to be very ill-conditioned.

Haegemans and Cools: Even small systems are hard to solve, despite the use of continuation techniques.

Comment: There was, here, some dispute on the value of continuation techniques for these problems. Keast was of the opinion that they hold great promise for the future, and that it will be eventually possible to solve even large moment systems by automatic techniques based on continuation.

Patterson (Comment): Even if we can solve the moment systems, is that the best approach? Is the algebraic approach necessarily the best way?

Haegemans: The problems themselves are interesting. It is nice to see the solutions.

Berntsen: I would say that even if the systems are hard to solve, and if that is not the way to go, it was good fun. One of the pitfalls is shown by a linearly consistent system arising in the cube, from a weight function w(x) = 1, which was inconsistent. This non-linear inconsistency is a harder thing to avoid.

Kahaner (Comment): I would like to make a remark on our concern with computation times. If we were physicists we would not worry about times, even if days of computing were necessary. If the problem can be solved we would do it. Most scientists would be happy with a code that works, almost regardless of the cost.

Kahaner (to de Doncker): Berntsen's tests show that QAGS is less efficient than QAG, except for end point singularities. Is that consistent with your experience?

de Doncker: No. But it is consistent when there are internal singularities. How well extrapolation works depends on the location of the singularity. If they are not at machine numbers QAGS is less efficient.

THURSDAY AUGUST 14

Keast (to Genz and de Doncker): An adaptive procedure run on a parallel machine cannot be truly global. Many subintervals may be divided simultaneously, resulting in the error requirement being over-satisfied.

de Doncker: Yes, but the error is still satisfied in the global sense. The use of several processors makes up for any inefficiency in obtaining an error that is too small.

Rabinowitz (Comments): The procedure should keep track of the number of times a subinterval is subdivided. Any one which is subdivided frequently should be marked for special treatment.

FRIDAY AUGUST 15

Gladwell (to Kahaner): A comment on avoiding COMMON in Fortran programs. This was to be a 'deprecated' feature in the 8X standard. Kahaner had suggested replacing

> FUNCTION F(X)
> COMMON /name/ A(10)

by   FUNCTION F(X,A)

where A is an indeterminate parameter vector. But if this is done we need also to pass the length of A. Thus we need

> FUNCTION F(X,A,N)

Kahaner also suggested the use of a parameter ISTOP to stop the process if the results were unsatisfactory, or out of range. One other possibility is to use 'reverse communication', a technique used in several NAG subroutines. This would necessitate the restructuring of codes, but certainly avoids COMMON.

Keast (Comment): But then it would be necessary to change the advice in NAG where the use of reverse communication is not urged.

Lyness (A comment on Carpentier's paper): If a zero is near the curve gamma things are difficult. It is hard to find how many zeros there are in a given region. Once that is known, the rest is easy.

Ioakimidis (to Lyness): Is there any easy way to find out how many zeros there are?

Lyness: The Delves/Lyness method assume that there are no poles on the contour. The Abd-Elall/Delves/Reid paper was published in a hard to get reference. It assumes there may be poles, but poles near the zeros give problems.

Baker (Comment): There was much discussion at breakfast about the publication of codes whose source is not available to researchers. Testing is difficult without the source.

Here there was general discussion on the difficulties that commercial libraries would face if the source was made available.

A brief discussion of principal problems followed. These included :

1. Indefinite integration.

2. Integration over product regions using methods which, when compounded, result in cancellation of some points.

3. Monte Carlo and Pseudo Monte Carlo methods, which seem to have been neglected recently.

LIST OF PARTICIPANTS

C. T. H. Baker
Dept. of Mathematics
Univ. of Manchester
Manchester
England   M13 9PL

B. G. S. Doman
Dept. of Stats. & Comp. Math.
University of Liverpool
PO Box 147 Liverpool
England L69 3BX

J. Berntsen
Department of Informatics
Universitetet I Bergen
Allegt. 55
N-5000 Bergen, Norway

T. O. Espelid
Department of Informatics
Universitetet I Bergen
Allegt. 55
N-5000 Bergen, Norway

M. P. Carpentier
Departamento de Matematica
Instituto Superior Tecnico
Av. Rovisco Pais
1096 Lisboa Codex, Portugal

G. Fairweather
Dept. of Mathematics
University of Kentucky
Lexington KY 40506
U.S.A.

R. Cools
University of Leuven
Computer Science Department
Celestynenlaan 200 A
3030 Heverlee Belgium

K. G. Foote
Institute of Marine Research
PO Box 1870
5011 Bergen
Norway

G. F. Corliss
Dept. of Math. Stat. & CS.
Marquette University
Milwaukee, WI 53233
U.S.A.

L. Gatteschi
Dipartimento di Matematica
Universita di Torino
Via Carlo Alberto 10
I-10123 Torino Italy

E. de Doncker-Kapenga
Computer Science Department
Western Michigan University
Kalamazoo, Michigan 49008
U.S.A.

A. C. Genz
Computer Science Department
Washington State University
Pullman WA 99164
U.S.A.

I. Gladwell
Dept. of Mathematics
Univ. of Manchester
Manchester
England M13 9PL

J. A. Kapenga
Computer Science Department
Western Michigan University
Kalamazoo, Michigan 49008
U.S.A.

S. Haber
National Bureau of Standards
Mathematical Analysis Division
Administration Building Room A302
Gaithersburg, MD 20899, U.S.A.

P. Keast
Math. Stats. & Comp. Sci.
Dalhousie University
Halifax, Nova Scotia
Canada B3H 3J5

A. Haegemans
University of Leuven
Computer Science Department
Celestynenlaan 200 A
3030 Heverlee Belgium

G. S. Krenz
Dept. of Math. Stat. & CS.
Marquette University
Milwaukee, WI 53233
U.S.A.

L.C. Hsu
Mathematics Department
Texas A & M University
College Station, TX 77843
U.S.A.

J. P. Lambert
Mathematical Sciences
University of Alaska,
Fairbanks, Alaska
U.S.A.   99775-1110

N. I. Ioakimidis
P.O. Box 1120
GR-261.10 Patras,
Greece

J. N. Lyness
Math and Computer Science Division
Argonne National Laboratory
Argonne IL 60439
U.S.A.

R. L. Johnston
3836 West Broadway
Vancouver B.C.
Canada V6R 2C3

F. Mantel
351 N. Stanley Ave.
Los Angeles CA 90036
U.S.A.

D. K. Kahaner
National Bureau of Standards
Technology Building Room A151
Gaithersburg, MD 20899
U.S.A.

H. M. Moeller
Fernuniversitat Hagen
Fachbereich Mathematik
Postfach 940 5800 Hagen
Federal Republic of Germany

P. H. Muir
Dept. of Math. and Comp. Sci.
St. Mary's University
Halifax, Nova Scotia
Canada B3H 3C3

H. J. Schmid
Mathematisches Institut
Univ. Erlangen-Nurnberg
Bismarckstrasse 1 1/2
8520 Erlangen FDR

T. N. L. Patterson
Dept. of Applied Mathematics
Queen's University
Belfast
Northern Ireland BT7 1NN

P. W. Smith
IMSL 2500 Park West, Tower 1
2500 City West Blvd
Houston   TX 77042-3020
U.S.A.

R. Piessens
Computer Science Department
University of Leuven
Celestijnenlaan 200 A
3030 Heverlee Belgium

T. Sorevik
Department of Informatics
Universitetet I Bergen
Allegt. 55
N-5000 Bergen, Norway

P. Rabinowitz
Dept.of Applied Mathematics
Weizmann Institute of Science
Rehovot
Israel

W. Squire
Dept. of Mech.& Aerospace Eng.
West Virginia University,
Morgantown, WV 26506-6101
U.S.A.

A. Scarpas
Research Fellow, NTUA,
132 Pendelis Avenue
Athens, 152-34,
Greece

A. H. Stroud
Mathematics Department
Texas A & M University
College Station   TX 77843
U.S.A.

LIST OF CONTRIBUTORS

An asterisk (*) denotes a joint author who did not attend the Workshop.

C. T. H. Baker
Dept. of Mathematics
Univ. of Manchester
Manchester
England M13 9PL

J. Berntsen
Department of Informatics
Universitetet I Bergen
Allegt. 55
N-5000 Bergen, Norway

*J. C. Butcher
Computer Science Department
University of Auckland
Auckland
New Zealand

M. P. Carpentier
Departamento de Matematica
Instituto Superior Tecnico
Av. Rovisco Pais
1096 Lisboa Codex, Portugal

R. Cools
University of Leuven
Computer Science Department
Celestynenlaan 200 A
3030 Heverlee Belgium

G. F. Corliss
Dept. of Math. Stat. & CS.
Marquette University
Milwaukee, WI 53233
U.S.A.

*L. M. Delves
Dept. of Stats. & Comp. Math.
University of Liverpool
PO Box 147 Liverpool
England L69 3BX

*M. S. Derakhshan
Dept. of Mathematics
Univ. of Manchester
Manchester
England M13 9PL

E. de Doncker-Kapenga
Computer Science Department
Western Michigan University
Kalamazoo, Michigan 49008
U.S.A.

B. G. S. Doman
Dept. of Stats. & Comp. Math.
University of Liverpool
PO Box 147 Liverpool
England L69 3BX

*S. Elhay
Computer Science Department
University of Adelaide
Box 498, Adelaide SA 5001
Australia

*D. M. Ennis
Phillip Morris Research Center
Commerce Road
Richmond, VA 23261
U.S.A.

T. O. Espelid
Department of Informatics
Universitetet I Bergen
Allegt. 55
N-5000 Bergen, Norway

L.C. Hsu
Mathematics Department
Texas A & M University
College Station, TX 77843
U.S.A.

K. G. Foote
Institute of Marine Research
PO Box 1870
5011 Bergen
Norway

N. I. Ioakimidis
P.O. Box 1120
GR-261.10 Patras,
Greece

L. Gatteschi
Dipartimento di Matematica
Universita di Torino
Via Carlo Alberto 10
I-10123 Torino Italy

D. K. Kahaner
National Bureau of Standards
Technology Building Room A151
Gaithersburg, MD 20899
U.S.A.

A. C. Genz
Computer Science Department
Washington State University
Pullman WA 99164
U.S.A.

J. A. Kapenga
Computer Science Department
Western Michigan University
Kalamazoo, Michigan 49008
U.S.A.

I. Gladwell
Dept. of Mathematics
Univ. of Manchester
Manchester
England M13 9PL

*J. Kautsky
School of Mathematical Sciences
Flinders University
Bedford Park SA5042
Australia

S. Haber
National Bureau of Standards
Mathematical Analysis Division
Administration Building Room A302
Gaithersburg, MD 20899, U.S.A.

G. S. Krenz
Dept. of Math. Stat. & CS.
Marquette University
Milwaukee, WI 53233
U.S.A.

A. Haegemans
University of Leuven
Computer Science Department
Celestynenlaan 200 A
3030 Heverlee Belgium

J. P. Lambert
Mathematical Sciences
University of Alaska,
Fairbanks, Alaska
U.S.A.   99775-1110

J. N. Lyness
Math and Computer Science Division
Argonne National Laboratory
Argonne IL 60439
U.S.A.

P. Rabinowitz
Dept.of Applied Mathematics
Weizmann Institute of Science
Rehovot
Israel

F. Mantel
351 N. Stanley Ave.
Los Angeles CA 90036
U.S.A.

H. J. Schmid
Mathematisches Institut
Univ. Erlangen-Nurnberg
Bismarckstrasse 1 1/2
8520 Erlangen FDR

H. M. Moeller
Fernuniversitat Hagen
Fachbereich Mathematik
Postfach 940 5800 Hagen
Federal Republic of Germany

T. Sorevik
Department of Informatics
Universitetet I Bergen
Allegt. 55
N-5000 Bergen, Norway

*K. Mullen
Dept. of Math. and Stat.
University of Guelph
Guelph, Ontario N1G 2W1
Canada

W. Squire
Dept. of Mech.& Aerospace Eng.
West Virginia University,
Morgantown, WV 26506-6101
U.S.A.

T. N. L. Patterson
Dept. of Applied Mathematics
Queen's University
Belfast
Northern Ireland BT7 1NN

A. H. Stroud
Mathematics Department
Texas A & M University
College Station   TX 77843
U.S.A.

R. Piessens
Computer Science Department
University of Leuven
Celestijnenlaan 200 A
3030 Heverlee Belgium

*Y. S. Zhou
Mathematics Department
Jilin University
Changchun
China

INDEX